Chernobyl – What Have We Learned?

The Successes and Failures to Mitigate Water Contamination over 20 Years

Edited by

Yasuo Onishi

Washington State University,
Richland, WA, USA

Oleg V. Voitsekhovich

Ukrainian Hydrometeorological Institute,
Kiev, Ukraine

and

Mark J. Zheleznyak

Ukrainian Center of Environmental and Water Projects,
Kiev, Ukraine

Springer

A C.I.P. Catalogue record for this book is available from the Library of Congress.

ISBN-10 1-4020-5348-7 (HB)
ISBN-13 978-1-4020-5348-1 (HB)
ISBN-10 1-4020-5349-5 (e-book)
ISBN-13 978-1-4020-5349-8 (e-book)

Published by Springer,
P.O. Box 17, 3300 AA Dordrecht, The Netherlands.

www.springer.com

Cover Photograph: Chernobyl Shelter (formerly called the sarcophagus) consisting of damaged Chernobyl Nuclear Plant Unit 4 structures and newer structures built after the accident. It is a temporary building.

Printed on acid-free paper

Chernobyl – What Have We Learned?

ENVIRONMENTAL POLLUTION

VOLUME 12

The titles published in this series are listed at the end of this volume.

Table of Contents

The Editors

Yasuo Onishi is an adjunct full professor of civil and environmental engineering at Washington State University and was Chief Scientist of fluid and computational engineering at Pacific Northwest National Laboratory in Richland, Washington. He has over 30 years of experience in fluid mechanics and environmental exposure and risk assessment. Dr Onishi pioneered surface water contaminant transport modeling accounting for interactions between sediment and contaminants. He developed many state-of-the-art sediment-contaminant-aquatic risk assessment and non-Newtonian fluid reactive transport codes. He is an adjunct member of the National Council of Radiation Protection and Measurements and was a member of the National Academy of Sciences' oil spill committee. He was U.S. Coordinator of the US/USSR Joint Coordinating Committee on Civilian Nuclear Reactor Safety on Chernobyl radionuclide behavior in soil-water systems and an IAEA advisor.

Oleg Voitsekhovich is the head of environmental radiation monitoring at the Ukrainian Hydrometeorological Institute. He has over 20 years of experience in hydrological sciences and environmental impact assessment. Since the Chernobyl accident, he has been involved in radiation monitoring, studying radionuclide behavior and its transport in the aquatic systems. He led development of water remedial action plans for the Chernobyl Exclusion Zone and has led many national and international research projects on aquatic ecosystems affected by the Chernobyl accident. He is an IAEA expert in radiation monitoring and environmental impact assessment for rivers and the Black Sea, and he co-authored the Chernobyl Forum Compendium in 2005. He also assesses uranium residue waste management in Ukraine and Central Asia.

Mark J. Zheleznyak is Research Director of the Ukrainian Center of Environmental and Water Projects in Kiev, Ukraine. He has more than 20 years of experience in developing and applying surface water transport modeling, especially related to Chernobyl accident aquatic assessments. He has directed a series of water resource, contaminant exposure, and risk assessment studies, including developing the aquatic assessment portion of a European emergency response and management computer system for radionuclide releases to the environment. He has collaborated with many U.S., European, Asian, and international organizations, including the IAEA.

Preface

Twenty years ago, on April 26, 1986, Unit 4 of the Chernobyl Nuclear Power Plant experienced a nuclear reactor accident and released six times more radionuclides into the atmosphere than the Hiroshima atomic bomb. The accident significantly damaged the environment and is expected to result in as many as 4000 deaths due to leukemia and other cancers.

The cause of the accident was specific to a particular Soviet-built reactor; but the environmental and human consequences are applicable to Western nuclear plant accidents. Contrary to public perception, the former Soviet Union did what they could to protect the public from the Chernobyl accident within funding restrictions. They involved the highest level of political and scientific leaders and even conducted public surveys to identify concerns and fears. However, the secrecy with which the information was treated caused mistrust of the Chernobyl assessment and contributed to the widespread mental health problems associated with the accident. Some Chernobyl countermeasures were successful, but many were ineffective or made things worse. We hope that the details presented in this book may be useful for Western countries in preparing for possible accidents or potential terrorist attacks with nuclear or chemical agents.

The assessment of the Chernobyl accident, its impacts, and its countermeasures and remediation has been published widely in many Western nations, but much of the research is still known only in Ukraine and Russia. Our original intention was to disseminate study results presented in the Ukrainian National Academy of Sciences' 1997-1998 publication, *Forecasts of the Water Contamination, Water Use Risk, and Water Protection Countermeasures Efficiency Assessments for the Aquatic Ecosystems of the Chernobyl NPP Accident Influence Outside Ukraine*. But instead, we decided to prepare a new book, which reflects a new understanding of the Chernobyl impacts and reports significant new activities and planning such as construction of the New Safe Confinement (NSC) over the Chernobyl Shelter (formerly called the sarcophagus). Thus, although this book includes parts of the above Ukrainian publication, it presents actual Chernobyl experience in assessing aquatic pathways and selecting and implementing water countermeasures and remediation activities over the last twenty years.

The Chernobyl Nuclear Plant is located on the Pripyat River, a tributary of the Dnieper River, which flows into the Black Sea. Water pathways make up a small part of the total dose received by the population from all dose-forming factors. However, the Dnieper River system has transported nuclides to twenty million people through aquatic pathways such as drinking water, irrigated produce, and milk and meat from cattle fed with contaminated feed and water. Thus, many research and remediation efforts have been focused on minimizing radionuclide transfer from the Chernobyl Exclusion Zone to the nearby Dnieper River to reduce radiological risks from water use.

The four main conclusions of this book are (1) scientifically defensive assessment tools and data must be developed and applied, (2) counter-measures and remediation selections must be based on a cost-risk analysis that directly connects the main physical and chemical processes to environmental and human health risks and costs, (3) decision makers must be knowledgeable of the phenomena being evaluated, and (4) the facts must be communicated quickly and honestly to the affected public.

A major upcoming event at Chernobyl is construction of the billion-dollar NSC, which is intended to reduce radionuclide releases to the environment, provide safer conditions for Shelter dismantling and waste management workers, and generate economic and social benefits. Normal operation of the NSC will reduce atmospheric fallout from Shelter collapse by about 15 to 20 times. Thus it will reduce radionuclide concentrations in groundwater and rivers. Even without the NSC there will be no measurable harmful effects to the Pripyat River environment from strontium, cesium, and plutonium.

The biggest public health effect of the Chernobyl accident involves mental health (e.g., hopelessness and anxiety). A major socio-economic problem was caused by the lack of a vital local economy, lower living standards, high unemployment, and increased poverty. Thus, regardless of the specific remedial benefits of the NSC, its construction will provide employment and build a high-quality labor force with high self esteem and high manufacturing/construction capabilities. We truly hope that it will revitalize the economy and generate hope, thus addressing the most critical Chernobyl issues, mental health and socio-economic problems.

The authors would like to express their sincere appreciation to all of the scientific and engineering staff who summarized the results and data, as well as those who were involved in preparation of the material in this book.

Spring 2006

Yasuo Onishi, Richland, Washington, USA

Oleg V. Voitsekhovich and Mark J. Zheleznyak, Kiev, Ukraine

Chapter 1
Soviet-Built Nuclear Plants and Their Safety

Yasuo Onishi
Washington State University, Richland, WA, USA
Formerly Pacific Northwest National Laboratory, Richland, WA, USA

Oleg V. Voitsekhovich
Ukrainian Hydrometeorological Institute, Kiev, Ukraine

Mark J. Zheleznyak
Ukrainian Center of Environmental and Water Projects and
National Academy of Sciences of Ukraine, Kiev, Ukraine

Twenty years ago, on April 26, 1986, Unit 4 of the Chernobyl Nuclear Power Plant suffered a core meltdown. Human error was a direct cause of this accident. Plant operators had dismounted automatic emergency shutdown devices to test safety equipment, and then made a mistake on plant shutdown with poorly designed safety rods, causing a steam explosion and graphite fire (DOE 1987, Marples and Young 1997). The accident released six times more radionuclides into the atmosphere than were released by the Hiroshima atomic bomb. This was the most devastating nuclear accident in history (IAEA 1991, Dreicer et al. 1996). The largest amounts of radionuclides released from the Chernobyl Plant were the short-lived (with half-lives of just hours to several days) xenon (^{133}Xe), iodine (^{133}I and ^{131}I), technetium (^{132}Tc), and neptunium (^{239}Np). The main radionuclides affecting human health are the longer-lived cesium (^{137}Cs), strontium (^{90}Sr), plutonium (^{238,239,240}Pu), and americium (^{241}Am), in addition to iodine (^{131}I). About 120,000 people were evacuated from the area within 30 km of the plant.

The Chernobyl plant is located on the Pripyat River in Ukraine. The Pripyat flows into the Dnieper River, which ends at the Black Sea. Figure 1.1 depicts the Dnieper River system with the Chernobyl plant shown at the upper left corner. The Pripyat and Dnieper rivers are the main aquatic pathways for the radionuclides to reach the people of Ukraine. The pathways include drinking water, eating fish, consuming irrigated produce; and drinking milk and eating the meat of cattle fed with contaminated feed and water, as well as swimming and other water recreational activities. Approximately three million Kiev residents drink water from the Dnieper River, and as many as 20 million Ukrainians eat foods irrigated with Dnieper River water.

Y. Onishi, O.V. Voitsekhovich and M.J. Zheleznyak (eds.), Chernobyl – What Have We Learned?, 1–7.

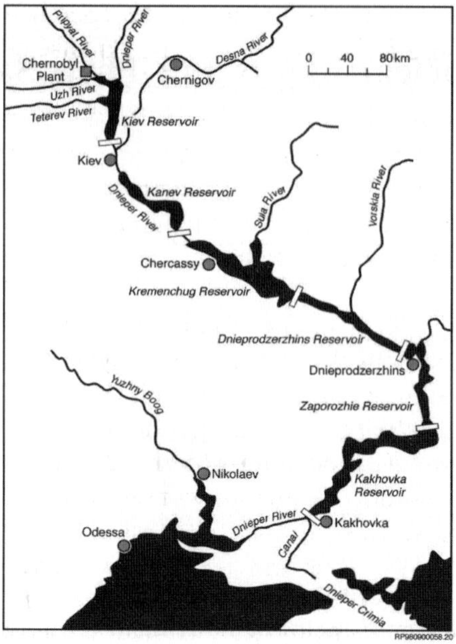

Figure 1.1. Dnieper River system and its cascade of six reservoirs

This book, published on the 20th anniversary of the Chernobyl accident, focuses on the aquatic environment that was affected by the accident. It describes what happened, who decided what actions to take, which water protection measures worked to protect the environment and human health and which ones did not, what the current aquatic environmental conditions and associated health risks are, what the future plan is for Chernobyl cleanup, and what lessons we have learned.

1.1 Civilian Nuclear Plants in the Former Soviet Union

Before we look into the Chernobyl accident and its effects on the aquatic environment, we briefly discuss why the former Soviet Union needed nuclear plants and how safe they were. The USSR had some of the largest oil, gas, and coal reserves and production in the world (Marples and Young 1997), as well as a vast expanse of forests. These fuels were, for the most part, adequate to supply the country's energy needs during the time when nuclear physics, thus nuclear bombs and nuclear plants, were being developed.

There may be both ideological and practical reasons for building nuclear power plants. In about 1920, a science fiction novelist, Aleksandr Bogdanov, considered that socialism needed atomic energy as a highly concentrated energy source for the people (Kramish 1959). The costs of fossil fuels were rising, and most of these fuels exist in the vast expanse of land east of the Ural Mountains and far from the energy-consuming areas of the western Soviet Union. Moreover, production of these fuels was not always stable. However, nuclear power plant development was not coordinated with these energy considerations. Atomic weapons work proceeded without the central control and coordination of fuel and energy production. This weapons work was the foundation of the Soviet nuclear power plant development and operation.

Scientists Peter L. Kapitsa (in 1921) and V. I. Vernadiskii (in 1922 ~ 1939), in close contact with Western nuclear physicists, initiated the work on atomic science in the Soviet Union. Vernadiskii, the founder of the Radium Institute, was aware of the potential of atomic power as both a useful energy source and a powerful weapon (Kramish 1959). He also emphasized that scientists should be responsible for the consequences of their work. These early years of atomic science in the 1920s produced some significant work; for instance, Dimitrii V. Skobeltsyn first observed nuclear reactions caused by bombardment of cosmic particles in a laboratory, and Vitalii G. Khlopin worked on extracting radium from uranium ores (Kramish 1959).

Nuclear physics took a great leap in the 1930s, promoted by physicist Abram F. Ioffe. He allegedly connected nuclear physics to the communist doctrine of Marx and Engels, implying that atomic theory was a basis for the materialistic interpretation of history (dialectic materialism—an important building block of the Soviet system) (Kramish 1959). He also thought of using atomic energy as a power source for industry. His arguments secured state funding of nuclear physics for a decade, until the early 1940s, when World War II brought a halt to Soviet nuclear physics research.

There were significant advances in atomic science in the late 1930s to early 1940s, both in the West and in the Soviet Union. These included nuclear fission, first discovered in Germany, a neutron a splitting uranium atom, and chain reactions. The discoveries showed fission as both energy and explosive, and Soviet scientists were fully aware of the possibilities. Soviet scientists were using a cyclotron, the first in Europe, to smash atoms to study the fission process, and produced fission products such as radioactive iodine from the fission of thorium and uranium and separation of uranium isotopes. The Soviet nuclear physics was at the world level. However, all Soviet atomic science work was stopped during World War II after the German invasion. Igor V. Kurchatov, the leader of Soviet nuclear research, was sent to the Black Sea to protect Soviet ships from German mines, and some nuclear physicists worked on nuclear medicine (Kramish 1959). Meanwhile, the West accelerated their nuclear work, culminating in the U.S. Manhattan Project to develop the bombs dropped on Hiroshima and Nagasaki in 1945.

Early Soviet reaction to U.S. nuclear activities was muted because Stalin and other Soviet leaders were unaware of the importance of atomic bombs as a weapon and their denial of the advancement of U.S. nuclear physics (Kramish 1959). However, by opposing the U.S. monopoly on nuclear technology as its policy, the Soviets renewed their efforts and built their first nuclear reactor in 1946–1947 and made their first atomic bomb in 1949 (Kramish 1959).

In 1954, the Soviet Union developed a civilian nuclear station they claimed was the first in the world, at Obninsk, a nuclear and physics research town about 100 km southwest of Moscow (Marples and Young 1997). It was a graphite-moderated reactor that generated 5 megawatts of electricity. The success of the first power plant was the foundation for building large nuclear power plants starting in 1958 under the sixth Five-Year Plan. The pace was accelerated in the 1970s to meet energy needs and provide a large amount of inexpensive, dependable electricity.

1.2 Soviet-Built Reactors

The two main types of Soviet reactors are the Reaktor Bolshoi Moschnoshi Kanalynyi (RBMK) and Vodo-Vodyanoi Energeticheskii Reaktor (VVER) (Marples and Young 1997, Egorov et al. 2000). There are also some Reaktor na Bystrykh Neytronakh (BN) fast breeder reactors (Bradley 1997). The RBMK is a boiling-water, graphite-moderated pressure tube reactor. The fuel is contained in 1,690 vertical tubes (channels) in a large graphite core. This reactor generates 1,000 or 1,500 megawatts of electricity. Unit 4 of the Chernobyl plant was an RBMK-1000. The RBMK does not meet international safety standards, having deficiencies in emergency core cooling, fire protection, instrumentation, and control systems (DOE 1996). It tends to be unstable at low power and has a questionable control rod function to control nuclear reactions. Unlike most Western reactors, no containment building covers the reactor. Some scientists regarded the first generation of this reactor as the most dangerous in the world (Marples and Young 1997).

The VVER is a pressurized water-cooled and water-moderated reactor similar to a Western pressurized water reactor. The first VVER was built in 1964 (Marples and Young 1997). The generating capacity was expanded to 1,000 megawatts in 1980 (VVER-1000). VVERs were also built in other East European countries. There are three main models in operation: the VVER-1000 and two models of VVER-440. The VVER-1000 is the largest and newest VVER and produces 1,000 megawatts of electricity. It uses safety systems common to Western nuclear power plants, including an emergency core cooling system and a containment building (DOE 1996). It meets most international safety standards. This is the most common Soviet reactor.

The VVER-440/230 reactor was an initial civilian model of the pressurized water reactor. Similar to Western reactors, it uses low-enriched uranium oxide fuel placed in thin-metal-clad rods. Although it is somewhat similar to Western reactors, it lacks many safety features such as reactor cooling and fire protection systems and has no containment building (DOE 1996). The VVER-440/213 is an improved version of the VVER-440/230. It has an emergency cooling system and a bubble-condenser tower that acts as containment to reduce off-site radionuclide release in the event of an accident.

By 1991, five years after the Chernobyl accident and the year the Soviet Union collapsed, 17 nuclear power plants with 59 power units were operating and producing approximately 40 gigawatts of electricity (DOE 1996, Egorov et al. 2000), as shown in Figure 1.2 (GAO 2000). There were 14 RBMK reactors, 25 VVER-440 reactors, and 20 VVER-1000 reactors. Unit 1 of the Chernobyl plant had one RBMK-1000 reactor at that time that is no longer operating.

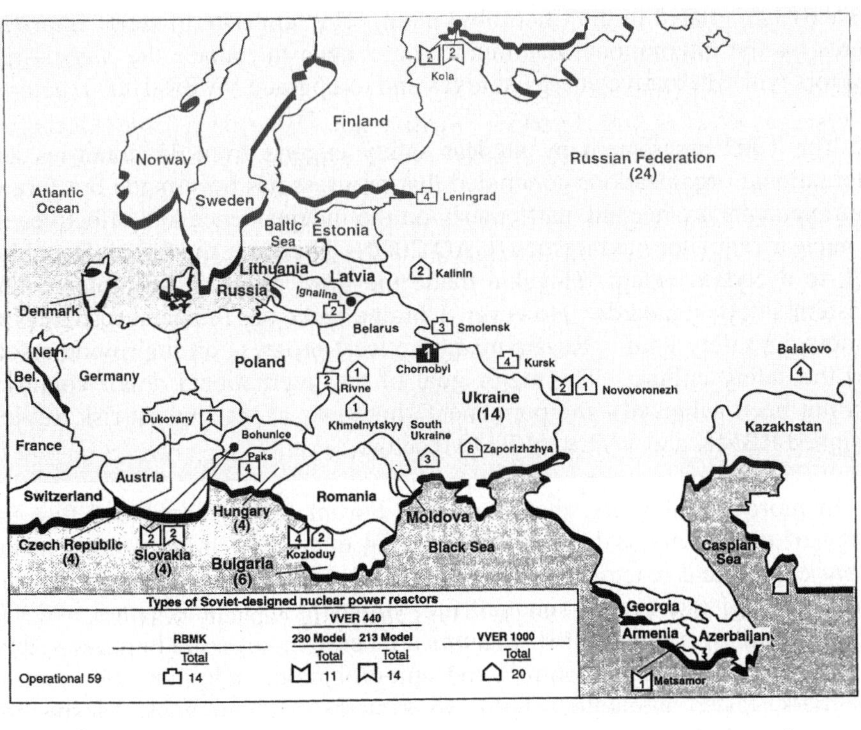

Figure 1.2. Nuclear power plants in Former Soviet Union and nearby countries

Today nuclear power provides about 13 percent of Russia's electricity and roughly 33 percent of Ukraine's electricity. Bulgaria, Czech Republic, Hungary, Lithuania, and Slovakia use nuclear power for 37, 29, 43, 88, and 54%, respectively, of their electricity needs (DOE 1996). Thus, nuclear energy provides a significant portion of the electricity in these countries. Thirteen RBMK Chernobyl-style reactors and 11 VVER 440/230 reactors are of greatest safety concern because they fall below Western safety standards and cannot be upgraded economically (GAO 2000).

The 1986 Chernobyl accident heightened public concern about the safety of the Soviet nuclear plants. To respond to the intensified public mistrust, the Soviet Union implemented several safety features to RBMK reactors, such as more fixed absorber rods, increased fuel enrichment, faster insertion times, and more containment (Bradley 1997). Training and radionuclide monitoring were also increased. In addition, since 1992, more than 20 countries and international organizations have contributed about $2 billion dollars to the seven former Soviet-block nations to reduce the risks associated with these reactors (DOE 1996). This support includes (1) management and operational safety, (2) engineering and technology, (3) plant safety and evaluation, (4) fuel cycle safety, (5) nuclear safety legislative and regulatory framework, and (6) activities related to the Chernobyl plant. The longer-term safety improvements by the international community have been to replace the highest-risk reactors with alternative energy sources and to upgrade VVER-1000 reactors.

The 1999 assessment by nuclear safety experts from 32 countries and international organizations concluded that progress has been made but further improvements are needed, particularly on the independence and effectiveness of nuclear regulatory authorities (GAO 2000). Hungary, the Czech Republic, and, to a certain extent, Slovakia made the most progress in implementing Western safety standards. However, Ukraine lacks the financial resources to achieve its safety goals. Russia made the least progress on improving safety and the safety culture. The major goal of the international donor countries had not been achieved—the permanent shutdown of the highest-risk Soviet-designed RBMK and VVER/440/230 reactors.

In more recent years, the Chernobyl Forum (2005) concluded that the safety of these reactors has vastly increased due to the improvement in the safety culture and reactors. All remaining Chernobyl units were shut down; Unit 2 in 1991 due to a turbine hall fire, Unit 1 in 1997, and Unit 3 in 2000. Moreover, all operating RBMK reactors are being modified to improve safety. Improvements include control rod modification, addition of neutron absorbers, faster automated shutdown systems, and automated inspection equipment. The Chernobyl Forum stated that these improvements make a repeat of the Chernobyl accident virtually impossible.

1.3 References

Bradley DJ. 1997. *Behind the Nuclear Curtain: Radioactive Waste management in the Former Soviet Union.* Battelle Press, Columbus, Ohio.

Chernobyl Forum. 2005. *Chernobyl legacy: Health, Environmental, and Socio-economic Impacts: and Recommendations to the Governments of Belarus, the Russian Federation and Ukraine.* International Atomic Energy Agency, Vienna.

Egorov NN, VM Novikov, FL Parker, and VK Popov (Eds.). 2000. *The Radiation Legacy of the Soviet Nuclear Complex.* Earthsan Publications Ltd., London.

Josephson PR. 2000. *Red Atom, Russia's Nuclear Power Program from Stalin to Today.* W.H. Freeman and Company. New York.

Kramish A. 1959. *Atomic Energy in the Soviet Union.* Stanford University Press, Palo Alto, California.

Marples DR and MJ Young (Eds.). 1997. *Nuclear Energy and Security in the Former Soviet Union.* Westview Press, Boulder, Colorado.

U.S. Department of Energy (DOE). 1987. *Health and Environmental Consequences of the Chernobyl Nuclear Power Plant Accident.* Prepared by the Interlaboratory Task Group on Health and Environmental Aspects of the Soviet Nuclear Accident, DOE Office of Health and Environmental Research, Washington, D.C.

U.S. Department of Energy (DOE). 1996. *Status Report: Improving the Safety of Soviet-Designed Nuclear Power Plants.* International Nuclear Safety Program, Office of Nuclear Energy, Science and Technology, DOE, Washington, D.C.

U.S. General Accountability Office (GAO). 2000. "Nuclear Safety: Concerns with the Continuing Operation of Soviet-Designed Nuclear Power Reactors," GAO/RCED-00-97, Submitted to the Subcommittee on Energy and Water Development, Committee on Appropriations, House of Representatives, Washington D.C.

Chapter 2

The Chernobyl Accident and Its Aquatic Impacts on the Surrounding Area

Oleg V. Voitsekhovich, Vladimir Kanivets
Ukrainian Hydrometeorological Institute, Kiev, Ukraine

Yasuo Onishi
Washington State University, Richland, WA, USA
Formerly of Pacific Northwest National Laboratory, Richland, WA, USA

Scientific and nuclear experts consider the Chernobyl accident unprecedented in the world's nuclear era. Among its consequences were significant health effects to the people living in the contaminated areas. The accident also had significant environmental impacts that continue to this day in the affected ecosystems. This book focuses on the *aquatic ecosystem* and on the international experience obtained in the decades since the accident.

The purpose of the research and the practical effort was to develop a strategy to minimize radionuclide transfer from the Chernobyl exclusion zone (CEZ) to the nearby Dnieper River so that radiological risks from water use would be reduced. The Dnieper River and its chain of reservoirs extend more than 1000 km, from the border of the exclusion zone (upper part of the Kiev Reservoir) to the Black Sea. This study demonstrates several achievements in the special tools and applications developed based on the monitoring, data analysis, and radiological protection principles applied to control radionuclide migration through the surface and groundwater pathways in the environment.

In the 20 years since the accident, the concern of the public, professionals and the media and the attention focused on radioactive contamination of the water, the aquatic ecosystem, and the associated habitats have clearly shown that this is a *very critical problem*. On one hand, the aquatic ecosystems of the CEZ such as nearby lakes, the Chernobyl nuclear power plant cooling pond, and river bays were most affected by direct radioactive fallout. Radiological effects were observed in some representative organisms living in the most contaminated water bodies.

Radiological effects that were obvious in the CEZ were not easily observed in water bodies farther away. A secondary radionuclide transfer from the contaminated CEZ to other areas has been a key factor in expanding

Y. Onishi, O.V. Voitsekhovich and M.J. Zheleznyak (eds.), Chernobyl – What Have We Learned?, 9–47.

contamination—the process of aquatic ecosystems and surface water runoff transporting radioactive substances from contaminated watersheds and waste sites to rivers downstream. This was a significant effect of the accident and one cause of stress-dependent health risks for the people living in the areas near the contaminated water bodies.

The radioecological assessment needs correct information based on reliable data. Specific required data are provided to state-of-the-art mathematical models for emergency response and short- and long-term predictions of the pathways of radionuclides in the environment. The main goals of this book are to (1) present as comprehensive an analysis as possible based on the studies of the migration of Chernobyl radionuclides in the environment and (2) describe major research achievements that can provide the scientific basis for justification and optimization of water protection and health risk analyses through studying the Chernobyl radionuclides as they migrate through the aquatic pathways.

2.1 Radioactive Release and Fallout due to Accident and Emergency Actions at the Chernobyl Nuclear Plant Site

The accident at reactor Unit 4 of the Chernobyl nuclear power plant took place shortly after midnight on April 26, 1986. Before that, the reactor had operated for many hours in non-design configurations preparing for an experiment on recovering the energy in the turbine in case of an unplanned shutdown. The course of the accident was rather complicated, but it can be considered a runaway surge in power level that caused the water coolant to vaporize inside the reactor. This in turn caused a further increase in power level with a resulting steam explosion that tore the reactor apart. After the initial explosion, the graphite in the reactor caught fire. Despite heroic efforts to control this fire, the graphite burned for many days, and releases of radioactive materials continued until May 10.

The details of the accident and the immediate consequences were discussed first at the Post-Accident Review Meeting (UN 2000) and then at many international forums (e.g., Jonssons 1996) and the United Nations Scientific Committee on the Effects of Atomic Radiation (UNSCEAR 1988, 2000). Reports appeared in the scientific literature (e.g., Izrael et al. 1990; Belyaev et al. 1996; Dreicer 1996; Borovoi et al. 1998; Likhtarev et al. 2002) and in other publications (DOE 1987). According to these studies, radioactivity was released to the surrounding area and the environment after the accident in the following four stages:

- **Stage 1** Immediate ejection of dispersed fuel and fragments from the core, fuel assembly and fuel rod fragments, reactor graphite, and reactor construction materials due to the explosion. Most of the ejected materials were concentrated as ruins near the reactor block walls and on the roofs of adjacent buildings. Fragments from the damaged reactor were ejected as far as 100 m (Figure 2.1a).
- **Stage 2** (April 26 to May 5 1986): Fine dispersed fuel discharged from the reactor entrained in a flow of hot gases (air and graphite combustion products) from the burning graphite. The level of fuel discharged was reduced gradually due to the mitigative actions taken to control and extinguish the graphite fire. During this period 40 tons of boron compounds, B_4C (reduces criticality), 800 tons of dolomite, $MgCa(CO_3)_2$, and 1800 tons of clay and sand (suppresses fire and creates filtering layer to reduce radioactive discharges), and 2400 tons of lead (Pb) (absorbs heat and provides a layer to seal and shield the top of the core vault) were added to the damaged reactor.
- **Stage 3** (May 6 to June 5, 1986): Increased discharge of fission products from the reactor block due to residual heat causing the fuel temperature to rise above 1700°C. Because of the temperature-dependent migration of fission products and the chemical transformation of the uranium oxide fuel, a discharge of fission products from the fuel matrix and subsequent aerosol (products of graphite combustion) occurred.

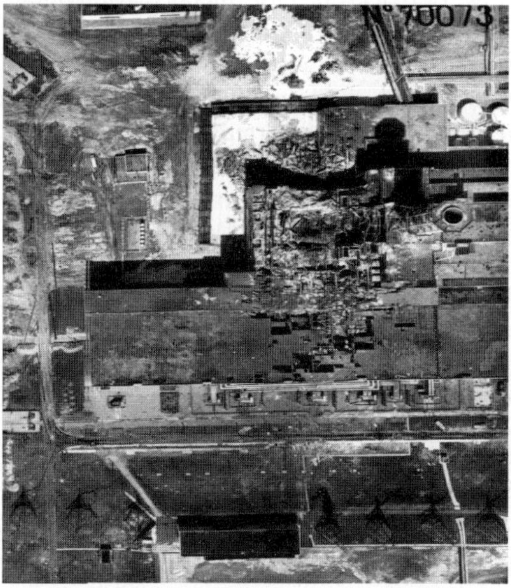

Figure 2.1a. The Industrial Site before Shelter construction

- **Stage 4** (after June 5, 1986): A fast reduction in the radioactive discharge rate from the damaged reactor attributed to the mitigating measures taken to reduce the fuel temperature, such as adding approximately 2400 tons of lead to the reactor central hall and feeding liquid nitrogen into the space beneath the reactor vault.

Measurements were taken by several organizations to establish the level and extent of contamination around the damaged reactor and the entire Chernobyl site. The first actions to prevent expansion of the scale and consequences of the accidental release are reported in the following paragraphs.

During **Stage 1**, polymeric coatings were applied to fix surface contamination and reduce the dose potential from inhalation and ingestion. The radiation conditions at Unit 3 were improved by installing a ferro-concrete plate under the foundation of Unit 4. The area immediately north of Unit 4 was prepared for construction of the Pioneer Wall around the reactor block and turbine hall. It was reported that approximately 5000 m^3 of radioactive waste was generated by the ground preparation work. Radiation levels in some areas remained high (up to 3 R/h) after this preparation was completed. At the end of Stage 1 it was decided to build the Chernobyl Shelter.

Stage 2 involved removing the contaminated soil and materials from around the damaged reactor to allow construction of the Shelter. Heavily shielded bulldozers and remotely operated equipment were used. Radioactive materials and waste were placed in 1-m^3 metal waste containers; those with a surface dose rate in excess of 100 R/h were entombed in the Pioneer Wall. After the contaminated top cover was removed, a 0.5-m concrete layer was laid over the treated ground, reducing the dose rate by a factor of 3 to 5. Construction of the Pioneer Wall reportedly reduced surface dose rate levels by a factor of 10 to 20. The Pioneer Wall and the concomitant reduction in radiation levels created a platform to support the erection of the Shelter.

During **Stage 3**, the Shelter was constructed and the final man-made layer formed at the Chernobyl site. The Cascade Wall was erected over the ruins on the north side of the damaged reactor. This involved pouring 133,500 m^3 of concrete. Some difficulties were reported with erection of the western wall; however, the erection of the Shelter was finished in November 1986. Conditions at the site before and during the erection of the Shelter are shown in Figures 2.1a and 2.1b, respectively.

Soils in the Chernobyl Industrial Zone were contaminated by radioactive materials released during the accident at Unit 4. Despite comprehensive decontamination activities between 1986 and 1988, the area still has high residual contamination.

Figure 2.1b. The Industrial Site during Shelter construction (June–December 1986)

2.2 Radionuclide Source Term in the Watershed

An early estimate based on measurements of contamination in the former Soviet Union and of the amount of ^{137}Cs released by the accident and deposited there was about 40 PBq (one million curies). A later estimate by international experts reported the radioactive deposition as 85 PBq ^{137}Cs, 54 PBq ^{134}Cs, 1760 PBq ^{131}I, 10 PBq ^{90}Sr, and 0.07 PBq $^{239,240}Pu$, along with many shorter-lived radionuclides of lesser radioecological significance (UNSCEAR 2000).[a] Major releases occurred for 10 days, during which the wind direction changed several times. Consequently, fallout was deposited over most of Europe. But the largest areas of contamination were in Russia, Ukraine, and Belarus. Much of the fallout occurred in the Dnieper River Basin. Figures 2.2 and 2.3 show the ^{137}Cs distributions, while Figures 2.4 and 2.5 depict distributions of ^{90}Sr and $^{239-240}Pu$. The greatest concentrations of ^{137}Cs are in the upper Dnieper watershed and the entire Pripyat watershed.

(a) Radioactivity is measured in becquerel (Bq), a unit equal to one disintegration per second. This is a very small unit, so radioactivity is often expressed as kBq (1000 Bq), MBq = 10^6 Bq, GBq =10^9 Bq, TBq = 10^{12}Bq and PBq = 10^{15} Bq. The non-metric unit curie (Ci) is also used; 1 Ci = 3.7 x 10^{10} Bq.

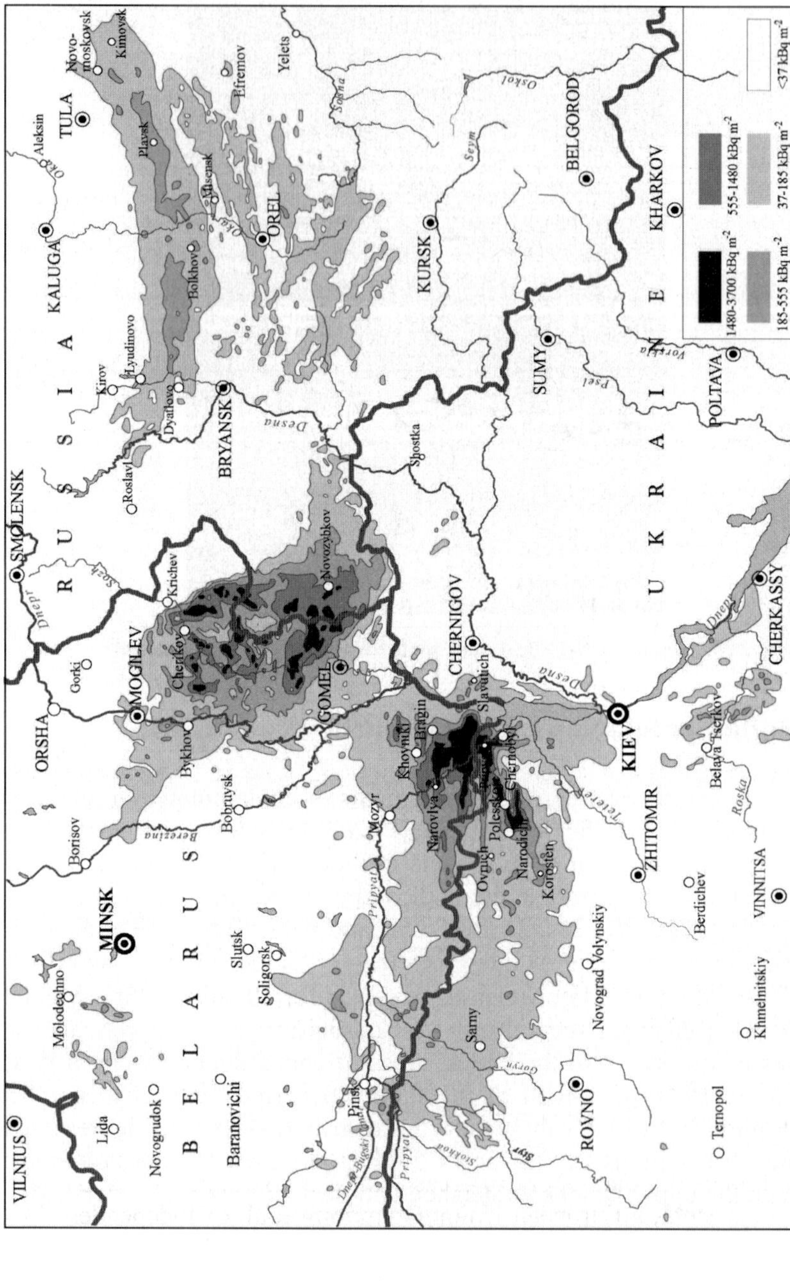

Figure 2.2. Distribution of deposited ^{137}Cs in the most contaminated areas of the Dnieper Basin, December 1989 (from UNSCEAR 2000)

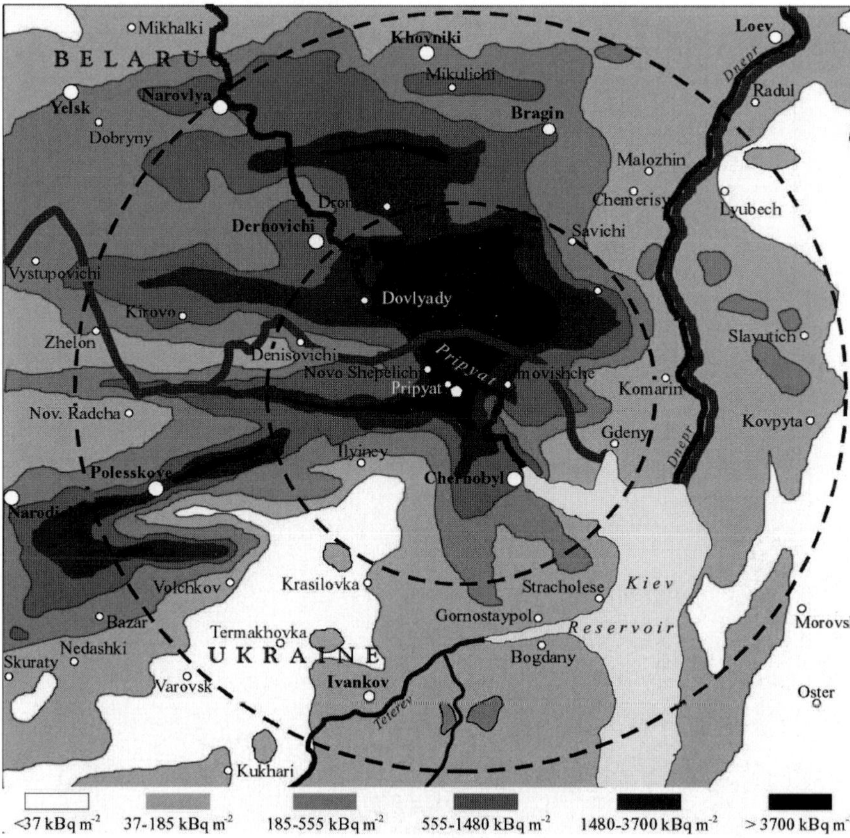

Figure 2.3. Distribution of ^{137}Cs within Chernobyl Exclusion Zone, 1986 (from UNSCEAR 2000)

Approximately 30% of the radioactivity deposited in the Dnieper Basin was in the CEZ, 30% in the far zone of the Belarus and Ukrainian sections of the Pripyat Basin, and about 40% in the Sozh and Iput river basins, which are the Gomel and Bryansk-Tula "hot spots" in Russia.

After the Chernobyl accident, the responsible agencies in many countries gathered data on soil contamination in their territories. These data were published as reviews, maps, and lists of contamination density in populated areas. Many of the data apply to areas within the Dnieper River Basin. In 1992 through 1995, a European Union/Commonwealth of Independent States (EU/CIS) program studied the consequences of the Chernobyl accident, and data on soil contamination density were gathered, processed, and published as an Atlas (DeCort et al. 1998), including a CD version in 2001.

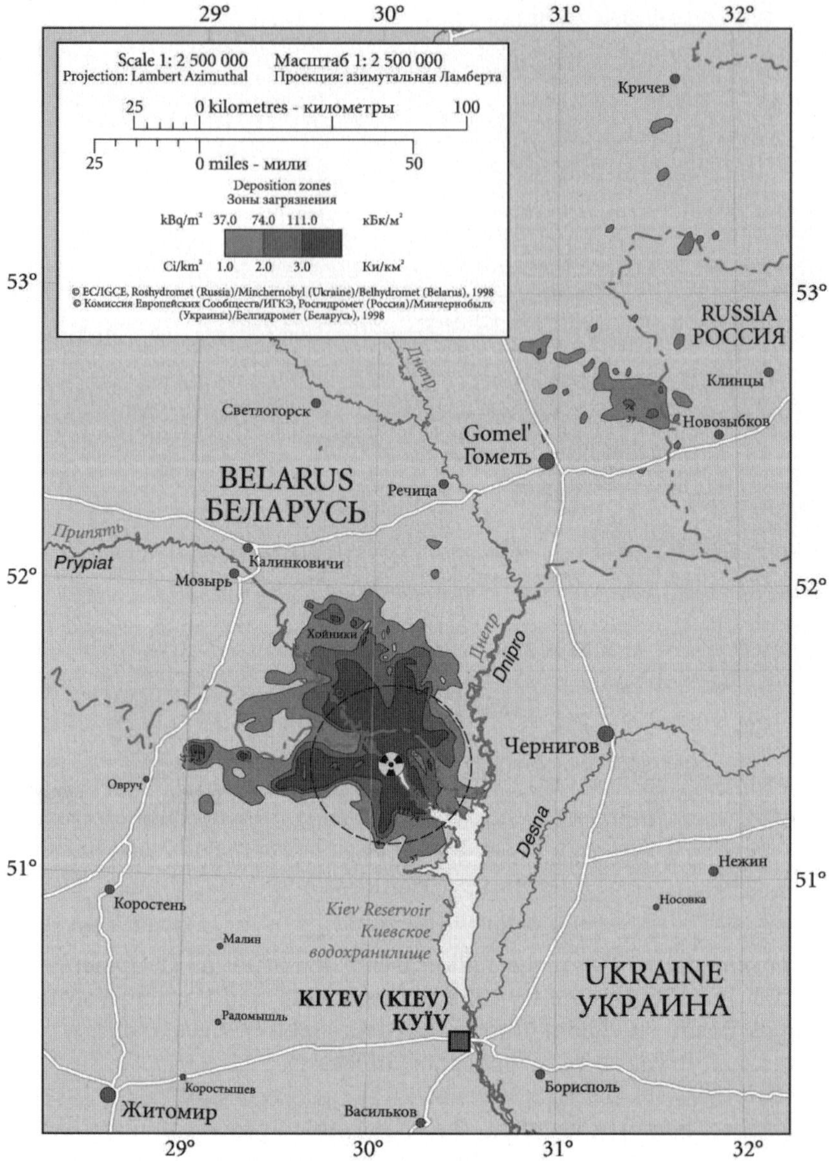

Figure 2.4. Distribution of deposited ^{90}Sr (December 1989) (from EC 2001)

Estimates of the releases have been refined over the years, and now the total release is thought to be about twice the earlier estimate for the former Soviet Union. Current estimates of the amounts of the more important radionuclides released are shown in Table 2.1. During the early period after the accident, the radionuclide of most concern was ^{131}I. Later, and over the next

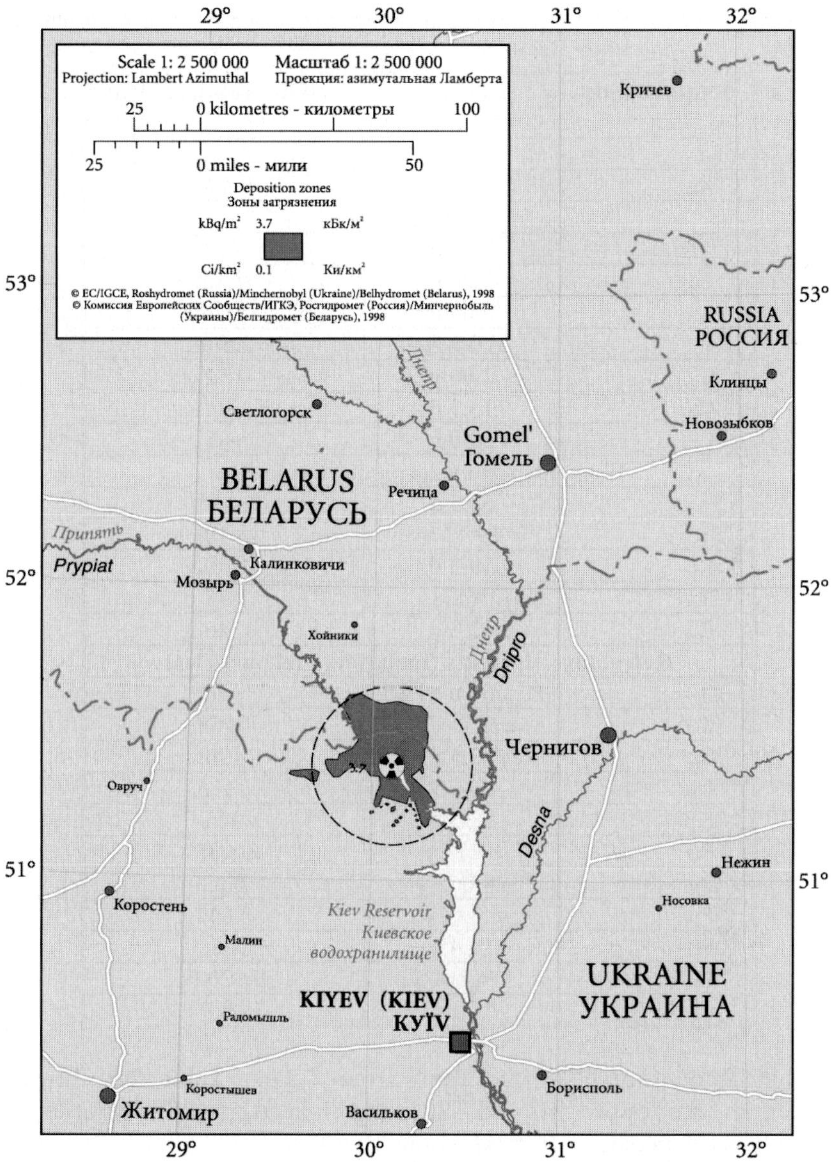

Figure 2.5. Distribution of deposited $^{239,240}Pu$ in December 1989 (from EC 2001)

two decades, the interest in radionuclide migration via the aquatic pathways was and continues to be on ^{137}Cs and ^{90}Sr. Over the long term (100s to 1000s of years) the only radionuclide anticipated to be of interest is plutonium. The only radionuclide expected to increase in the coming years is ^{241}Am. This radionuclide arises from the decay of ^{241}Pu. It takes about 100 years for the

Table 2.1. Revised estimates of principal radionuclides released during the accident (from Dreicer et al. 1996; UNSCEAR 2000)

Radionuclide	Half life	Activity released, PBq
Inert gases		
^{85}Kr	10.72 y	33
^{133}Xe	5.25 d	6,500
Volatile elements		
^{129m}Te	33.6 d	240
^{132}Te	3.26 d	~1,150
^{131}I	8.04 d	~1,760
^{133}I	20.8 h	2,500
^{134}Cs	2.06 y	~54
^{136}Cs	13.1 d	36
^{137}Cs	30.0 y	~85
Elements with intermediate volatility		
^{89}Sr	50.5 d	~115
^{90}Sr	29.12 y	~10
^{103}Ru	39.3 d	>168
^{106}Ru	368 d	>73
^{140}Ba	12.7 d	240
Refractory elements (including fuel particles)		
^{95}Zr	64.0 d	196
^{99}Mo	2.75 d	>168
^{141}Ce	32.5 d	196
^{144}Ce	284 d	~116
^{239}Np	2.35 d	945
^{238}Pu	87.74 y	0.035
^{239}Pu	24,065 y	0.03
^{240}Pu	6,537 y	0.042
^{241}Pu	14.4 y	~6
^{242}Pu	376,000 y	0.00009
^{242}Cm	18.1 y	~0.9

maximum amount of ^{241}Am to form from ^{241}Pu. Table 2.2 shows the estimated inventories of ^{137}Cs and ^{90}Sr in the main rivers and tributaries of the Dnieper River watershed. In total, about 19.6 PBq (530 kCi) of ^{137}Cs and 2.3 PBq (62 kCi) of ^{90}Sr were deposited within the Dnieper River system.

The extent to which the source term contributes to contamination of water bodies depends not only on the initial inventory but also on factors affecting the dynamics and availability of each radionuclide, including:

- The extent of radioactive decay
- Local and climatic factors affecting water runoff and soil erosion

Table 2.2. Estimated inventories (1986) of ^{137}Cs and ^{90}Sr in soils of Dnieper River Basin upstream of Kiev (from Voitsekhovich et al. 1994)

River	**Basin, km²**		**Inventory, PBq**	
	Total	**Activity** >37 kBq/m²	^{137}**Cs**	^{90}**Sr**
Upper Dnieper, (upstream of Kiev Reservoir)	105,000	29,000	10.2	0.22
Pripyat, mouth	115,000	27,000	6.7	1.5
Braginka and area between Dnieper and Pripyat Rivers	2,000	2,000	2.1	0.44
Desna	89,000	61,000	0.29	0.037

- Reduction of radionuclide activity in the upper layer of the soil as the result of vertical migration
- Transformation of initial physical and chemical forms of radionuclides in the soils of the watersheds and floodplains.

In the first hours to weeks after the accident, fallout of short-lived radionuclides (especially ^{131}I) dominated the dose rate. After the first weeks and for a few years, ^{106}Ru, ^{144}Ce, and ^{134}Cs were important. From a few years to a few hundred years, secondary transportation processes increase in importance and the dominant radionuclides are ^{137}Cs and ^{90}Sr. The dose rate is reduced by radioactive decay, so as of September 2003 the activity of ^{137}Cs and ^{90}Sr is about 33 percent less than it was in 1986. After a few hundred years, most of the ^{137}Cs and ^{90}Sr will have decayed, and other radionuclides such as ^{241}Am and ^{239}Pu will become more important, though their overall impact will be much less.

Because radionuclides are transferred into surface runoff mainly from the upper layer of soil, the extent of contamination of waterways depends on the vertical distribution of radionuclides in the soil. Depending on the chemical properties of the radionuclide, soil type, landscape, and geochemical factors, the activity has migrated to a greater or lesser extent into the soil over the past 20 years. Thus, the inventory of radionuclides available for migration by surface transport decreases with time.

Major studies on radionuclide migration in the watersheds and water bodies are reported by Borsilov et al. (1988), Bobovnikova et al. (1991), Konoplev et al. (1990, 2002), Konoplev (1992), Bulgakov et al. (1996, 2004), Kashparov et al. (1999), and Sobotovitch et al. (2002). Their findings are discussed in Chapter 3.

2.3 Physical and Chemical Forms of Released Materials

Radionuclides were released from the stricken reactor in the form of gases, condensed particles, and fuel particles. The presence of the latter was an important characteristic of the accident. Less-oxidized fuel particles from the initial explosion were released primarily toward the west. More-oxidized and soluble particles prevailed in the fallout on other parts of the Chernobyl area. During oxidation and dispersal of the fuel, some radionuclides became volatile. After the initial cloud cooled, the more volatile radionuclides remained as gases and the less volatile condensed on particles of soot and dust. Thus the chemical and physical forms of radionuclides in the Chernobyl release were determined by volatility as well as conditions inside the reactor. Radioactive compounds with relatively high vapor pressure (primarily isotopes of inert gases and iodine) were released in the gaseous phase; isotopes of refractory elements (e.g., cerium, zirconium, niobium, plutonium) were released into the atmosphere primarily as fuel particles. Other radionuclides (e.g., isotopes of cesium, tellurium, and antimony) were found as both fuel and condensed particles. Fuel particles were the most important part of the fallout near the source. Their chemical and radionuclide composition was like that of irradiated nuclear fuel with less volatile radionuclides, a higher oxidation state of uranium, and the presence of various admixtures, especially in the surface layer. In contrast, the chemical and radionuclide composition of condensed particles varied widely. The specific activity of radionuclides in these particles was determined by the length of the condensation process and the process temperature as well as particle characteristics. The activity of some of the particles was contributed almost entirely by one or two nuclides such as $^{103,106}Ru$ or $^{140}Ba/^{140}La$ (Sandals et al. 1993).

The form of a radionuclide in the release determined the distance of its atmospheric transport. Even the smallest fuel particles consisting of a single grain of nuclear fuel were relatively large (5 to 10 μm). Because of their size, they traveled only tens of kilometers. Larger aggregates of particles were found only several kilometers from the plant. The deposition of refractory radionuclides strongly decreased with distance from the reactor, and only traces of refractory elements could be found outside the immediate vicinity of the power plant. In contrast, significant deposition of gaseous radionuclides and submicron condensed particles took place thousands of kilometers from Chernobyl (Borovoi et al. 1998).

Another important characteristic of fallout is its solubility in aqueous solutions, which determines the mobility and bioavailability of radionuclides in soils and surface waters during the initial period after deposition. There were many investigations of dissolution of particles separated from soils near the Chernobyl reactor (e.g., Konoplev et al. 1992; Kashparov et al. 1999).

The chemical and physical forces that bind radionuclides to the soil determine the extent of their transfer to water. The nature of the binding depends on both the characteristics of the radionuclide and the properties of the soil. The traditional method of experimentally determining the type of binding is sequential extraction, where the water-soluble form is extracted from the soil using distilled water. The exchangeable form is then extracted using a concentrated solution of an electrolyte, usually 1 M CH_3COONH_4. The portion remaining in the solid phase after extraction is considered to be nonexchangeable (Konoplev et al. 1990, 2002; Osipov et al. 1996; Bulgakov et al. 1996b, 2003; Sobotovitch et al. 2003).

In fallout sampled at the Chernobyl meteorological station from April 26 to May 5, 1986, the sum of water-soluble and exchangeable forms of ^{137}Cs varied from 5 to more than 30 percent (Bobovnikova et al. 1991). The water-soluble and exchangeable forms of ^{90}Sr on April 26 accounted for about 1 percent, increasing to 5–10 percent in the following days. The low solubility of ^{137}Cs and ^{90}Sr indicates that fuel particles were the major part of fallout even 20 km from the source. Closer in, the smaller proportion of water-soluble and exchangeable forms of ^{137}Cs and ^{90}Sr was due to the presence of larger particles; at longer distances the fraction of soluble condensed particles increased. Table 2.3 shows data for sequential extraction of ^{137}Cs and ^{90}Sr from soils in the Gomel and Bryansk region determined by the Scientific Production Association, Typhoon, of Russia. The data show that, independent of soil type, the exchangeable fraction of ^{137}Cs is much less than that of ^{90}Sr. Data on the form of ^{137}Cs in soils of the CEZ are presented in Table 2.4.

Migration of ^{137}Cs and ^{90}Sr can be understood by appreciating the chemistry of their respective elements, cesium and strontium. Cesium has an affinity for clay minerals that frequently occur in natural soils. Binding of cesium onto soil retards its lateral and vertical migration. Strontium is less firmly bound onto minerals and is consequently more mobile in the environment. The soils of the CEZ are heavily contaminated with ^{90}Sr; some washes off during floods when low-lying areas become inundated.

Absorption of radionuclide fallout into watershed soils and river and lake sediments plays an important role in determining subsequent transport in aquatic systems. The fraction of a radionuclide that is absorbed by suspended particles in surface waters strongly influences both its transport and bio-accumulation. In fresh waters, the fraction of a radionuclide that is attached to the solid phase varies considerably. In general, almost all ^{90}Sr is found in the dissolved phase, with only 0.05 to 5 percent in the solid phase. In the Chernobyl near zone, however, a significant proportion of strontium fallout was in the form of fuel particles.

Table 2.3. ^{137}Cs and ^{90}Sr content in soils in the Bryansk region in the mid-1990s (from IAEA-UNDP 2003)

Soil	Soil Type[(a)]	Water Extract		Extract 1 M CH_3COONH_4	
		^{137}Cs	^{90}Sr	^{137}Cs	^{90}Sr
Sod-podsol[(b)]	1	0.2±0.2	6.1±1.4	2.8±0.7	67.1±2.8
Sandy[(b)]	2	0.2±0.1	1.9±0.2	2.6±0.4	51.4±4.2
Loamy sand[(b)]	2	0.5±0.1	5.2±3.1	8.3±0.2	69.0±0.5
Sod-podsol loamy sand[(c)]	3	0.2±0.1	1.7±0.8	4.2±0.2	71.9±3.4
Sod-podsol[(b)]	1	0.5±0.5	8.3±6.7	23.2±2.4	57.3±9.3
On moraine[(b)]	2	0.7±0.1	1.7±0.1	25.6±1.4	57.4±3.8
Sod-podsol[(b)]	1	0.4±0.3	5.8±7.6	7.2±1.4	67.3±9.6
On sand[(b)]	2	0.5	1.1	23.3	64.4
Sod-podsol gley loamy sand[(d)]	1	1.4±0.6	1.9±1.2	16.1±2.0	60.5±6.9
Grey forest soil[(b)]	1	0.8	1.9	10.3	58.7
Grey forest soils[(b)]	1	0.9	1.5	3.2	42.9
Marsh low-lying[(b)]	1	0.3±0.1	2.2±1.7	1.0±0.2	40.8±3.8
Marsh low-lying humus-peaty[(d)]	3	0.1±0.02	1.3±0.5	1.9±0.7	53.2±7.9
Boggy soils[(b)]	1	7.4±1.2	5.6±3.8	16.7±3.0	57.5±6.6
Peaty- podsol-gley[(e)]	4			9.4[(f)]	
Peaty- sod- podsol[(e)]	4			18.1[(f)]	
Humus-peaty[(e)]	4			24[(f)]	
Alluvial sod acid loamy sand[(c)]	1	0.1±0.05	2.8±1.1	0.8±0.2	68.8±7.8

(a) 1, non-arable soil; 2, agricultural; 3, arable lands; 4, forest.
(b) Osipov 1996.
(c) Konoplev et al. 2002.
(d) Bulgakov et al. 1996b.
(e) Konoplev 1999.
(f) Sum of water-soluble and exchangeable forms.

Table 2.4. Exchangeable forms of ^{137}Cs in CEZ (from Sobotovitch et al. 2003)

Type of soil	Sector, Distance from Chernobyl	Percent Exchangeable Form
Automorphic	Northern, 2–15 km	6–15
Automorphic	Northern, 15–50 km	15–30
Hydromorphic	Northern, 2–15 km	2–9
Hydromorphic	Northern, 15–50 km	2–28
Peat-bog soils	Northern, 15–50 km	6–9
Podzol-sandy soils	Northern, 3–4 km	2–13
Peat-podzol soils	Western, 3–4 km	1–10
Podzol-sandy soils	Western, 4–5 km	3–6

In the Pripyat River in the first decade after the accident, 40 to 60 percent of radiocesium was found in the particulate phase (Voitsekhovich et al. 1997), but estimates in other systems vary from season to season, depending on the composition and concentration of suspended particles and water chemistry. Fine clay and silt particles absorb more radioactivity than larger, less-reactive sand particles. Sandy river beds, even close to the reactor, were relatively uncontaminated, but fine particles could transport radioactivity relatively long distances. Settling of fine particles in the deep parts of the Kiev Reservoir, for example, led to high levels of contamination of bed sediments (Voitsekhovich 1991; Kanivets 1997).

Analysis of these data allows a better understanding of why, ten or more years after the accident, different rates of release exist depending on the specific radionuclide, soil type, and region. For instance, even when the most contaminated lowland of the CEZ was inundated during the highest flood of 1999, the flux of ^{137}Cs into the Pripyat River was much less than ^{90}Sr. At present the percentage of mobile cesium in the upper contact layer of the 30-km zone soils does not exceed 2 or 3 percent. The same data and the nature of ^{90}Sr distribution (Figure 2.4) demonstrate that the only significant source of ^{90}Sr washoff at present is the flood plain territories and waters crossing the near zone of the Chernobyl plant.

2.4 Radionuclides in Surface Waters

Initial concentrations of radioactivity in rivers in parts of Ukraine, Russia, and Belarus were relatively high from direct fallout onto the river surfaces and washoff of contamination from the watershed. During the first few weeks after the accident, however, concentrations in river waters declined rapidly because the short-lived isotopes physically decayed and radionuclide deposits absorbed into watershed soils and bottom sediments. In the longer term, relatively long-lived ^{137}Cs and ^{90}Sr formed the major component of contamination of aquatic ecosystems. Though long-term levels of these isotopes in rivers were low, temporary increases during flooding of the Pripyat River caused serious concern in areas using water from the Dnieper reservoirs. During high flood periods the heavily contaminated flood plains and river bays were the main sources of radioactive contamination in the Pripyat River (Figure 2.6).

Small amounts of radionuclides migrate from the soil to rivers, lakes, and eventually the marine system through erosion of surface particles and washoff in the dissolved phase. Estimates of rates of ^{137}Cs removal from catchments affected by Chernobyl fallout show that losses of ^{137}Cs from the watershed per

Figure 2.6. During the spring floods the heavily contaminated water bodies in the CEZ become significant sources of Pripyat River secondary contamination (April 1999)

year (values normalized to annual runoff) ranged from 1 to $5x10^{-5}$ mm^{-1} (Borsilov et al. 1988; Bulgakov et al. 1990; Smith et al. 1999; Konoplev et al. 2002). Thus, in the long term, washoff does not reduce the amount in the terrestrial system significantly, though it continues (at a low level) to contaminate rivers and lakes.

Lakes and reservoirs were contaminated by fallout on the water surface and transfer of radionuclides from the surrounding contaminated lands. Radioactivity concentrations in water declined relatively rapidly in reservoirs and in lakes with significant inflows and outflows ("open" lake systems). In some cases, however, levels of cesium in lakes remained relatively high due to runoff of radioactivity from organic soils in the watershed. In addition, internal cycling of cesium in "closed" lake systems (little inflow and outflow) led to much higher concentrations in the water and aquatic biota than were seen in open lakes and rivers.

2.4.1 Radioactivity in the Most Affected Rivers

In all water bodies, the maximum radioactivity concentrations in water were observed during and shortly after the accident (Table 2.5). Levels declined rapidly during the first few weeks. Initial contamination was primarily due to deposition of radionuclides directly on the water surface and washoff from watersheds. Since then, relatively long-lived ^{90}Sr and ^{137}Cs retained in watershed soils are slowly transferred to river water by erosion of soil particles and (in the dissolved phase) desorption from soils. The rates of transfer are influenced by the extent of soil erosion, the strength of radionuclide binding to soils, and migration down the soil profile. An example time series of ^{90}Sr and ^{137}Cs observation in the Pripyat River near Chernobyl is shown in Figure 2.7.

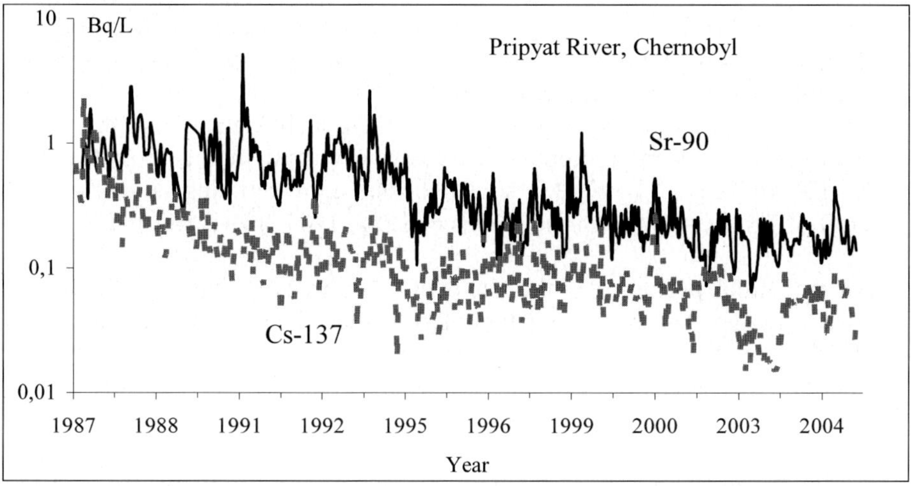

Figure 2.7. ^{90}Sr and ^{137}Cs in the Pripyat River (dissolved phase); monthly averaged data

Initial concentrations of radioactivity in the Pripyat, Teterev, Irpen, and Dnieper rivers were largely due to direct fallout on the river surface. The highest concentrations were observed in the Pripyat River at Chernobyl, where ^{131}I activity was up to 4440 Bq/L (Table 2.5).

Since the Chernobyl accident, there have been no long high water periods or flooding in the contaminated areas. The spring flow of the Pripyat River, with discharges of 800 to 2200 m^3/s, has not exceeded normal conditions; the

Table 2.5. Maximum radionuclide levels (dissolved phase) measured in the Pripyat River at Chernobyl Site (from Vakulovsky et al. 1990; Vakulovsky et al. 1994; Kryshev 1995)

Radionuclide	**Max. Concentration in Pripyat River (Bq/L)**	**Radionuclide**	**Max. Concentration in Pripyat River (Bq/L)**
^{137}Cs	1591	^{106}Ru	271(a)
^{134}Cs	827(b)	^{144}Ce	380
^{131}I	4440	^{141}Ce	400
^{90}Sr	30	^{95}Zr	1554
^{140}Ba	1400	^{95}Nb	420
^{99}Mo	670	^{241}Pu	33(c)
^{103}Ru	814	$^{239+240}Pu$	0.4

(a) From ^{103}Ru measurement assuming $^{103}Ru/^{106}Ru$ ratio (~3) for fallout.
(b) From ^{137}Cs measurement and $^{134}Cs/^{137}Cs$ ratio ~ 0.52.
(c) From $^{239,240}Pu$ measurement and $^{241}Pu/^{239,240}Pu$ ratio (~82) for fallout.

maximum discharge can exceed 5000 m^3/s (as occurred in 1979). After flooding in 1988 and 1991, it became clear that, unless mitigative actions were implemented in the contaminated area, the floodplain would remain a potential hazard well into the future. For this reason, remedial actions focused on preventing further significant movement of radionuclides from the flood plain. Major flooding took place in 1994 and the spring of 1999, when maximum discharge reached about 3000 m^3/s. In 1999, the heavily contaminated terraces of the flood plain near the Chernobyl plant were inundated for about two weeks and radionuclide releases occurred, mainly from the right bank (Figure 2.6). Figure 2.7 shows peaks corresponding to substantial washoff of ^{90}Sr during the spring floods of 1988, 1991, 1994, and 1999.

In the initial period after the accident, ^{90}Sr washoff from the far watersheds was higher than from the Chernobyl Nuclear Plant accident zone. This can be explained by the relatively low mobility of ^{90}Sr in fuel particles during that period. Subsequently, as fuel particles disintegrated, the mobility of ^{90}Sr increased. In contrast, ^{137}Cs quickly became fixed onto the soil minerals (especially clays). Over time, the relative contribution of ^{90}Sr increased, and it is now the most important radionuclide in terms of contamination of the Dnieper River system. Figure 2.8 shows the ratio of dissolved ^{90}Sr to dissolved ^{137}Cs as a function of time. The floods in 1991, 1994 and 1999 are clearly discernible as sharp peaks when the flood plains became inundated (Voitsekhovich et al. 2001).

A significant reduction in ^{90}Sr release from the CEZ is expected after completion of flood control (water protective) measures on the southwest (right) bank of the river, drying up the cooling pond, and completion of runoff

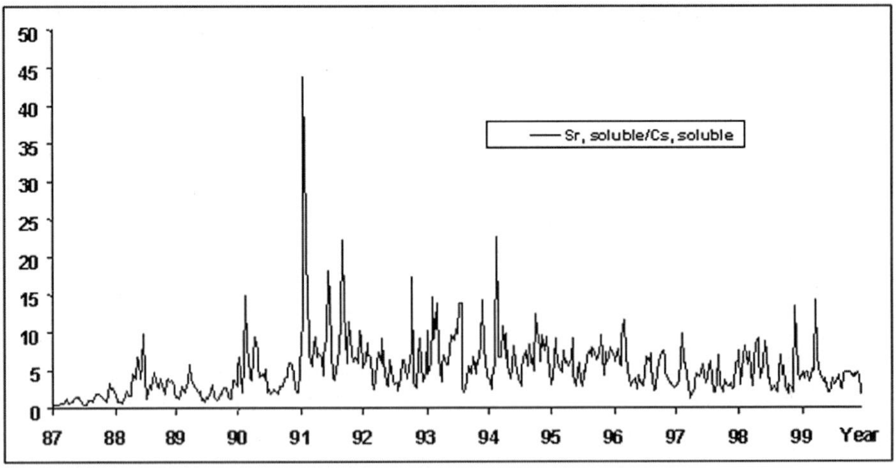

Figure 2.8. Relative contribution of ^{90}Sr and ^{137}Cs in the Pripyat River at Chernobyl Town

control on the northeast (left)-bank polder. In 2001, 10 km of canals on the east bank polder systems (for reclamation) were dredged, and flow gates and drainages were restored and repaired. This allowed better control of runoff from the polder system.

After the Chernobyl accident, water monitoring stations were established in the CEZ and along major rivers to determine the concentration and total flux of radionuclides. Measurements from these stations allow estimates of ^{90}Sr and ^{137}Cs flux into and out of the CEZ. The migration of ^{137}Cs has decreased markedly with time and changes little from upstream to downstream of the CEZ (Figure 2.9). On the other hand, the migration of ^{90}Sr has fluctuated from year to year depending on the extent of flooding along the banks of the Pripyat River (lower chart in Figure 2.9). For ^{90}Sr, there is also a significant flux from the CEZ. Fluxes downstream (observation point near

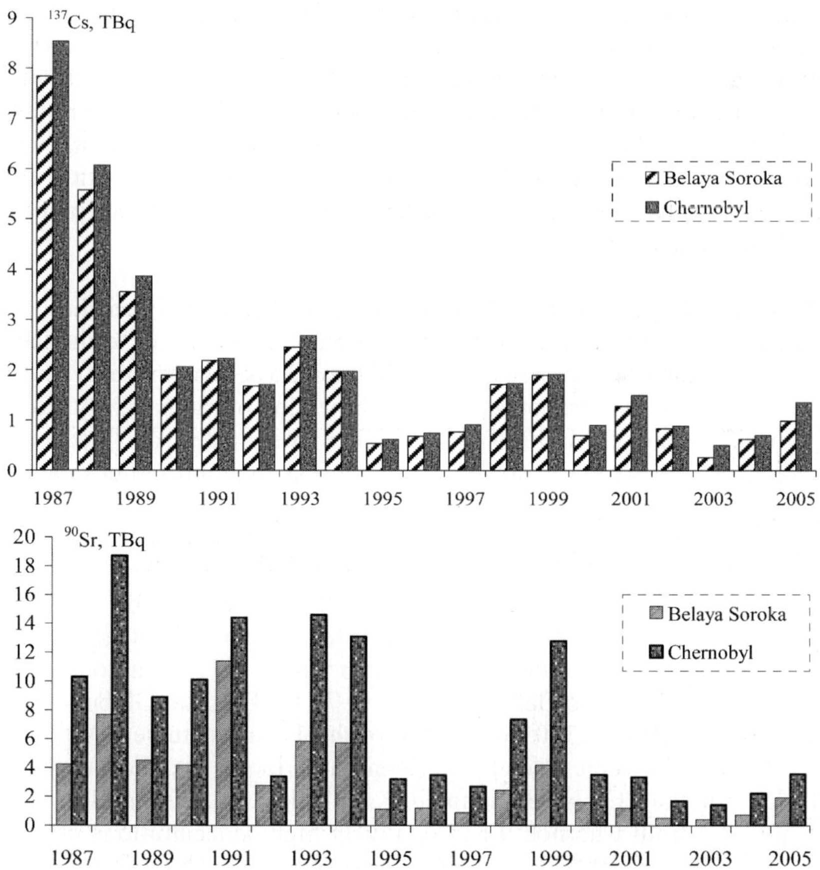

Figure 2.9. Annual fluxes of ^{137}Cs and ^{90}Sr in the Pripyat River at Belaya Soroka, near the Belarus-Ukraine border (inlet to CEZ) and downstream of Chernobyl (outlet of CEZ)

Chernobyl) are much higher than those upstream (Belaya Soroka and Dovlyadi). The extent of washout of radionuclides by the river system, however, is only a small percentage of the total inventory in the watershed.

The monitoring data shown in Figure 2.9, mainly collected by the Chernobyl Ecocenter and the Ukrainian Hydrometeorological Institute, were used to develop and modify remedial activities after the initial contamination and to report on the self-restoration abilities of the natural landscape in terms of the watershed and floodplain in the CEZ and as a source term for the secondary contamination due to runoff and radionuclide wash off process (IAEA-UNDP 2003).

2.4.2 Radioactivity in Lakes and Reservoirs

In the affected areas of Ukraine, Russia, and Belarus, many lakes were extensively contaminated by fallout. In most, radionuclides were well mixed in the water during the first days and weeks after the fallout (Kudelsky et al. 2002; Smith et al. 2000). However, according to Santschi et al. (1990) and Zibold et al. (2002), deep lakes like Lake Zurich (143 m deep) or Bodesee Lake in Germany (about 200 m deep) took several months for full vertical mixing. In some areas of Northern Europe, lakes were covered with ice at the time of the accident, so maximum concentrations in the water were observed after the ice melted.

Radionuclides deposited in lakes or reservoirs are removed through the outflow and transferred to bed sediments. Radionuclide inputs to the lake are from contaminated watershed soils. Studies have shown that long-term contamination can also be caused by remobilization of radioactivity from bed sediments (Comans et al. 1989). Thus, like rivers, lakes declined in ^{137}Cs concentrations fairly rapidly during the first weeks to months after fallout.

In some shallow lakes where there are no significant surface inflows or outflows, the bed sediments play a major role in controlling radionuclide concentrations. These are called "closed" lakes (Vakulovsky et al. 1994; Bulgakov et al. 2002). The most contaminated waters in the Chernobyl area are the closed lakes of the Pripyat floodplain within the 30-km zone. During 1991, ^{137}Cs levels in these lakes were up to 74 Bq/L (Lake Glubokoye), and ^{90}Sr activity was 100 to 370 Bq/L in six of the 17 lakes studied (Vakulovsky et al. 1994). In the years since the accident, high levels of radioactivity in the closed lakes in the CEZ have remained (Kuzmenko et al. 2001). Some closed lakes quite far from Chernobyl had relatively high concentrations of cesium; for example, during 1996, Lakes Kozhanovskoe and Svyatoe in the Bryansk region of Russia (approximately 400 km from Chernobyl), contained 0.6 to 1.5 Bq/L of ^{90}Sr and 10–20 Bq/L of ^{137}Cs (Figure 2.10).

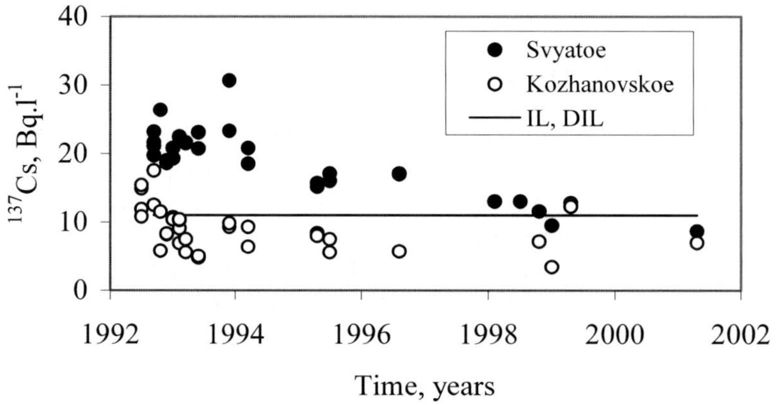

Figure 2.10. Dynamics of ^{137}Cs concentration in the water of Lakes Svyatoe and Kozhanovskoe (Russia), approximately 400 km from Chernobyl (from Bulgakov et al. 2001)

Activity concentrations in water were higher than in most lakes near Chernobyl because of remobilization from sediments in these "closed" lakes. The Russian intervention level (IL) and the derived intervention level (DIL) of Basic Safety Standard (IAEA 1996) of 11 Bq/L for ^{137}Cs are shown for comparison in Figure 2.10 (IAEA-UNDP 2003).

The Chernobyl cooling pond covers an area of approximately 22 km^2 and contains approximately 145 million m^3 of water. It is situated between the Chernobyl nuclear plant and the Pripyat River. The total inventory of radionuclides in the pond is estimated to be in excess of 200 TBq (about 80% is ^{137}Cs, 10% ^{90}Sr, 10% ^{241}Pu, and less than 0.5% each of ^{238}Pu, ^{239}Pu, ^{240}Pu, and ^{241}Am) with the deep sediments containing most of the radioactivity. The ^{90}Sr annual flux from the cooling pond to the Pripyat River via groundwater was estimated in a recent study (Bucley et al. 2002) as 0.37 TBq. This is a factor of 10–30 less than the total annual ^{90}Sr fluxes via the Pripyat River during recent years. Thus, the cooling pond is not a significant source of ^{90}Sr contamination for the Pripyat. Radioactivity concentrations in cooling pond water (Figure 2.11) are at present relatively low, about 1–2 Bq/L. Seasonal variation of ^{137}Cs in the water masses of the cooling pond is an algae and phytoplankton biomass seasonal dynamic (Nasvit et al. 2002).

2.4.3 Radioactivity in Reservoirs of the Dnieper Cascade

The reservoirs in the Dnieper River were significantly affected by both atmospheric fallout and input from the contaminated zone (see Figures 2.2 through 2.5). In the long term, radioactivity in the Dnieper reservoirs was dominated by ^{137}Cs and ^{90}Sr. In spite of low mixing and dilution, radionuclide

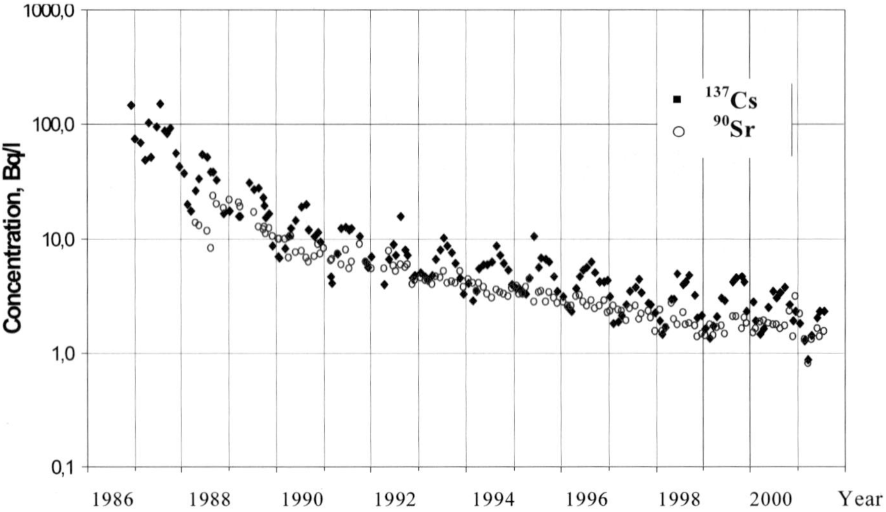

Figure 2.11. ^{137}Cs and ^{90}Sr in the water of the Chernobyl cooling pond (from Nasvit 2002)

concentrations in Kiev and downstream reservoirs declined rather quickly (Figure 2.12), and since the end of the initial post-accident period the contamination level has not exceeded maximum permissible levels for safe water use (Radiation Safety Standard of Ukraine 1997).

The different affinities of these radionuclides for suspended matter influenced their transport through the Dnieper system. Cesium-137 tends to become fixed onto clay sediments that are deposited in the deeper sections of the reservoirs, particularly in Kiev Reservoir (Figure 2.13). Because of this

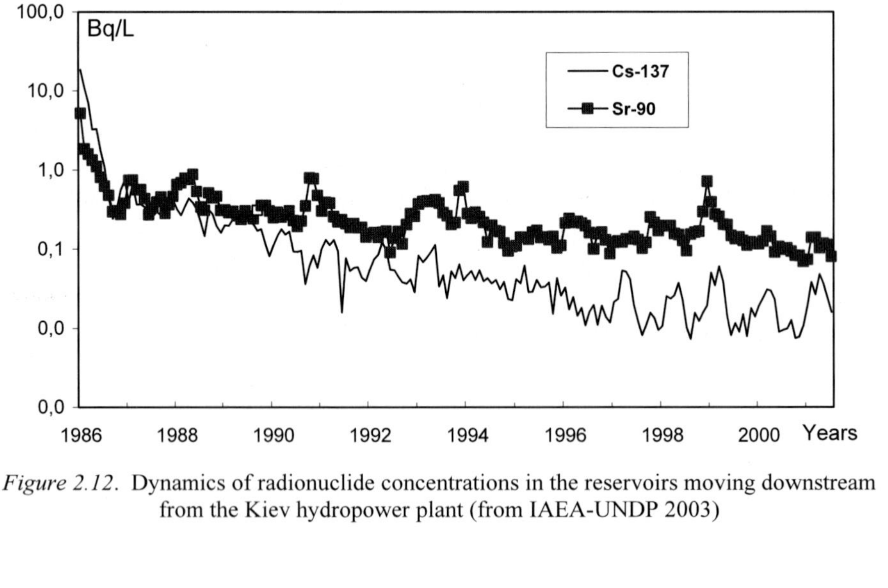

Figure 2.12. Dynamics of radionuclide concentrations in the reservoirs moving downstream from the Kiev hydropower plant (from IAEA-UNDP 2003)

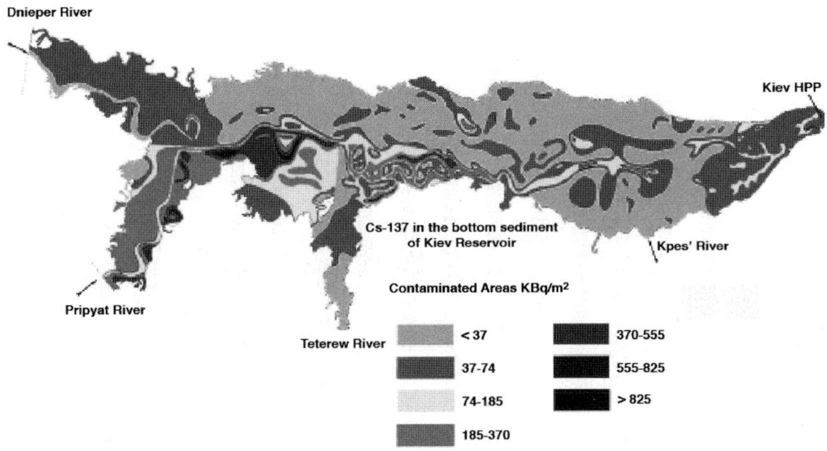

Figure 2.13. ^{137}Cs in bottom sediment of Kiev Reservoir (from Kanivets 1996; Voitsekhovich et al. 2001)

process, very little ^{137}Cs flows through the cascade of reservoirs, and the present concentration entering the Black Sea is indistinguishable from background. On the other hand, although ^{90}Sr concentration decreases with distance from the source (mainly due to dilution), about 40 to 60% passes through the cascade and reaches the Black Sea. Figure 2.14 depicts the trend in average annual ^{90}Sr concentration in the Dnieper reservoirs since the accident. Figure 2.13 shows ^{137}Cs trapped by sediments in the reservoir system, so activity in the lower part of the system is orders of magnitude less than in Kiev Reservoir. Sr-90 is not strongly bound by sediments, so concentrations in the lower part of the reservoir system are similar to those measured in Kiev Reservoir.

The peaks of ^{90}Sr activity in the reservoirs of the Dnieper cascade (Figure 2.14) were caused by flooding of the most contaminated floodplains in the Chernobyl Exclusion Zone. For instance, flooding of the Pripyat caused by ice dams in the river in the winter of 1991 led to temporary but significant increases in ^{90}Sr concentrations but did not affect ^{137}Cs concentrations (Voitsekhovich et al. 1994). Concentrations of ^{90}Sr increased from about 1 to 8 Bq/L for a 5–10 day period during the winter of 1991 (Vakulovsky et al. 1994). Similar events took place during the summer rainfall in July 1993, the winter flood of 1994 (Voitsekhovich et al. 1997), and the high spring flood in 1999 (Voitsekhovich et al. 2001).

In the Dnieper Basin are more than 1.8 million hectares of irrigated agricultural land, almost 72% of which is irrigated with water from the Kakhovka and other Dnieper reservoirs. Radionuclides can accumulate in plants through

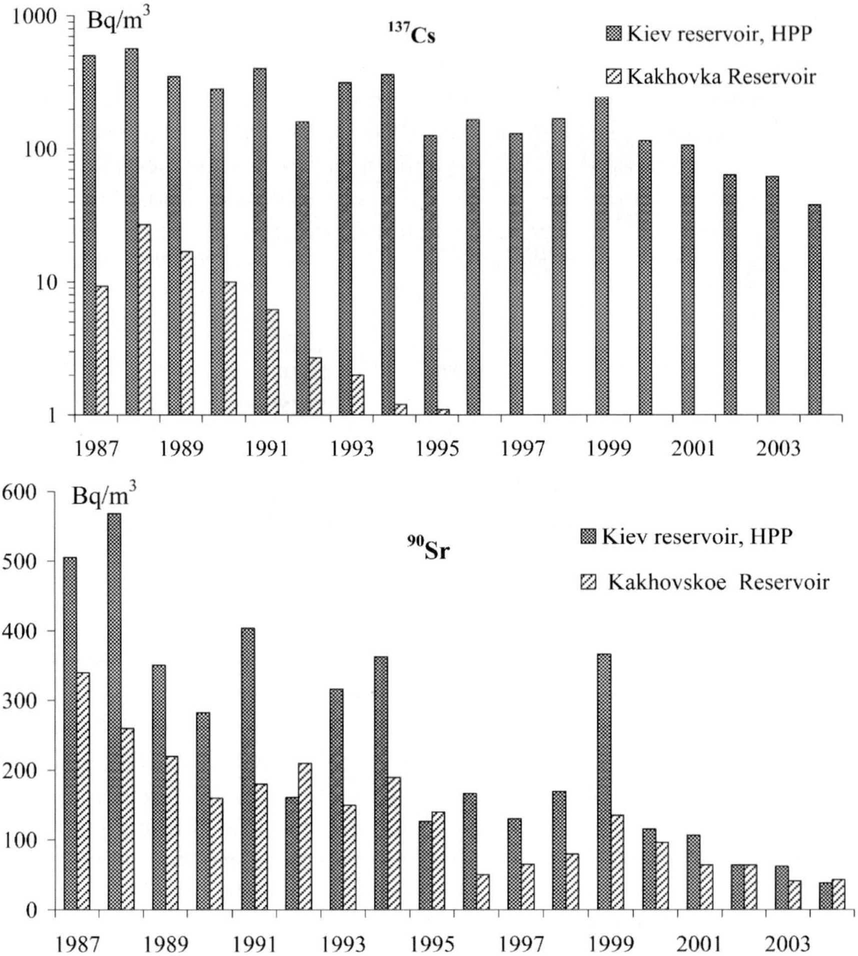

Figure 2.14. Annually averaged content of ^{137}Cs and ^{90}Sr in Kiev (Vishgorod) (near the dam) and Kakhovka reservoirs of the Dnieper cascade

root uptake of contaminated irrigation water and direct incorporation through leaves due to sprinkling. However, recent studies (COMETES 2002) show that in southern Ukraine, irrigation water did not add significant radioactivity to crops, compared with that initially deposited by fallout and subsequently taken up from the soil.

2.4.4 Radioactivity in Fresh Water Sediments

In spite of high levels of fallout onto the water, the bottom sediments of the streams and rivers self-cleaned soon after the initial contamination. During spring floods, the average velocity in the Pripyat and Dnieper rivers can exceed 1 m/s. The ripple and dune mechanisms of bottom sediment transport are typical for the Pripyat, Dnieper, and other rivers near Chernobyl. The turbulence and other processes have resuspended relatively small clay and silt particles with adsorbed radioactivity into the flow, which delivers them to the Kiev Reservoir and other downstream areas with low velocity where they are deposited. This explains why the bottom sediment of the rivers remains uncontaminated even a short distance from the reactor. Information about radioactive contamination in the bottom sediments of water bodies near Chernobyl is given in Nosov et al. (1989), Voitsekhovich et al. (1991, 1997), Kanivets et al. (2000), and Bulgakov et al. (2003).

The bottom sediment of floodplain lakes and old river channels in the CEZ, even a fairly long distance from Chernobyl, received significant radioactive fallout. However, in contrast to the rivers, the bed sediments of lakes and reservoirs are an important long-term sink for radionuclides. Radionuclides attach to suspended particles that settle into the sediment at the bottom of the lake. Radioactivity also diffuses into bed sediments (see Figure 2.13). Radionuclides can be removed from lake water by self-cleaning (Santschi et al. 1990).

In the Chernobyl cooling pond about one month after the accident, most of the radioactivity was found in the bed sediments (Kanivets et al. 2002). If there is no significant secondary influx of radionuclides from the terrestrial ecosystem into the lakes, approximately 95 to 99.9 percent of the ^{137}Cs and 90 percent of the ^{90}Sr will be found in the bed sediments (Sansone et al. 1996).

In the rapidly accumulating sediments of Kiev Reservoir, the layer of maximum radioactivity is now buried several tens of centimeters below the sediment surface (Figure 2.15). In sediments with slower accumulation, however, the peak cesium activity remains near the sediment surface. Peaks in sediment layer contamination in 1988 and 1993 reflect the consequences of high summer rainfall, floods, and soil erosion in the Pripyat River Basin.

Closer to Chernobyl, a high proportion of radioactivity was deposited as fuel particles. Radionuclides deposited as fuel particles are generally less mobile than those deposited in dissolved form. In Glubokoye Lake in 1993, most fuel particles remained in the top 5 cm of sediment (Sansone et al. 1996; Smith et al. 2002). Konoplev et al. (1992, 1996) and Kashparov et al. (2001) observed that fuel particles broke down slower in lake sediments than in soils. Studies in the cooling pond have shown that the half-life of fuel particles in sediments is approximately 35 years. In other words, it is estimated that by 2056 (70 years after the accident) one-fourth of the radioactivity deposited as fuel particles in the cooling pond will remain in fuel particle form.

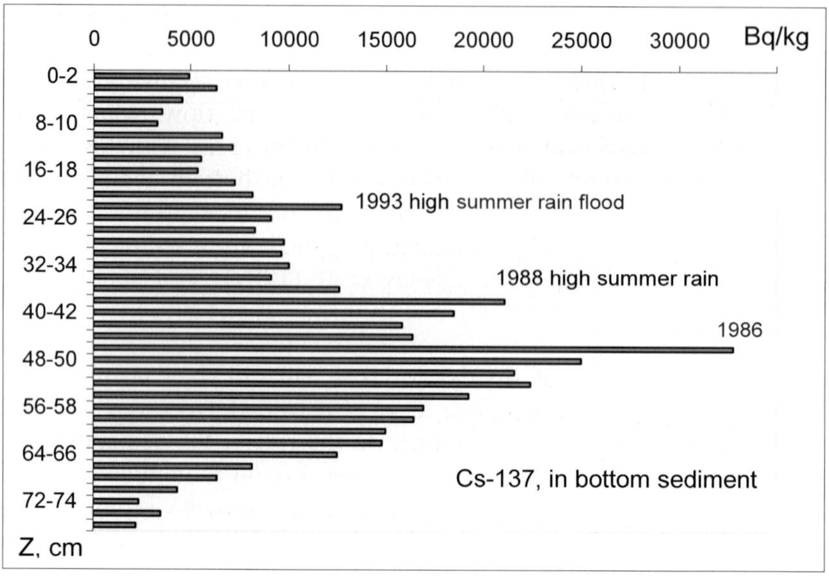

Figure 2.15. ^{137}Cs in the deep silt deposits in the upper part of Kiev Reservoir, 1998

Natural attenuation processes such as radionuclide fixation on the solid particles and sedimentation in the reservoir helps the ecosystem recover from radiological contamination. In spite of the large amount of radionuclides that accumulated in the Kiev and other reservoirs due to annual deposition and radionuclide vertical migration, recent core measurements revealed that the upper layer of the bottom sediment has a relatively low level of radioactive contamination. Benthic organisms, zooplankton and others, which are an important food chain component for the fish in most reservoirs of the Dnieper cascade, were shown to have rather low levels of radioactive contamination in 2000–2002 (TRANSAQUA 2003).

2.4.5 Radionuclide Transport Modeling in Pripyat and Dnieper Rivers

A series of mathematical modeling has been performed to assess aquatic environmental impacts and possible remediation in the Pripyat and Dnieper rivers (Zheleznyak and Voitsekhovich 1990, Zheleznyak et al. 1992). As a part of this effort, the CHARIMA (Holly et al. 1990) and TODAM (Onishi et al. 1983) codes were applied to these rivers from the Chernobyl nuclear plant to the Kakhovka Dam (see Figure 1.1). CHARIMA is a one-dimensional code but can handle networks of rivers connected in multiple ways. CHARIMA was used to predict river velocity and changes in water depth affected by the six dams on the Dnieper River. TODAM is an unsteady, one-dimensional, finite element code that applies to multiple networks of rivers and estuaries. It simulates:

- migration (transport, deposition, resuspension) of sand, silt, and clay
- dissolved contaminant transport with adsorption/desorption of both suspended and bottom sediments and chemical/biological degradation and radionuclide decay
- particulate contaminant migration and contaminant adsorption and desorption by both suspended and bottom sediments of sand, silt, and clay fractions
- bed elevation change and bed sediment size distributions
- particulate contaminant concentration distributions associated with sand, silt, and clay within the bed.

TODAM simulated migration of (1) sand, silt, clay, (2) dissolved radionuclides of ^{90}Sr and ^{137}Cs, and (3) sand-, silt-, and clay-sorbed radionuclides in both the water and the bed of the Pripyat and Dnieper rivers between the Chernobyl plant and Kakhovka Dam. Figure 2.16 shows good agreement between predicted and measured ^{90}Sr concentrations at the ends of Kiev and Kanev reservoirs. Like the Pripyat River, the Dnieper reservoirs registered the highest contamination levels in the first few days after the accident.

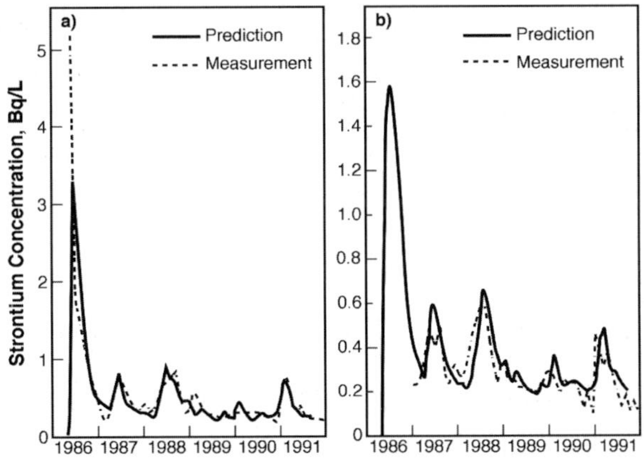

Figure 2.16. Predicted and measured ^{90}Sr concentrations at (a) Kiev and (b) Kanev reservoirs

2.5 Radioactivity in Marine Systems

Marine systems were not seriously affected by fallout from Chernobyl. The seas nearest the reactor are the Black (about 520 km) and the Baltic (about 750 km). The primary path of contamination of these seas was atmospheric fallout, with smaller influx from river transport in the years after the accident. Surface deposition of ^{137}Cs on the Black Sea was estimated to be 1.8 to 2.0 PBq by Vakulovsky et al. (1994) and 2.8 PBq by Eremeev et al. (1995). The estimated ^{137}Cs deposition over the Baltic Sea was about 3 PBq (Vakulovsky et al. 1994). The average ^{137}Cs activity in the Baltic Sea in the initial period after deposition was 50 to 60 Bq/m^3, with the maximum concentration two to four times greater. In addition to ^{137}Cs, short-lived radionuclides such as ^{144}Ce and ^{106}Ru were observed in both the Black and Baltic Seas (Aarkrog 1988; Polikarpov et al. 1991).

Radioactive fallout onto the surface of the Black Sea was not uniform and occurred mainly during early May (Makhonko et al. 1987; Eremeev et al. 1993; Vakulovsky et al. 1994). On the surface, concentrations of ^{137}Cs ranged from 15 to 503 Bq/m^3 in June and July of 1986, but by 1989 horizontal mixing of surface waters resulted in relatively uniform concentrations of 41 to 78 Bq/m^3 (Vakulovsky et al. 1994). By 2000 the levels declined to between 20 and 35 Bq/m^3 (IAEA 2003). In addition to ^{137}Cs, short-lived radionuclides such as ^{144}Ce and ^{106}Ru were observed (Aarkrog 1988; Polikarpov et al. 1991). The inventory of ^{137}Cs in the water of the Black Sea due to Chernobyl deposition doubled the existing inventory from atmospheric nuclear weapons tests (Vakulovsky et al. 1994) to approximately 85,000 Ci (3.1×10^{15} Bq). The amount of ^{90}Sr increased 19 percent over the pre-Chernobyl period and was estimated to be about 1760 TBq (Vakulovsky et al. 1994; IAEA 2003).

Vertical mixing of surface-deposited radioactivity reduced the maximum observed over the months and years after fallout. Removal of radioactivity to deeper waters steadily reduced ^{137}Cs activity in the surface layer (0 to 50 m) of the Black Sea (Figure 2.17). However, nearly 20 years after the initial contamination, ^{137}Cs from Chernobyl remains in the upper 200–250 m of the sea (IAEA 2004). The main mechanism of radionuclide transport from the surface layers to the bottom layers is sedimentation in pellets and other particulates through the complex geochemical transformation occurring in the sea water column.

The radionuclide influx to the Black Sea from rivers was much less significant than direct atmospheric fallout on the surface. Between 1986 and 2000, influx for ^{137}Cs was only 3 to 5 percent and influx for ^{90}Sr was approximately 15 to 25 percent of atmospheric deposition (Kanivets et al. 1999; IAEA-2003).

In the Baltic Sea, riverine inputs were at a level similar to atmospheric fallout for both ^{137}Cs and ^{90}Sr (Nies and Nielsen 1996). The greater relative riverine input of ^{90}Sr is due to its weaker adsorption to watershed soil and lake

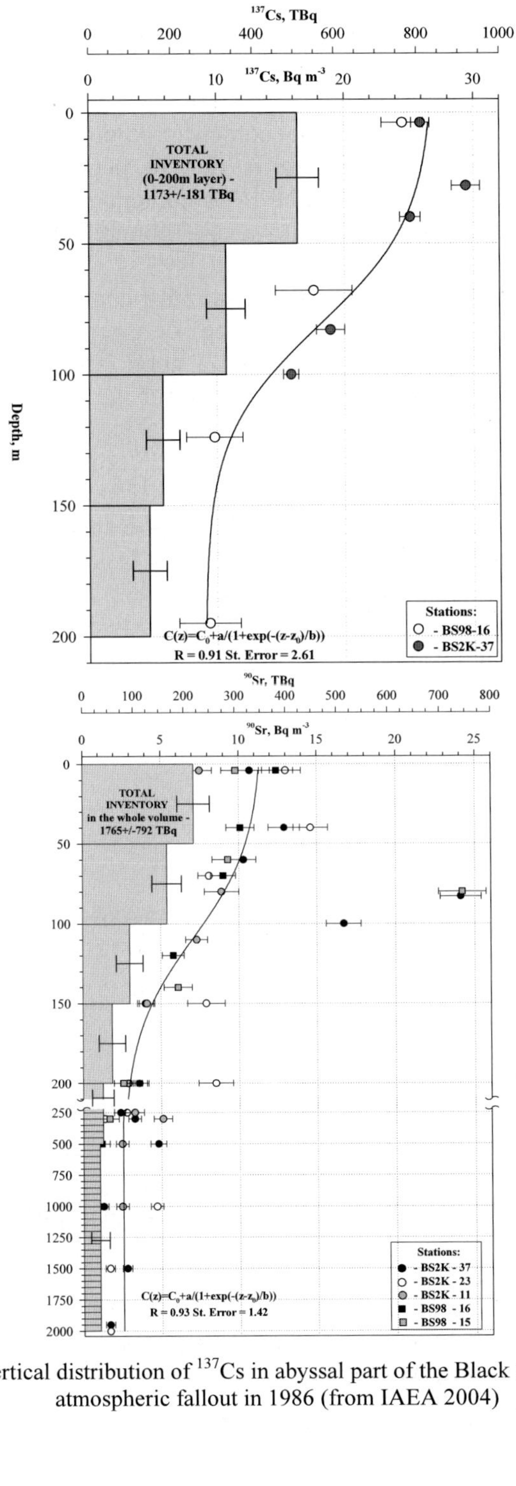

Figure 2.17. Vertical distribution of ^{137}Cs in abyssal part of the Black Sea since Chernobyl atmospheric fallout in 1986 (from IAEA 2004)

and river sediments and to lower ^{90}Sr atmospheric fallout longer distances from the reactor. Like the fresh water environment, sedimentation processes are important in self-cleaning the marine environment.

The relatively high radionuclide content traditionally observed in the northwestern part of the Black Sea is due to accumulation of riverine sediment from the Danube River. The ^{137}Cs content in the upper part of the sediment on the shelf ranged from 40 to several hundred Bq/kg^2 (IAEA 2004).

Shortly after the initial fallout, the maximum contamination in the deep abyssal sediment was 1,000 to 2,000 Bq/kg^2 or more. The relatively high ^{137}Cs content in these layers is considered a good isotope marker for sedimentation rate studies. The sedimentation rate for the Black Sea is relatively low, estimated at about 40 to 60 g-m^2/y for the abyssal sediment and one to two orders of magnitude higher for the shelf area (Voitsekhovich et al. 2002).

The data in Figure 2.18 demonstrate that in the central deep basin of the Black Sea the Chernobyl deposition is covered by less than 1 cm of sediment since the accident (Voitsekhovich et al. 2002, IAEA 2004). Due to dilution and sedimentation, the concentration of ^{137}Cs quickly declined by two to four times that observed in the summer of 1986.

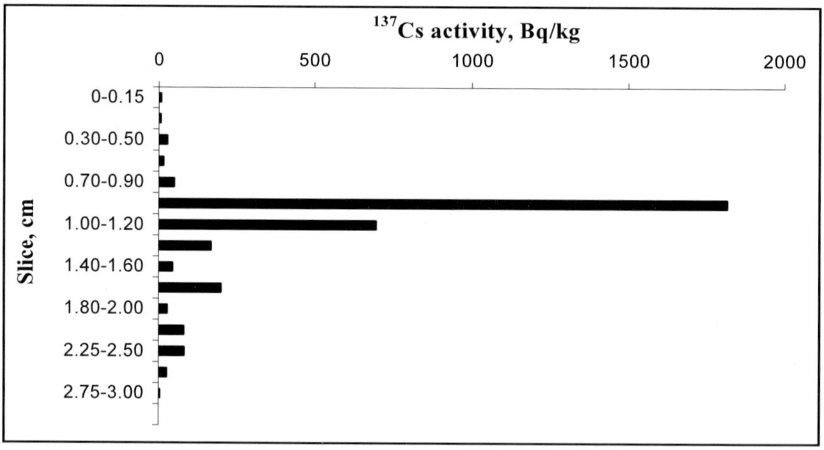

Figure 2.18. ^{137}Cs profile in bottom sediment (Core BS-23/2000) taken during an IAEA Black Sea expedition in 2000 (from IAEA 2004)

According to a comprehensive study by researchers from Ukraine, Romania, Turkey, and others, no significant radioecological impacts to the marine biota have materialized due to radionuclides from the Chernobyl accident (IAEA 2004).

2.6 Radionuclides in Groundwater in the CEZ

Sampling of groundwater in the affected areas showed that radionuclides were transferred from surface soils to groundwater. However, the level of groundwater contamination in most areas from radioactive waste storage and the Chernobyl industrial site is very low. The rate of migration from the soil to groundwater is very low as well. Some relatively fast radionuclide migration was found in the aquifers in areas with morphological depressions (Shestopalov et al. 2002). Horizontal fluxes of radionuclides in groundwater are low because of the slow flow of groundwater and high retardation of radionuclides (Bugai et al. 1998).

The only significant transfer of radionuclides to groundwater occurs in the CEZ. In some wells during the last 10 years, ^{137}Cs concentration has declined while ^{90}Sr has increased in shallow groundwater (Figure 2.19). Radionuclides were transferred to groundwater from radioactive waste disposal sites in the exclusion zone. After the accident, fuel-containing masses and radioactive debris were temporarily stored at the plant and in areas near the Pripyat River floodplain. In addition, trees from the Red Forest were buried in shallow, unlined trenches. At these disposal sites, ^{90}Sr concentrations in groundwater are as high as 1000 Bq/L (Voitsekhovich et al. 1996).

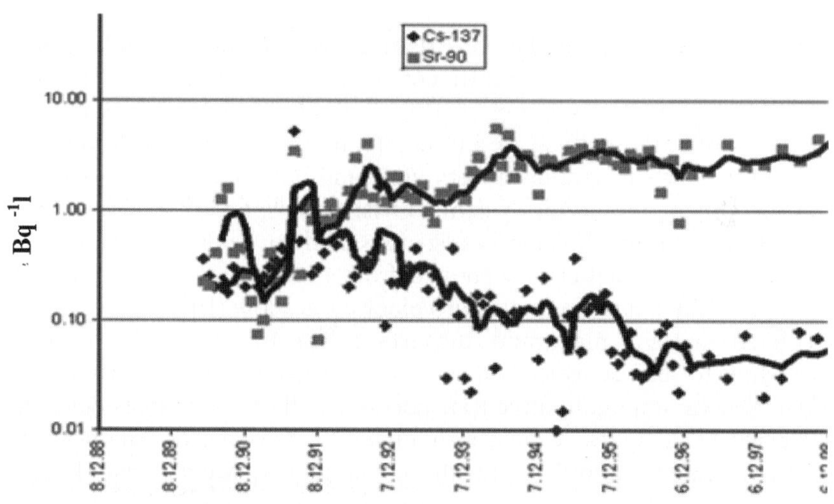

Figure 2.19. ^{137}Cs and ^{90}Sr in shallow groundwater at the Red Forest area near the Chernobyl industrial site (data from CEZ Ecological Centre)

The most contaminated groundwater is in the area near the Chernobyl Shelter. Radionuclide contamination of the groundwater at the Shelter is much higher than other areas and varies from place to place within the site. In most recent studies, the primary contamination sources of groundwater are (1) precipitation that has accumulated inside the Shelter's underground rooms, (2) groundwater that has accumulated near the Pioneer Wall, where there is no drainage system, and (3) other water infiltrating from the nuclear power plant site and (4) radionuclides being carried into the aquifer.

In some places, ^{137}Cs in groundwater near the Shelter reaches 100 Bq/L and even 3–5x10^3 Bq/L. However, in most areas around the site, ^{137}Cs concentrations in groundwater range from 1 to 10Bq/L (Bugai et al. 1998; Shestopalov et al. 2002).

Typical amounts of ^{90}Sr in groundwater around the Shelter range from 2 to 160 Bq/L. In the last five years, the maximum concentrations in groundwater range from 1–3x10^3 Bq/L. Concentrations of transuranic elements in this area vary widely from 3x10^{-3} Bq/L to 3– Bq/L for ^{238}Pu and $^{239+240}Pu$ and 1x10^{-3} to 8–10 Bq/L for ^{241}Am.

Groundwater contamination around the main waste disposal sites is essentially constant. The groundwater is moving toward the river at rates of about 3 to 30 m/yr; accounting for the retardation capacity of the surrounding soils, the corresponding transport of radionuclides is estimated to be 10 to 100 times slower than the groundwater velocity.

Although there is potential for offsite transfer of radionuclides from the disposal sites, it will not be significant compared with washout of surface-deposited radioactivity (Bugai et al. 1996). Studies have shown that groundwater fluxes of radionuclides are in the direction of the Pripyat River, but the rate of radionuclide migration is very slow and does not present a significant risk to the Dnieper reservoir system. Offsite transport of groundwater contamination around the Shelter is also expected to be insignificant because radioactivity in the Shelter is separated from groundwater by an unsaturated zone 5 to 6 m thick and groundwater velocities are low (Bugai et al. 1996). It is predicted that the maximum subsurface ^{90}Sr transport rate from waste disposal sites to surface water bodies will occur from 33 to 145 years after the accident. Maximum cumulative transport from all of the sources described is estimated to be 130 GBq/yr in approximately 100 years, or 0.02 percent per year of the total inventory within the contaminated watersheds. Integrated radionuclide transport for a 300-year period is estimated as 15 TBq, 3 percent of the total initial inventory of radioactivity in the watersheds.

More detailed analyses of groundwater contamination of the Chernobyl area and model predictions are presented in Chapter 8.

2.7 References

Aarkrog A. 1998. "A retrospect of anthropogenic radioactivity in the global marine environment." *Radiation Protection Dosimetry*, Vol. 75, pp. 23-31.

Bechtel, Battelle, and EDF. 2003. *Environmental Impact Assessment for NSC.* Bechtel Inc., San Francisco.

Belyaev ST, VF Demin, VA Kutkov, VG Bariakhtar, and EP Petriaev. March 1996. "Characteristics of the Development of the Radiological Situation Resulting from the Accident, Intervention Levels and Countermeasures." *The radiological consequences of the Chernobyl accident, proceedings of the first international conference,* A Karaoglou et al. (Eds.). EUR 16544 EN, Luxemburg, pp. 19-28.

Bobovnikova TI, KP Makhonko, AA Siverina, FA Rabotnova, and AA Volokitin. 1991. "Physical chemical forms of radionuclides in atmospheric fallout after Chernobyl accident and their transformation in soil." *Atomnaya Energiya*, Vol. 5, pp. 449–454 (in Russian).

Borovoy AA et al. 1998. *The Shelter current safety analysis and situation development forecasts: Final Report.* TACIS, European Commission.

Borsilov VA, AV Konoplev, CK Revina, TI Bobovnikova, PM Liutik, YV Shveikin, and AV Sherbak. 1988. "Experimental study of the radionuclides washed out as result of Chernobyl Accident." *Meteorology and Hydrology,* Vol. 11, pp. 43-53 (in Russian).

Brittain JE, A Storruste, and E Larsen. 1991. "Radiocaesium in Brown Trout (*Salmo trutta*) from a subalpine lake ecosystem after the Chernobyl reactor accident." *J. Environmental Radioactivity,* Vol. 14, pp. 181-191.

Bucley MJ, D Bugai, L Dutton, M Gerchikov, V Kashparov, A Ledenev, O Voitsekhovich, D Weiss, and M Zheleznyak. 2002. *Drawing up and Evaluating Remediation Strategies for the Chernobyl Cooling Pond.* C6476/TR/001/2002, Final Report, NNC Ltd.

Bugai DA, RD Waters, SP Dzhepo, and AS Skalskiy. 1996. "Risks from radionuclide migration to groundwater in the Chernobyl 30-km zone." *Health Physics,* Vol. 71, pp. 9-18.

Bugai DA, SP Dzhepo, and AS Skalskiy. 1998. "Prognoses of geomigration, risk assessment and water protection principles to be developed for mitigation of groundwater radioactive contamination at the CEZ." *Radiogeoecology of the water bodies at the Chernobyl affected areas,* Chapter 6, Vol. 2, pp. 218-252.

Bulgakov AA, AV Konoplev, VE Popov, and AV Stcherbak. 1990. "Dynamic of the long-lived radionuclide wash-off from soils at the area around Chernobyl NPP." *Pochvovedenie,* Vol. 4, pp. 47-54 (in Russian).

Bulgakov AA and AV Konoplev. 1996. "Basic transformation processes of chemical species in soil and bottom sediments." *Modeling and study of the mechanisms of the transfer of radioactive material from terrestrial ecosystems to and in water bodies around Chernobyl,* U Sansone and O Voitsekhovich (Eds.). CEC-CIC Joint Programme on the Consequences of the Chernobyl Accident, Experimental Collaboration Project No 3, Final Report. Luxemburg, pp. 122-135.

Bulgakov AA, AV Konoplev, VV Kanivets, and OV Voitsekhovich. 2001. "Modeling the long-term dynamics of radionuclides in rivers." *ECORAD-2001*, Aix-de-Provence, 2:649-654.

Bulgakov AA, AV Konoplev, JT Smith, J Hilton, RNJ Comans, GV Laptev, and BF Christyuk. 2002. "Modeling the long-term dynamics of radiocaesium in a closed lake system." *J. Environmental Radioactivity,* Vol. 61, pp. 41-53.

Bulgakov AA, AV Konoplev, EV Samokhvalova, and GV Laptev. 2004. "Parameters of radiocesium sorption by bottom sediments in Kozhanovskoe and Svyatoe lakes in the Bryansk region of Russian Federation." *Meteorology and Hydrology,* Vol. 2, pp. 64-71 (in Russian).

Comans RNJ and DE Hockley. 1992. "Kinetics of caesium sorption on illite." *Geochimica et Cosmochimica Acta,* Vol. 56, pp. 1157-1164.

COMETES. 2002. *Implementing computerised methodologies to evaluate the effectiveness of countermeasures for restoring radionuclide contaminated fresh water ecosystems,* L Monte (Ed.). Project Report ERB IC15-CT98-203, COMETES. Italy.

Dreicer M, A Aarkrog, R Alexakhih, L Anspauh, NP Arkhipov, and KJ Jonsson. 1996. "Consequences of the Chernobyl accident for the natural and human environments." *Proc. International Conference on One Decade after Chernobyl: Summing up the Consequences of the Accident,* International Atomic Energy Agency, Vienna, pp. 319-361.

Eremeev VN, TV Chudinovskih, and GF Batrakov. 1993. "Artificial radioactivity of the Black Sea." *UNESCO Marine Science Series*, Vol. 59, p. 95.

Eremeev VN, LM Ivanov, AD Kirwan, and TM Margolina. 1995. "Amount of ^{137}Cs and ^{134}Cs radionuclides in the Black Sea produced by the Chernobyl accident." *Journal of Environmental Radioactivity*, Vol. 27, pp. 49-63.

Elliott JM, J Hilton, E Rigg, PA Tullett, DJ Swift, DP Leonard. 1992. "Sources of variation in post-Chernobyl radiocaesium in fish from two Cumbrian lakes (northwest England)." *J Applied Ecology,* Vol. 29, pp. 108-119.

Fleishaman DG, VA Nikiforov, AA Saulus, and VT Kotov. 1994. "^{137}Cs in fish of some lakes and rivers of the Bryansk region and Northwest Russia in 1990-1992." *J. Environmental Radioactivity*, Vol. 24, pp. 145-158.

Hadderingh RH, GM Van Aerssen, IN Ryabov, OA Koulikov, and N Belova. 1997. "Contamination of fish with ^{137}Cs in Kiev reservoir and old river bed of Pripyat near Chernobyl." *Freshwater and Estuarine Radioecology,* G Desmet, R Blust, RNJ Comans, J Fernandez, J Hilton, and A de Bettencourt (Eds.). Elsevier Studies in Environmental Science, Amsterdam, Vol. 68, pp. 339-351.

Håkanson L. 1991. "Radioactive caesium in fish in Swedish lakes after Chernobyl–Geographical distributions, trends, models, and remedial measures." *The Chernobyl Fallout in Sweden–Results from a Research Programme on Environmental Radiology,* L Moberg (Ed.). Swedish Radiation Protection Institute, pp. 239-281

Håkanson L and T Andersson. 1992. "Remedial measures against radioactive caesium in Swedish lake fish after Chernobyl. *Aquatic Sciences,* Vol. 54, pp. 141-164.

Hansen HJM and A Aarkrog. 1990. "A different surface geology in Denmark, the Faroe Islands, and Greenland influences the radiological contamination of drinking water." *Water Research,* Vol. 24, pp. 1137-1141.

Holly FM Jr, JC Yong, P Schwarz, J Schaefer, SH Hsu, and R Einhelig. 1990. "Numerical Simulation of Unsteady Water and Sediment Movement in Multiply Connected Networks of Mobile-Bed Channels." Report No. 343, Iowa Institute of Hydraulic Research, University of Iowa, Iowa City.

IAEA. 2004. "Marine Environmental Assessment of the Black Sea. Working Material." Final Report, Regional Technical Co. Operation Project RER/2/003, IAEA, Vienna.

Izrael YA, SN Vakulovskoy, VA Vetrov, VN Petrov, FY Rovinskiy, and ED Stukin. 1990. "Chernobyl: Radioactive contamination of the environment." *Hydrometeoizdat*, pp. 356.

Jonssons KJ. 1996. "Consequence of the Chernobyl accident for the natural and human environments." *Proc. International Conference on One Decade after Chernobyl: Summing up the Consequences of the Accident.*" IAEA, Vienna, pp. 319–361.

Kanivets VV. 1996. "Analysis of main tendencies in radiation state development at the Dnieper's aquatic system." *Visnik agrarian science,* Vol. 4, pp. 40-56 (in Russian).

Kanivets VV, OV Voitsekhovich, VG Simov, and ZA Golubeva. 1999. "The post-Chernobyl amount of ^{137}Cs and ^{90}Sr in the Black Sea. *J. Env. Research*, Vol. 43, pp. 121-135.

Kanivets VV and OV Voitsekhovich. 2000. "Radioactive contamination of the bottom sediment of the Chernobyl cooling pond." *Proc of UHMIH,* Vol. 248, pp. 154-171, Kiev, Ukraine (in Russian).

Kashparov VA, DH Oughton, VP Protsak, SI Zvarisch, and SE Levchuk. 1999. "Kinetics of fuel particle weathering and ^{90}Sr mobility in the Chernobyl 30 km exclusion zone." *Health Physics,* Vol. 76, pp. 251-259.

Kashparov VA, SM Lundin, and YV Khomutinin. 2001. "Soil contamination with ^{90}Sr in the near zone of the Chernobyl accident." *J. Env. Radioactivity,* Vol. 56, pp. 285-298.

Konoplev AV and TI Bobovnikova. 1990. "Comparative analysis of chemical forms of long-lived radionuclides and their migration and transformation in the environment following the Kyshtym and Chernobyl accidents." *Proc. of Seminar on Comparative Assessment of the Environmental Impact of Radionuclides Released during Three Major Nuclear Accidents: Kyshtym, Windscale, Chernobyl.* CEC, Luxemburg, pp. 371-395.

Konoplev AV. 1992. "Behaviour of long-lived radionuclides in a soil-water system." *Analyst,* Vol. 117, pp. 1041-1047.

Konoplev AV, R Comans, J Hilton, J Madruga, A Bulgakov, O Voitsekhovich. and J Smith. March 1996. "Phisico-Chemical and hydraulic mechanisms of radionuclides mobilisation in aquatic systems." *The radiological consequences of Chernobyl accident. Proceedings of the first international conference,* A Karaoglou et al. (Eds.). European Commission, Luxemburg, pp. 137-146.

Konoplev AV, AA Bulgakov, VG Zhirnov, et al. 1998. "Study of the behaviour of ^{137}Cs and ^{90}Sr in Svyatoe and Kozhanovskoe lakes in the Bryansk region." *Meteorology and Hydrology,* Vol. 11 (in Russian).

Konoplev AV, G Deville-Cavelin, OV Voitsekhovich, et al. 2002. "Transfer of Chernobyl ^{137}Cs and ^{90}Sr by surface runoff." *Proc. of ECORAD-2001*, Aix-de-Provence, Vol. 1, pp. 315-318.

Koulikov AO and IN Ryabov. 1992. "Specific caesium activity in freshwater fish and the size effect." *Science of the Total Environment*, Vol. 112, pp. 125-142.

Kryshev II and TG Sazykina. 1994. "Accumulation factors and biogeochemical aspects of migration of radionuclides in aquatic ecosystems in the areas impacted by the Chernobyl accident." *Radiochimica Acta,* Vol. 66/67, pp. 381-384.

Kryshev II. 1995. "Radioactive contamination of aquatic ecosystems following the Chernobyl accident." *J. Environmental Radioactivity,* Vol. 27, pp. 207-219.

Kudelsky AV, JT Smith, SV Ovsiannikova, and J Hilton. 1996. "Mobility of Chernobyl–derived ^{137}Cs in a peatbog system within the catchment of the Pripyat River, Belarus." *Science of the Total Environment*, Vol. 188, pp. 101-113.

Kudelsky AV, JT Smith, and AA Petrovich. 2002. "An experiment to test the addition of potassium to a non-draining lake as a countermeasure to ^{137}Cs accumulation in fish." *Radioprotection – Colloques,* Vol. 37, pp. 621-626.

Kuzmenko MI, VD Romanenko, VV Derevets, OM Volkova, et al. 2001. *Impact of radionuclide contamination on to hydrobionts of the Chernobyl Exclusion Zone*, p. 318.

Likhtarev IA, LN Kovgan, P Jacob, and LR Anspaugh. 2002. "Chernobyl accident: Retrospective and prospective estimates of external dose of the population of Ukraine." *Health Phys.*, Vol. 82, pp. 290–303.

Nasvit OI, MI Fomovskiy, and VG Klenus. 1997. "Radionuclides in hydrobionts of the water bodies in the CEZ." *Radiogeoecology of water bodies at the Chernobyl Exclusion zone. K: Chernobylinterinform, Vol. 1. Monitoring of radioactive contamination of the natural water in Ukraine,* pp. 215-222.

Nasvit OI. 2002. "Radioecological Situation in the Cooling Pond of Chernobyl NPP." *Recent Research Activity about Chernobyl NPP Accident in Belarus, Ukraine and Russia.* KURAI-KR-79, 74-85, Kyoto University, Japan.

Nies H and SP Nielsen. 1996. "Radioactivity in the Baltic Sea." *Radionuclides in the oceans. Inputs and Inventories.* IPSN, pp. 219-231.

NRBU-97. 2001. Radiation Safety Standards of Ukraine.

Onishi Y, G Whelan, and RL Skaggs. 1983. *Development of a Multimedia Radionuclide Exposure Assessment Methodology for Low-Level Waste Management.* PNL-3370, Pacific Northwest National Laboratory, Richland, Washington.

Polikarpov GG, VI Timoschuk, LG Kulebakina, and NA Stokozov. 1991. "^{90}Sr and ^{137}Cs in surface water of the Dnieper River, Black Sea, and Aegean Sea in 1987–1988." *J. Environment Radioactivity*, Vol. 13, pp. 25-38.

Romanenko VD, MI Kuzmenko, and NY Evtushenko. 1992. "Radioactive and chemical contamination of the Dnieper and its reservoirs after Chernobyl accident." Naukova Dumka, Kiev, p. 316.

Ryabov IN. 1992. "Analysis of countermeasures to prevent intake of radionuclides via consumption of fish from the region affected by the Chernobyl accident." *Proc. Int. Seminar on*

Intervention Levels and Countermeasures for Nuclear Accidents. EUR 14469, Luxemburg: European Commission, pp. 379-390.

Ryabov I, N Belova, L Pelgunova, et al. 1996. "Radiological phenomena of the Kojanovskoe Lake." *The Radiological Consequences of the Chernobyl Accident, Proceedings of the First International Conference,* A Karaoglou, G Desmet, GN Kelly, et al. (Eds.). EUR 16544, pp. 213-216.

Sandalls FI, MG Segal, and NV Victorova. 1993. "Hot particles from Chernobyl: A review." *J. Environmental Radioactivity,* Vol. 18, pp. 5-22.

Sansone U and OV Voitsekhovich. 1996. *Modeling and study of the mechanisms of the transfer of radioactive material from terrestrial ecosystems to and in water bodies around Chernobyl.* Final Report, Experimental Collaboration Project No. 3, EUR 16529, European Commission, DG XII, Brussels-Luxemburg.

Santschi PH, S Bollhalder, S Zingg, A Luck, and K Farrenkother. 1990. "The self-cleaning capacity of surface water after radionuclide fallout. Evidence from European lakes after Chernobyl 1986-88." *Environ. Sci. Technol.*, Vol. 24, pp. 519-527.

Shestopalov VN. 2002. *Chernobyl disaster and groundwater.* Balkema Publishers, p. 289.

Smith JT, RNJ Comans, and DG Elder. 1999. "Radiocaesium removal from European lakes and reservoirs: key processes determined from 16 Chernobyl contaminated lakes." *Water Research,* Vol. 33, pp. 3762-3774.

Smith JT, RT Clarke, and R Saxén. 2000. "Time dependent behaviour of radiocaesium: a new method to compare the mobility of weapons test and Chernobyl derived fallout." *J. Environmental Radioactivity*, Vol. 49, pp. 65-83.

Smith JT, AV Kudelsky AV, IN Ryabov, and RH Hadderingh. 2000. "Radiocaesium concentration factors of Chernobyl-contaminated fish: a study of the influence of potassium, and 'blind' testing of a previously developed model." *J. Environ. Radioactivity*, 48:359-369.

Smith JT, AV Konoplev, AA Bulgakov, RNJ Comans, MA Cross, B Khristuk, A De Koning, AV Kudelsky, MJ Madruga, OV Voitsekhovich, and G Zibold. 2001. *Dorset Centre for Ecology and Hydrology; AQUASCOPE Final Report.* European Commission project IC15-CT98-0205, p. 114.

Smith JT, AV Kudelsky, IN Ryabov, SE Daire, L Boyer, RJ Blust, JA Fernandez, RH Hadderingh, and OV Voitsekhovich. 2002. "Uptake and elimination of radiocaesium in fish and the "size effect." *J. Environmental Radioactivity*, Vol. 62, pp. 145-164.

Sobotovitch EV, GN Bondarenko, LV Kononenko, et al. 2002. *Geochemistry of technogenic radionuclides.* Naukova Dumka, Kiev, p. 332 (in Russian).

TRANSAQUA. 2003. "Transfer in Aquatic Ecosystems." Final report of the French-German Initiative for Chernobyl, CD database, IRSN, Cadarach.

United Nations. 2000. *Sources and Effects of Ionizing Radiation.* Report to the General Assembly by the Committee on Effects of Atomic Radiation, New York, II:451-566.

UNDP. 2001. *Water Quality Evaluation and Prediction in the Areas Affected by the Chernobyl Accident. Bryansk Region.* Final Project Report, p. 64.

UNDP. 2003. *Radioactive contamination of the Dnieper River basin: Contribution to the transboundary diagnostic analysis and recommendation for the strategic action plan.* Summary report, Regional technical cooperation project RER/9/072, IAEA, Vienna, p. 24.

UNSCEAR. 1988. *Sources, Effects and Risks of Ionizing Radiation.* Report to the General Assembly by the UN Scientific Committee on Effects of Atomic Radiation, New York, pp. 309–374.

UNSCEAR. 2000. *Sources and Effects of Ionizing Radiation.* Report to the General Assembly, (with Scientific Annexes). UN Scientific Committee on the Effects of Atomic Radiation, Vol. II, pp. 451-566.

U.S. Department of Energy (DOE). 1987. *Health and Environmental Consequences of the Chernobyl Nuclear Power Plant Accident.* Prepared by the Interlaboratory Task Group on Health and Environmental Aspects of the Soviet Nuclear Accident for the DOE Office of Health and Environmental Research, Washington, D.C.

Vakulovskij SM, OV Voitsekhovich, IY Katrich, and VI Medinets. 1990. "Radioactive contamination of water system in the area affected by releases from Chernobyl accident." *Proc. of the IAEA, UNEP, FAO, WHO Symposium,* Vol. 1, pp. 231-246. Vienna.

Vakulovsky SM, AI Nikitin, VB Chumichev, IY Katrich, OV Voitsekhovich, VI Medinets, VV Pisarev, LA Bovkum, and ES Khersonsky. 1994. "Caesium-137 and Strontium-90 contamination of water bodies in areas affected by releases from Chernobyl Nuclear Power Plant accident: an overview." *J. Environmental Radioactivity,* Vol. 23, pp. 103-122.

Voitsekhovich OV. 1991. "Radioactive contamination of the Dnieper basin." *Nature,* Vol. 5, pp. 52-56 (in Russian).

Voitsekhovich OV, VA Borzilov, and AV Konoplev. 1991. "Hydrological Aspects of Radionuclide Migration in Water Bodies Following the Chernobyl Accident." *Proc. of Seminar on Comparative Assessment of the Environmental Impact of Radionuclides Released during Three Major Nuclear Accidents: Kyshtym, Windscale, Chernobyl*, Vol. 2. Commission of European Communities, Radiation Protection-53, EUR 13574, Luxemburg, pp. 527-548.

Voitsekhovich OV, MJ Zheleznyak, and Y Onishi. 1994. *Chernobyl nuclear accident hydrological analyses and emergency evaluation of radionuclide distribution in the Dnieper River, Ukraine during the 1993 summer flood.* PNL-9980, Pacific Northwest National Laboratory, Richland, Washington.

Voitsekhovich OV, U Sansone, MJ Zheleznyak, and DA Bugai. 1996. "Water quality management of contaminated areas and its effect on doses from aquatic pathways." *The radiological consequences of Chernobyl accident. Proceedings of the first international conference,* A Karaoglou et al. (Eds.). European Commission, Luxemburg, pp. 401-410.

Voitsekhovich OV et al. 1997. *Radiogeoecology of water bodies at the Chernobyl affected zone. Vol. 1. Monitoring of radioactive contamination of the natural water in Ukraine*, p. 297.

Voitsekhovich OV, VM Shestopalov, AS Skalskiy, and VV Kanivets. 2001. *Monitoring of surface and groundwater contamination after Chernobyl accident.* Kiev, p. 147.

Voitsekhovich OV, GV Laptev, OB Kostezh, and VV Kanivets. 2002. "Assessment of the rate of sedimentation in the Black Sea using isotope dating techniques." *Proc of UHMI*, pp. 221-231 (in Russian).

Zarubin OL and OL Zalisskiy. 2000. "Fish radioactive contamination in the Chernobyl cooling pond." *Bulletin of the ecological state in the CEZ*, Vol. 16, pp. 39-43 (in Russian).

Zarubin OL and OL Zalisskiy. 2002. "Radioactive contamination of algae and aquatic plants in the Pripyat River." *Bulletin of the ecological state in the CEZ*, Vol. 19, pp. 39-45 (in Russian).

Zheleznyak MJ and OV Voitsekhovich. 1990. "Mathematical Modeling of Radionuclide Dispersion in Surface Waters after the Chernobyl Accident to Evaluate the Effectiveness of Water Protection Measures." *Proc. Seminar on Comparative Assessment of the Environmental Impacts of Radionuclides Released During Three Major Accidents: Kyshtym, Windscale, and Chernobyl*, Vol. 2. Radiation Protection 53, EUR 13574, CEC, Luxembourg, 401-410.

Zheleznyak MJ, R Demchenko, S Khursin, Y Kuzmenko, P Tkalich, and N Vitjuk. 1992. "Mathematical modeling of radionuclide dispersion in the Pripyat-Dnieper aquatic system after the Chernobyl accident." *The Science of the Total Environment*, 112:89-114.

Zibold G, A Forschner, J Drissner, E Klement, R Miller, and M Walser. 1998. "Time-dependency of the caesium-137 concentration factor in pike from Lake Vorsee." *Proceedings of the UIR Topical Meeting*, Mol, Belgium.

Zibold G, S Kaminski, E Klemt, and JT Smith. 2002. "Time-dependency of the ^{137}Cs activity concentration in freshwater lakes, measurement and prediction." *Radioprotection – Colloques*, Vol. 37, pp. 75-80.

Chapter 3
Radioecological Aspects of Water Use

O. V. Voitsekhovich
Ukrainian Hydrometeorological Institute, Kiev, Ukraine

G. Perepelyatnikov
Ukrainian Radiological Educational Center, Kiev, Ukraine

O. Nasvit, V. Klenus, V. Belyaev, O. Volkova, L. Yurchuk
Institute of Hydrobiology, Ukrainian National Academy of Science, Kiev, Ukraine

I. Ryabov
Institute of Animal Biology/Evolutional Morphology, Russian Academy of Sciences, Moscow

This chapter examines radionuclide transfer to the human body through consumption of irrigated crops and fish. These are two important ecological food chains in the aquatic system of Dnieper reservoirs. The first part of this chapter describes radioecological studies performed on Ukrainian irrigated lands during 1986–1996. Included are the main mechanisms of migration of biologically significant radionuclides in "water-plant" and "water-soil-plant" systems. The long-term dynamics of radionuclides in the main irrigated crops are discussed, and the parameters of radionuclide transport into the plants are quantified. As part of this evaluation, contamination of rice by radionuclides in the Dnieper River irrigation water and the main controlling factors were also evaluated. Chernobyl studies conducted by various Ukrainian research institutes are synthesized and generalized to draw some conclusions. The second part of the chapter describes radionuclide transfer through fish trophic chains and radioactive contamination of the fish population. Radionuclide total distribution coefficients for fish and the effects of size on contamination are also discussed.

3.1 Radioecological Aspects of Irrigated Agriculture

G. Perepelyatnikov and O.V. Voitsekhovich

Agriculture in the southern Ukraine consists of growing produce on irrigated lands. The southern and southeastern regions of the Ukrainian steppe and the steppe part of the Crimea (an irrigated area) have the highest

Y. Onishi, O.V. Voitsekhovich and M.J. Zheleznyak (eds.), Chernobyl – What Have We Learned?, 49–85.

seasonal temperatures and the lowest air moisture. They have the lowest precipitation in Ukraine and a large number of dry, windy days. The extensive irrigation changes the microclimate there and optimizes the water regime of the upper layer of soil.

The Dnieper River, the largest river in Ukraine, supplies 64 percent of the country's water needs and is widely used for irrigation. About 30 percent of the water consumed in the Dnieper River Basin is used for agriculture (Palamarchuk 1992; Doroguntsov et al. 1992). The Dnieper irrigates about1.6 million of the 2.6 million hectares (ha) of land in Ukraine (mid-1990s). Irrigation water taken from the Dnieper is almost 4,700 million m^3 annually. These values were used for water consumption in this study.

These areas are a very important region of the country. The main irrigated lands (about 80 percent) are situated in five southern regions of Ukraine (Table 3.1). The top soil of the irrigated lands generally consists of mean and minor humus (often solonetz), ordinary chernozem (black earth), and dark chestnut solonetz soils. The composition of the typical soil varies from light to heavy loam.

The largest areas of irrigated lands are used for growing grain (rice, winter and spring wheat, and barley) and fodder (corn, peas, mangel beet, clover, and alfalfa). Vegetables (potatoes, tomatoes, cucumbers, cabbage, onions, carrots, red beets, vegetable marrow) and fruits are grown less in the irrigated areas (Table 3.1).

The Chernobyl nuclear power plant accident deposited a large amount of radionuclides in the Dnieper River and on its watershed. In addition, radionuclides have washed from the catchments into the river (runoff). Ten years after the 1986 accident, no significant reduction in ^{90}Sr contamination was observed in the Dnieper reservoirs.

Table 3.1. Irrigated land in main regions of Ukraine in 1992

Region	Amount of Irrigated Land (10^3 ha)			
	Total	Grain	Vegetable	Fodder
Crimea	330	120	20	170
Kherson	420	470	20	210
Zaporizhya	260	100	20	120
Dniepropetrovsk	140	50	8	65
Mikolayiv	130	50	8	53
Totals	1280	790	76	618

The Chernobyl accident resulted in the transfer of ^{137}Cs and ^{90}Sr from the Chernobyl plant to the Dnieper River and farther on to the irrigated lands. It is thus necessary to quantitatively estimate the radioactive contamination of the crops irrigated with the polluted Dnieper River water and to study the dynamics of this transfer process. It was especially important to determine long-term ^{90}Sr and ^{137}Cs concentrations in irrigated crops contaminated directly by irrigation water (especially in rice paddies) and indirectly by contaminated soil (Perepelyatnikov 1993).

The Chernobyl accident resulted in the transfer of ^{137}Cs and ^{90}Sr from the Chernobyl nuclear power plant to the Dnieper River and on to the irrigated lands. Thus, it is necessary to quantitatively estimate the radioactive contamination of the crops irrigated with the polluted Dnieper River water and to study the dynamics of this transfer process. It is especially important to determine long-term ^{90}Sr and ^{137}Cs concentrations in crops contaminated directly by irrigation water (especially in rice paddies) and indirectly by contaminated soil (Perepelyatnikov 1993).

Irrigation enhances radionuclide migration through trophic food chains more than dry land agriculture (Perepelyatnikov 1983) because irrigation provides both short (irrigated "water-terrestrial organs") and long (irrigated water-soil-root system-terrestrial organs) pathways. Irrigation also increases the mobility of radionuclides in soils and their accessibility to plants (Table 3.2). Under conditions of excessive moisture or inundation of soils, the physiological processes of mineral feeding of plants intensifies and results in assimilation of both stable and radioactive elements.

In southern Ukraine, there are two main irrigation practices: artificial rain (overhead sprinklers) and surface irrigation (furrow and flooding). Drip and under-soil irrigation is used to a small extent. During irrigation by artificial rain, most radioactive substances in the water fall onto the leaves, stems, flowers, and fruits and are directly absorbed by the plant. Thus, the sorption of radioactive substances, a key factor in soil contamination, is almost excluded from the process, which could be a barrier to radionuclide transfer to the soil.

Under furrow and free irrigation, water is supplied to the surface of the irrigated soil area, and radionuclide transfer into the plants occurs mainly via the roots. During irrigation by flooding (as on rice paddies) radionuclides enter the plants via roots that are in direct contact with the water, and thus are transferred to the plant through the roots. The type of irrigation is thus an important factor in estimating radioecological consequences of irrigation of agricultural lands from radioactively contaminated water sources.

Table 3.2. Net irrigating rates for leading agricultural crops

Culture	Irrigation rate *1000 m^3 ha^{-1}			Moisture supply *1000 m^3 ha^{-1}		
	p=50%	p=75%	p=95%	p=50%	p=75%	p=95%
Zaporozhye region						
Winter wheat	1.9	2.4	3.2	0.6	0.8	1.0
Spring wheat	1.3	1.7	2.3			
White beet	3.0	3.7	4.9			
Perennial herbs	4.0	4.7	5.6			
Corn for grain	2.4	2.9	3.6			
Stubble corn	2.4	3.0	3.8	0.4	0.6	0.8
Potatoes	1.9	2.4	3.1			
Tomato, vegetable	3.5	4.3	5.3	0.3	0.4	0.5
Nikolaev region						
Winter wheat	1.7	2.3	3.0	0.6	0.8	1.0
Spring wheat	1.1	1.6	2.0			
White beet	2.7	3.2	4.1			
Perennial herbs	3.5	4.1	5.1			
Corn for grain	1.9	2.5	3.1			
Stubble corn	2.2	2.7	3.5	0.4	0.6	0.8
Potatoes	1.6	2.1	2.6			
Tomato, vegetable	2.9	3.7	4.8	0.3	0.4	0.5

p = 50% - dry year with average of once in two years.
p = 75% - dry year with average of once in four years.
p = 95% - dry year with average of once every 20 years.

3.1.1 Vertical Migration of Radionuclides in Soils of Irrigated Lands

Data obtained by the Experimental Scientific-Research Station (SRS) of rice in the Skadovsk region provide an insight to ^{137}Cs and ^{90}Sr migration in the soil of irrigated rice paddies (Table 3.3). Measured ^{90}Sr distributions in the soil of these paddies are more vertically uniform than ^{137}Cs distributions in

Table 3.3. Vertical distribution of ^{137}Cs in soils of rice paddies in 1988 after three years of irrigation, Bq kg^{-1} of dry soil (data from Ukrainian Rice Research Station)

Layer, cm dry land	1986 1987 1988	Rice Alfalfa Fallow	Fallow Rice Alfalfa	Alfalfa Fallow Rice	Rice Fallow Rice	Rice Rice Rice
0-0.5	16.6	15.2	19.2	22.6	22.9	26.3
0.5-2	15.9	14.4	14.4	16.6	18.9	18.1
2-4	15.2	15.2	13.7	15.2	14.1	13.7
4-6	14.1	13.3	14.8	14.1	15.5	15.2
6-8	15.9	16.6	15.5	13.7	14.4	15.5
8-10	13.7	13.7	14.1	14.1	13.7	11.8
10-20	7.0	14.8	14.4	15.2	15.2	15.2

a dryland area. The upper 0.5 cm is enriched with ^{137}Cs by water for one growing period, and its concentration is 10 to 40 percent greater there than at 0.5 to 2.0 cm deep, as shown in Table 3.4. According to data obtained by the Rice Research Station in the Kherson region, ^{90}Sr distributions are more uniform because of its greater mobility. However, their data indicate that the top 0.5 cm of soil accumulated significantly more ^{90}Sr than the lower layers, as shown in Figure 3.1. But this reflects the radionuclide contamination (1-5 $Bq \cdot kg^{-1}$) due to global fallout.

Table 3.4. ^{90}Sr vertical distribution in wild solonetz soil measured in 1986-1988 ($Bq\ kg^{-1}$)

Layer, cm	1986 1987 1988	Rice Fallow Rice	Rice Rice Rice
0-0.5	2.55	4.07	4.44
0.5-2	2.41	4.07	4.07
2-4	2.22	3.71	4.51
4-20	2.15	3.71	4.25

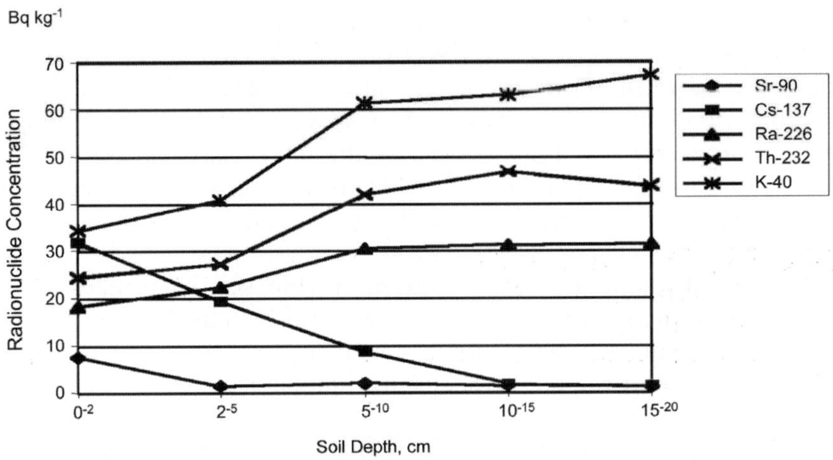

Figure 3.1. Measured vertical radionuclide distributions in wild solonetz soil

The Institute of Biology of the Southern Seas of the National Academy of Sciences of Ukraine and the Rice Research Station measured vertical distributions of radionuclides in wild sandy loamy and clay solonetz soils in southern Ukraine. The data are in good agreement with a vertical migration pattern of radionuclides in analogous soils in other areas of Ukraine. The data also indicate that the concentrations of some natural radionuclides in the upper layers of irrigated soils are reduced by being partially removed with the annual crops gathered from the irrigated lands.

The physical and chemical properties of radionuclides and their retention time in a soil layer also influence migration (see Figure 3.1). The largest amounts of ^{137}Cs and ^{90}Sr from fallout after the accident are concentrated in the top 10 cm of soil, as indicated in Table 3.5. The natural radionuclides have migrated into the deeper soils, and their highest specific activities occur in a layer 10–20 cm deep.

Table 3.5. Radionuclide vertical distributions in wild solonetz (Bq kg^{-1}) in 1994

Layer, cm	0-2	2-5	5-10	10-15	15-20
^{90}Sr	7.44	1.41	1.91	1.39	1.12
^{137}Cs	31.8	19.3	8.49	1.84	1.45
^{226}Ra	18.1	22.3	30.5	31.1	31.3
^{232}Th	24.4	27.4	42.0	46.8	43.7
^{40}K	343	408	613	629	671

Vertical distributions of ^{137}Cs and ^{90}Sr measured in 1994 in meadow chestnut, residual-solonetz surface-glysol mean-sandy loamy soil in Skadovsk of the Kherson region were nonuniform in rice paddies. The highest ^{137}Cs concentration in an arable layer of rice paddies at the end of the growing period occurred in a 1–2-cm layer and is 25–30 percent greater than those found in the upper and lower layers, as shown in Table 3.6. This shows that ^{137}Cs concentrations were reduced in irrigation water by one order of magnitude from 1987 to 1994. Thus, as the more contaminated irrigation water moves from the top 1-cm layer to the next 1-cm layer, the top layer is less contaminated.

The ^{90}Sr distribution is more uniform in the soil of paddies, but its concentration in the top 1 cm is 20–25 percent less than that in other layers of arable horizon. This is because more contaminated water was advected away to deeper layers. Plowing increases this distribution pattern.

Table 3.7 presents concentrations of ^{90}Sr and ^{137}Cs in plowed soils of both irrigated areas and drylands. These data do not show any definite distribution trends, and it is difficult to identify small differences in radionuclide concentrations in different soil layers. This may be because their distributions are not for paddy crops but for ordinary crops irrigated at 1 to 3x10^3 m^3 ha^{-1}.

Table 3.6. Radionuclide concentrations in soil of irrigated paddies in Kherson region, 1994 (Bq kg^{-1})

Layer, cm	0-1	1-2	2-5	5-10	10-20
^{90}Sr	4.7	6.0	7.0	6.1	5.6
^{137}Cs	24.8	34.4	25.9	27.4	23.3

Table 3.7. 1993 distributions of ^{137}Cs and ^{90}Sr in soils irrigated by Dnieper reservoirs (Bq kg^{-1})

Layer, cm	Dnieprodzerzhynsk Reservoir		Kakhovka Reservoir		Zaporozhye Reservoir			
	Borodaevka		Vyshetarasovka		Volosskoye		Mouth of Kil'chen River	
	Irrig	Dryland	Irrig	Dryland	Irrig	Dryland	Irrig	Dryland
				^{137}Cs				
0-5	12.0	14.0	15.0	9.0	17.0	9.0	17.0	26.0
5-20	14.0	12.0	9.5	9.0	14.0	8.0	11.0	20.0
20-35	7.7	11.0	8.0	4.7	12.0	10.0	18.0	13.0
				^{90}Sr				
0-5	0.46	0.67	0.46	0.37	0.40	0.50	0.66	0.48
5-20	0.74	0.50	0.39	0.31	0.42	0.40	0.18	0.51
20-35	0.54	0.51	0.51	0.50	0.46	0.66	0.58	0.49

Thus it is useful to evaluate radionuclide migration in irrigated fields and its effect on exposure pathways to a person using data obtained in southern Ukraine. A 10-year Ukrainian study concluded that irrigation significantly affects radionuclide distributions in the soil of rice paddies. In addition, radionuclide distributions in irrigated soils vary strongly in different periods after the accident.

An evaluation of observation data quality revealed that radionuclide distributions of irrigated vegetable or other arable crops (not rice) were not studied enough in the years immediately after the accident, and some observation data were questionable. It may be because the measuring methods of agro-ecological practice and preparations of soils and plant samples used were not adequate to evaluate low contamination levels.

3.1.2 Distribution of Radionuclides in Elements of Irrigated Network

Concentrations of ^{137}Cs in various parts of an irrigated system (main and irrigated channels, inlets and outlets of paddies, drainage channels) show an 80 percent decrease in the first year after the accident due to adsorption by soil and irrigation systems (Figure 3.2), according to the SRS of Rice and the Ukrainian Scientific Research Institute of Agricultural Radiology. The figure also shows ^{90}Sr concentrations. In 1993–1996, the amount of ^{137}Cs adsorbed by soils in paddies was reduced to 10 percent, but concentration in the water of the discharge channels increased due to washout (desorption) from soils, as indicated in Figure 3.3. Changes in ^{90}Sr concentration in the irrigation water show a similar trend, though not as pronounced as those of ^{137}Cs.

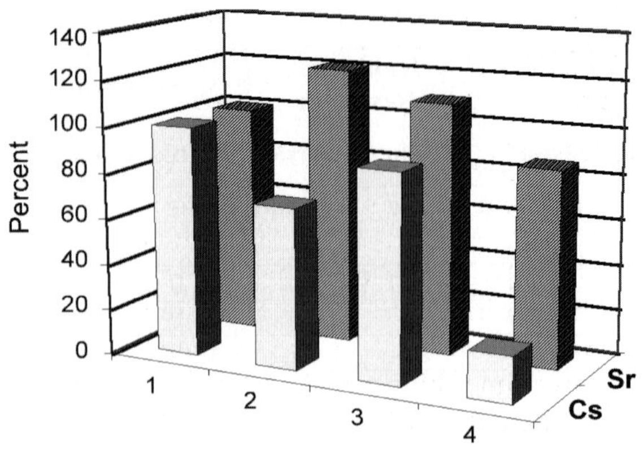

Figure 3.2. Relative radionuclide content in water of irrigation channels (1989) compared with Kakhovka Reservoir: 1 – main, 2 – irrigation, 3 – inlet, 4 –outlet

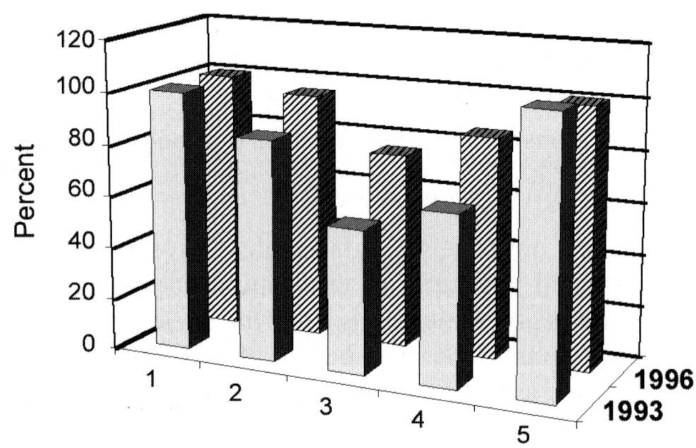

Figure 3.3. Relative ^{137}Cs content in irrigation system compared with Kakhovka Reservoir: 1 – main channel, 2 – irrigation channel, 3 – inlet 4– outlet, 5 – drainage channel

The main 161-km-long Northern-Crimea channel is mostly used for irrigation with Kakhovka Reservoir water. Along this channel, the ^{137}Cs concentration is reduced by 10 times due to sedimentation of contaminated suspended silt. In July 1995, the ^{137}Cs concentration coming from the channel into the Karkinitsky Gulf on the Black Sea was an order of magnitude less than it was near the upstream intake. The reduction is also due to deposition of some suspended silt to the channel bottom and other drainage and sorption losses confirmed by the Institute of Agroecology in 1992–1993 and the Institute of Biology of the Southern Seas of the National Academy of Sciences of Ukraine.

In addition to the sedimentation and sorption along the channel, soil and sediment wash into the drainage channel. Unlike ^{137}Cs, ^{90}Sr distributions in the irrigation system have not changed with time or channel distance.

3.1.3 Transfer of Radionuclides from Water into Agricultural Crops

During 1987 and 1988, immediately after the accident, ^{137}Cs transfer to crops was high when the source of irrigation was downstream of the CEZ (Kiev and Kanev reservoirs). The ^{137}Cs concentration in crops irrigated with Kanev Reservoir water (100 to 150 km downstream of the 30-km exclusion zone) was two to three times higher than Kakhovka Reservoir water (700–850 km downstream of Chernobyl) (Table 3.8) because contamination of irrigated crops with artificial rain (sprinklers) is directly proportional to radionuclide concentration in the water. Pollution of Kanev Reservoir was also two to three times greater than that of the Kakhovka Reservoir downstream.

As shown in Tables 3.8 and 3.9 and Figure 3.4, 1988 ^{137}Cs and ^{90}Sr concentrations in the crops from lands irrigated by Dnieper River water were five to ten times higher than those where other water sources were used (e.g., Kharkov and Donetsk regions). The influx of ^{137}Cs into crops in 1996 was not significantly different from 1988, probably because of increasing radionuclide root transport into the plants.

Table 3.8. ^{137}Cs Concentration in crops irrigated by Dnieper reservoirs (Bq kg^{-1} air-dry mass)

Agriculture		**Kanev Res.** 1987 1988	**Kremen-chug Res.**	**Dnieprod-zerjinsk Res.**	**Kakhovka Res.**	**Water sources(a)**	
						Kharkov Region	**Donetsk Region**
Winter wheat	grain	1.85	1.85	0.92	1.11	0.29	0.37
		1.11	1.48	0.37	1.11	0.37	0.37
Corn	grain	0.37	0.37	0.18	0.22	0.07	0.07
		0.37	0.18	0.22	0.11	0.04	0.07
Alfalfa	hay	22.2	22.2	13.7	11.8	2.96	3.70
		14.8	14.8	11.1	7.40	3.33	3.33
Cabbage	heads	0.22	0.26	0.11	0.11	0.04	0.04
		0.22	0.22	0.07	0.11	0.04	0.04
Tomato	fruits	0.74	0.74	0.37	0.37	0.22	0.18
		0.74	0.37	0.37	0.74	0.22	0.18
Tomato	veg. mass	25.9	29.6	18.5	14.8	5.92	9.25
		14.8	18.5	6.40	6.40	6.40	6.40
Cucumbers	fruits	1.48	1.48	0.74	0.74	0.37	0.37
		1.11	1.48	0.74	0.37	0.37	0.74
Cucumbers	veg. mass	44.4	59.2	29.6	22.2	12.9	11.1
		29.6	40.7	14.8	11.1	11.1	7.40

(a) Water sources for irrigation were not connected to the Dnieper reservoirs.

Table 3.9. Radionuclide concentrations in crops grown with irrigation water (1988, Kherson Region), Bq kg^{-1} of air-dry mass

Culture or Part Used		Kakhovka Reservoir		Source of irrigation water[(a)]	
		^{137}Cs	^{90}Sr	^{137}Cs	^{90}Sr
Winter wheat	grain	1.1–1.9	0.1-0.3	0.30–0.40	0.07–0.11
Corn	grain	0.1–0.4	0.07–0.19	0.04–0.07	0.01–0.03
Alfalfa	hay	11.1–22.8	3.7–11.1	1.85–3.70	0.07–0.37
Cabbage	heads	0.1–0.3	0.004–0.015	0.04–0.07	0.0004–0.004
Tomato	fruits	0.3–0.7	0.02–0.04	0.07–0.03	0.004–0.037
Cucumbers	fruits	0.6–1.5	0.37–1.48	0.26–0.37	0.07–0.01
Red beet	root	0.4–0.7	0.001–0.004	0.19–0.37	0.001–0.002
Carrot	root	0.37–0.74	0.07–0.22	0.15–0.30	0.0004–0.0007
Vegetable marrow	fruit	0.19–0.26	0.07–0.11	0.07–0.20	0.0002–0.0004
Onion	onions	0.74–1.11	0.01–0.11	0.30–0.56	0.0007–0.004
Fennel	herbs	1.11–1.85	0.01–0.02	0.37–0.74	0.002–0.006

(a) Source is not the Dnieper River, and contamination has not been significantly affected by the Chernobyl accident.

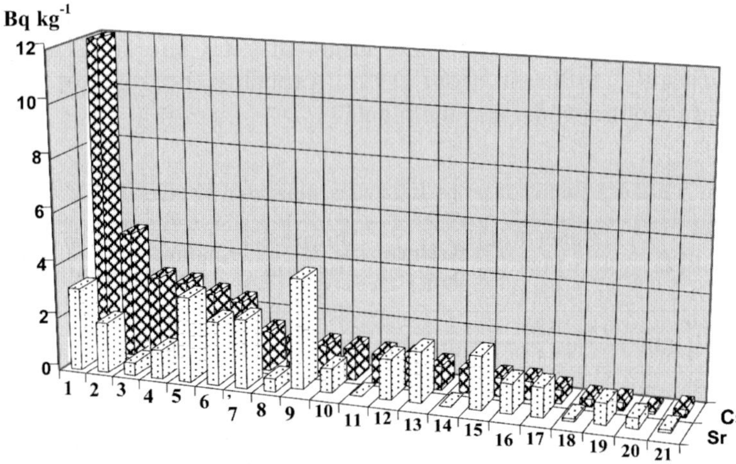

Figure 3.4. ^{137}Cs and ^{90}Sr content in the crops grown with irrigation water (1996, Kakhovka Reservoir, Kherson region), Bq kg-1 of air dry mass, where 1, alfalfa, hay; 2, rice, straw; 3, maize, silo; 4, cucumber, fruit; 5, head cabbage; 6, vegetable marrow, fruit; 7, mangel-wurzel, root; 8, pumpkin, fruit; 9, pea, grain; 10, winter wheat, straw; 11, barley, grain; 12, carrot, root; 13, tomato, fruit; 14, spring wheat, grain; 15, onion; 16, rice, grain; 17, winter wheat, grain; 18, maize, grain; 19, potatoes, tuber; 20, millet, grain; 21, paprika, fruit.

The transfer dynamics of ^{137}Cs and ^{90}Sr into rice, as shown in Figures 3.5 and 3.6, point to the steady increase in these radionuclides averaged over different sorts of rice species grown in Ukraine in post-accident years. The ^{90}Sr concentration in rice increased 18 times from 1986 to 1996, and it has not

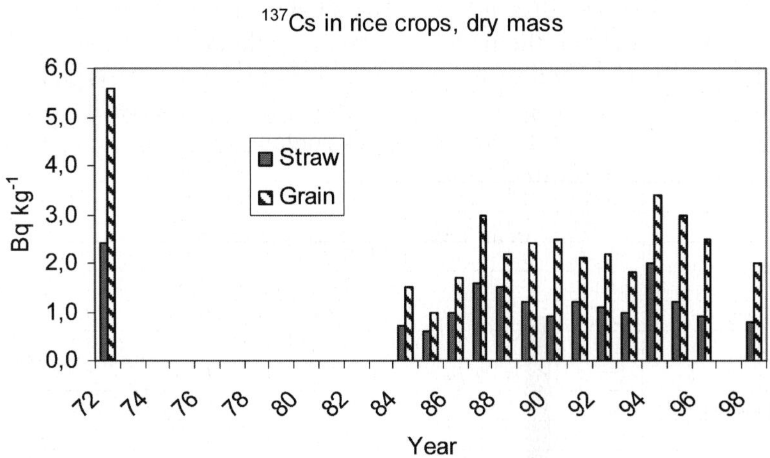

Figure 3.5. Dynamic of ^{137}Cs in rice in Bq kg^{-1} (dry mass)

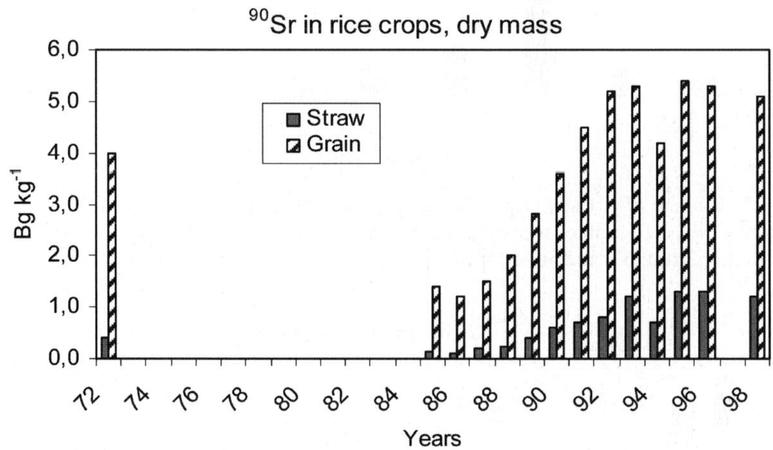

Figure 3.6. Dynamic of ^{90}Sr in rice in Bq kg^{-1} (dry mass)

yet reached steady state, while the ^{137}Cs concentration increased 7.5 times over the same period and did reach steady state. Table 3.10 shows ^{137}Cs and ^{90}Sr concentrations in different varieties of rice.

Table 3.10. Radionuclide concentrations in different varieties of rice grown in Kherson region in 1995 (Bq kg^{-1})

Sort	^{137}Cs		^{90}Sr	
	Grain	**Straw**	**Grain**	**Straw**
Mutant	3.33	2.10	1.50	3.70
Krasnodarsky-424	0.43	2.05	1.00	2.85

The 1992 levels of ^{137}Cs in the crops shown in Figure 3.7 are 4 to 80 times higher in the northern Kiev region without irrigation, and the agricultural products grown there, as shown in Figure 3.8, have 7 to17 times higher levels than the irrigated southern regions. The greater crop contamination in Kiev area is due to the higher contamination of the land (about 1 Ci/km^2), which is closer to Chernobyl.

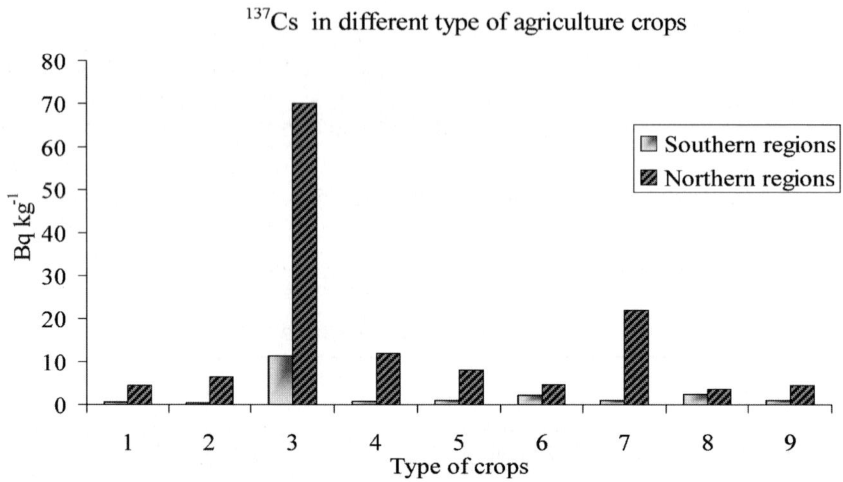

Figure 3.7. Content of ^{137}Cs in crops (Bq kg^{-1}) from southern (a) and northern (b) regions of Ukraine (1992); 1 – winter wheat, grain; 2 - maize, silo; 3 – alfalfa, hay; 4 – cabbage; 5 - tomato, fruit; 6 – cucumber, fruit; 7 - mangel-wurzel, root; 8 – onion; 9 – carrot

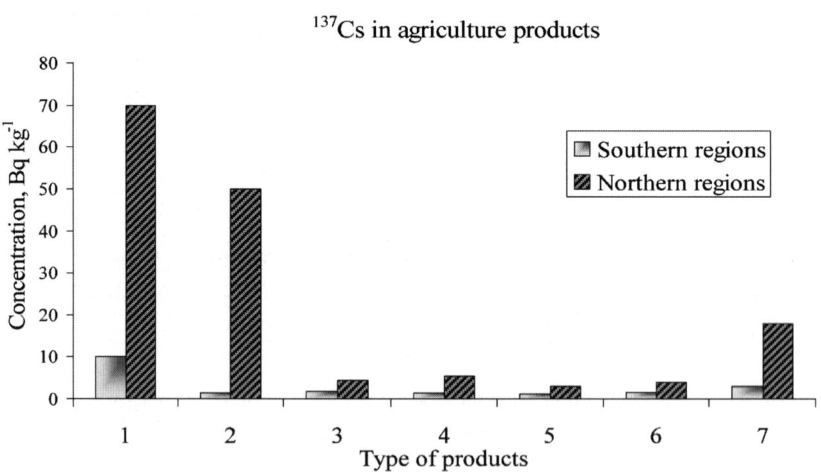

Figure 3.8. ^{137}Cs content in agricultural products (Bq kg^{-1}) grown in southern (a) and northern (b) regions of Ukraine (1992): 1 – roughage, 2 – rich fodder, 3 – food concentrates, 4 – vegetables, 5 – groats/bread, 6 – milk, 7 – meat

Thus the southern region is in a better situation to produce "cleaner" products. Table 3.11 shows the transfer values of ^{137}Cs and ^{90}Sr into irrigated crops in 1996 using overhead sprinklers. These values can be used to predict radionuclide transport into crops or radiation doses from the agricultural products grown on irrigated lands. Tables 3.12 and 3.13 present additional information for simulating radionuclide accumulation in irrigated crops.

During the first years after the accident (1987–1988), the ^{137}Cs contamination of crops was 5 to 10 times higher than ^{90}Sr contamination. By 1996, concentrations in crops were almost the same. This may be due to (1) increasing ^{90}Sr root transport from irrigated soils, (2) decreasing ^{137}Cs concentration in irrigation water by more than an order of magnitude, and (3) maintaining a relatively constant ^{90}Sr concentration in the Dnieper River. The accumulation of ^{137}Cs in rice was stabilized at 1 Bq kg^{-1} for the grain and 2 Bq kg^{-1} for straw. Rice accumulated ^{90}Sr steadily, at least up to 1997.

Table 3.11. Average transfer of ^{137}Cs and ^{90}Sr into 1999 crops irrigated by artificial rain

Culture	**Part**	**^{137}Cs**		**Sr^{90}**	
		TF[(a)]	**AF[(b)]**	**TF**	**AF**
Winter wheat	grain	0.11	24	1.33	2.90
	straw	0.53	110	4.42	11
Spring wheat	grain	0.15	32	-	-
Barley	grain	0.08	15	-	-
Peas	grain	0.19	38	1.37	3.1
Millet	grain	0.05	9	-	-
Corn	grain	0.25	47	0.42	0.9
	silage	1.25	290	1.33	2.7
Rice[(c)]	grain	0.13	80	0.55	13
	straw	0.88	530	2.65	51
Alfalfa	hay	3.80	920	21	64
Mangel beet	root crops	0.63	170	2.20	5.9
Carrot	root crops	0.18	37	0.39	0.8
Potato	tubers	0.05	9	0.33	0.7
Cucumbers	fruits	0.42	86	0.11	0.3
Tomato	fruits	0.28	39	0.17	0.4
Sweet pepper	fruits	0.13	27	0.22	0.7
Vegetable marrow	fruits	0.02	4	0.17	0.5
Pumpkin	fruits	0.06	10	0.17	0.5
Cabbage	heads	0.08	15	0.47	1.6
Onion	onions	0.27	37	1.86	5.8
Eggplant	fruits	0.11	23	0.25	-

(a) TF = transfer factor or coefficient of proportionality (Bq kg^{-1} of mass of crop of used humidity)/(kBq m^{-2}).
(b) AF = accumulation factor, or coefficient of accumulation (Bq/kg of mass of crop of used humidity)/(Bq l^{-1}).
(c) Irrigation of rice paddies by flooding.

Table 3.12. Yield (kg m^{-2}) and LAI (m^2 m^{-2}) as a function of season

Herbage					
Date	Jan 1	Mar 15	May 15	Oct 31	Nov 1
Yield	0.01	0.05	1.5	1.5	0.05
Winter wheat					
Date	Jan 1	Apr 20	Jun10	Aug 5	Aug 6
LAI[(a)]	0	1	7	1	0
Spring wheat					
Date	Apr 15	Jun 20	Aug 15	Aug 16	
LAI	0	6	1	0	
Winter barley					
Date	Jan 1	Jan 4	May 25	Jul 15	Jul 16
LAI	0	1	6	1	0
Oats					
Date	Apr 15	Jun 20	Aug 15	Aug 15	
LAI	0	5	1	0	
Rye					
Date	Jan 1	Mar 20	May 20	Aug 1	Aug 8
LAI	0	1	6	1	0
Corn					
Date	May 15	Jun 20	Aug 1	Oct 15	Oct 16
LAI	0	1	5	4	0
Beet					
Date	May 15	Jun 20	Aug 1	Nov 1	Nov 2
LAI	0	1	4	3	0
Potatoes					
Date	May 20	Jul 7	Aug 1	Sep 15	
LAI	0	4	4	0	
Root-plants, vegetables, fruits, berries					
Date	Apr 15	Jul 7	Oct 1	Nov 1	
LAI	0	5	5	0	

(a) LAI is surface of leafy part of plants per unit area of growth.

3.1.4 Soil Transfer of Radionuclides to Agricultural Plants in Irrigation

Irrigation of farmlands leads to radionuclide accumulation in soil, and in time uptake by roots becomes important. Eight to ten years after the accident, the ^{90}Sr uptake by vegetables and other plants through soil (roots) became the main transfer route (Voitsekhovich et al. 1997), but ^{137}Cs transfer still occurs primarily through the aquatic (non-root) pathway. The relative importance of root and non-root pathways of ^{137}Cs for irrigated lands remains the same until its concentration in water is reduced by two to three orders of magnitude.

In addition, the period for which equilibrium is established in relation to the amount of a radionuclide entering a plant by root and non-root pathways is different for various plants. The equilibrium period may be 14 to 16 years for

Table 3.13. Crop harvest time in Ukraine

Plant	Harvest Times for Crops	
	Recommended	Actual
Herbage	May 1–Oct 31	Apr 25–Oct 31
Winter wheat	Aug 5	Jul 25–Aug 5
Spring wheat	Aug 15	Aug 1–Aug 15
Winter barley	Jul 15	Jul 20–Jul 30
Spring barley	Aug 5	Jul 25–Aug 5
Oats	Aug 10	Aug 5–Aug 9
Rye	Jul 31	Jul 25–Aug 5
Corn, silage	Aug 15–Sep 15	Aug 15–Sep 15
grain	Oct 15	Sep 25–Oct 15
Beet	Sep 20–Oct 31	Sep 20–Oct 31
Mangel beet	Sep 20–Oct 31	Sep 20–Oct 31
Potatoes	Aug 15–Sep 24	Aug 15–Sep 24
Fruit and leaf vegetables	Aug 8–Oct 31	Jun 1–Oct 31
Vegetables-root-plants	Aug 1–Oct 31	Aug 1–Oct 31
Fruits	Jul 1–Oct 15	Jul 1–Oct 15
Berries	Jul 1–Oct 15	Jun 1–Sep 15

grains, eight to ten years for vegetables, and two to six years for alfalfa (Perepelyatnikov 1983; Sanzharova 1997). For corn, equilibrium for ^{90}Sr transfer may be reached after 16 to 20 years of irrigation. The period for ^{137}Cs is similar; ^{137}Cs concentrations will be reduced by two to three orders of magnitude in irrigated water from the 1986–1987 levels.

3.1.5 Dynamics of Radionuclide Contamination of Soils Through Irrigation with Radioactive Water

The radionuclide mass balance in irrigated lands consists of influx into the soil with irrigated water and then uptake by the plant, vertical migration away from the root layer of soil, and reduction by radioactive decay. Fifty-three to 85 percent of ^{137}Cs and ^{90}Sr concentrations in the irrigation water applied over a growing season are retained in the upper layer of soil (Perepelyatnikov 1993), and less than 10 percent is transferred to the plants.

During the first years after the disaster, most of the ^{137}Cs and ^{90}Sr introduced to the soil was from the air. For three years the ^{137}Cs concentration increased by 35 to 57 percent in the top 0.5-cm of soil and 14 to 20 percent in the next 1.5-cm compared with areas not irrigated. After three years of irrigation, the content of ^{90}Sr in rice paddies increased to twice that in dryland areas (Figure 3.9 shows ^{137}Cs and ^{90}Sr concentrations in tomatoes).

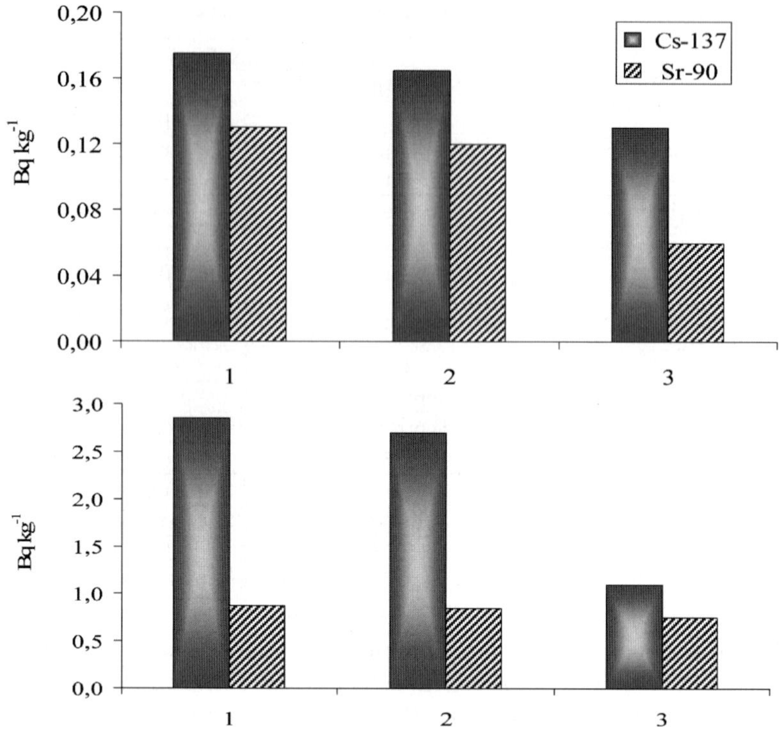

Figure 3.9. Effect of irrigation method on radionuclide accumulation in fruits of tomatoes (dry mass) (top) and tops of tomatoes (bottom); (1) sprinklers, (2) furrow, (3) drip

The amount of ^{137}Cs and ^{90}Sr in the soil of rice paddies increased 170 and 270 percent, respectively, from 1986 to 1997, as shown in Table 3.14. Though these low levels of radioactive contamination and the additional radioactivity

Table 3.14. Dynamics of radionuclide in soil of rice paddies (data from Skadovsk Rice Research Station)

Year	^{137}Cs		^{90}Sr	
	Bq kg^{-1}	**kBq m^{-2}**	**Bq kg^{-2}**	**kBq m^{-2}**
1987	13.7	4.1	2.2	0.7
1988	15.4	4.6	2.9	0.9
1990	18.6	5.6	5.2	1.6
1992	15.4	4.6	5.5	1.6
1993	18.5	5.6	5.4	1.6
1994	18.5	5.6	5.5	1.7
1995	18.7	5.6	5.5	1.7
1996	20.1	6.0	5.6	1.7
1997	22.7	6.8	6.0	1.8

from irrigated water are not major safety concerns, the increased concentrations require detailed analysis and continuous monitoring. These significant increases in ^{90}Sr concentration in soils of rice paddies are difficult to explain by the introduction of radioactivity from the Dnieper River alone. There must be other factors to explain the data in Table 3.14. One must assess the possibility of methodical errors in experimental data. Nevertheless, these measured values and conclusions should be used in the long-term radioecological assessments.

3.1.6 Principal Factors Limiting Radionuclide Migration in Irrigated Biogeocenosis

The methods and rates of irrigation and quality of water affect crop contamination from radionuclides in the irrigation water. Among irrigation methods being applied in Ukraine, the two that transfer the greatest amount of radionuclides are artificial rain (sprinklers) and flooding of paddies. On tomatoes (fruits and tops), artificial rain (overhead sprinklers) generates 1.3 to 2 times more ^{137}Cs contamination than drip irrigation. Artificial rain increased ^{90}Sr contamination up to 2.2 times for the fruit of tomatoes but not for the tops, as shown in Figure 3.9.

Where Dnieper River water is used, the rate of irrigation is the most important factor affecting the transport of radionuclides into soil. Current irrigation rates for rice paddies are from 15,000 to 30,000 m^3 ha^{-1}. Thus, to minimize crop contamination, irrigation technologies must be optimized.

The quality of irrigation water (e.g., hydrochemical composition and mineralization) can influence the mobility of radionuclides and their accessibility to the plants due to formation of complex compounds with different dissolution and biological accessibilities. Experiments at the Ukrainian Institute of Irrigation Land Use demonstrated that in crops irrigated with artificial rain water of varying hydrochemical composition and mineralization, the ^{137}Cs contamination varies more than two to five times (see Table 3.15).

Table 3.15. Effect of mineralization of irrigated water on contamination of crop with ^{137}Cs, (Bq kg^{-1} of air-dry mass)

Sources of irrigation water	Water mineralization (mg/L^{-1})	Winter wheat		Alfalfa	Corn
		grain	straw	hay	silage
Dnieper River	480	0.27	0.55	0.97	0.16
Dnieper 20%, Inguletz 80%	590	0.13	0.24	0.57	0.12
Dnieper 80%,Inguletz 20%	510	0.16	0.78	1.10	0.32
Artesian water	230	0.60	0.38	0.60	0.24

The smallest accumulation of radionuclides was seen in irrigation with mineralized water. For example, a mixture of 80 percent Inguletz and 20 percent Dnieper River water had mineralization of up to 590 mg l^{-1}. This water resulted in the smallest ^{137}Cs accumulation in the crops, reducing the buildup by two to five times. These results agree well with the data of Aleksahin et al. (1985).

3.2 Radionuclide Contamination of Fish

O. Nasvit, V. Klenus, V. Belyaev, O. Volkova, L. Yurchuk, and O. Ryabov

3.2.1 Radionuclide Transfer by Food Trophic Chains

This section discusses radionuclide transfer from the river to various commercial and choice fish. Food consumption is the main pathway of radionuclide uptake by fish. Even at its larva stage, fish take up radionuclides through feeding on the active grab of their food subjects and the feed organism of the fish. According to the prevailing feeding pattern, the fish are customarily subdivided into the following ecological groups:

- planktophages feeding on plankton organisms in a water column
- benthophages feeding on benthic organisms (dwelling on the surface of the bottom or in its thickness) at the bottom
- ichthyophages or predators feeding on other fish.

According to foods they consume under natural conditions, fish may be divided into the following three trophic groups:

- vegetarians feeding on plants
- carnivores feeding on invertebrates
- predators feeding on other fish.

These divisions mainly refer to adult fish. Young fish, irrespective of kind, mostly eat similar food—rotifera, nauplius glochidium, and small zooplankton. Such division may also be called conventional because it characterizes only the prevailing type of feeding.

The food selections of fish are usually subdivided into four categories by ichthyologists. The first is the basic food. The secondary food is that always found in the stomach or bowel but in small amounts. The third category, casual food, is seldom found in the bowels. The fourth category is the food fish are forced to consume because of a shortage of their basic food.

Different hydrobionts (aquatic biota) and abiotic components of water ecosystems accumulate radionuclides differently, creating varying amounts of radioactivity in the foodstuffs depending on the ration and conditions of radionuclide accumulation in the fish of different trophic groups. It was shown that the ^{90}Sr concentration in fish is less in predators than in vegetarians. This is usually associated with the varying efficiency of assimilation of radionuclides present in the food with different ways of feeding. In particular, the main source of ^{90}Sr intake in the food for a predator is the bony tissues of the victim, because it contains up to 90 percent of the radionuclides. However, the bones are only slightly digested in the bowels of fish and then excreted. By contrast, ^{90}Sr is assimilated well by the vegetative tissues, which are the food of vegetarian fish, and taken up by the organism and accumulated in the bones.

Unlike ^{90}Sr, ^{137}Cs in fresh water increases in concentration through the food chain. Thus, its concentration in predators is higher than in their victims. Vegetarian and carnivorous fish as a rule have higher concentrations than their food. In fresh water, ^{137}Cs concentrations are one to three times higher in benthic fish and three to five times higher in predators than in planktophages (Ilchenko 1968). The higher content of ^{137}Cs in benthophages is apparently connected to the fact that up to 90 percent of this radionuclide is deposited in the bottom sediments.

For estimating radionuclide accumulation in fish (and, in general, hydrobionts) in the water radioecology, a coefficient of accumulation (C_a), or bioaccumulation factor, is commonly used. It is the ratio of radionuclide concentration in an organism to that of the water in its habitat. The coefficient of accumulation is an integral feature and depends on ecological conditions, particularly hydrochemical factors, temperature, and structure of food chains.

The other important factor is the coefficient of travel (C_t). It is determined as the ratio of the radionuclide concentration in the body of the fish to the concentration in its food. The coefficient of travel, to a lesser extent than the coefficient of accumulation, is subjected to the effect of ecological conditions and is mainly determined by physiological features of the type and composition of food objects.

The accumulation of radionuclides in fish occurs over time. Thus, coefficients of accumulation and travel will retain their values in a steady-state condition; i.e., the radionuclide concentration in fish correlates to that in water and fish food only under equilibrium conditions. In a steady radiological situation in a water body, C_t is considered to be above 1 for ^{137}Cs.

In nonequilibrium water ecosystem conditions, values of C_t may not reach or exceed equilibrium values depending on the phase of the process (Koulikov and Ryabov 1992). The cooling pond at Chernobyl is a suitable test ground for studying transfer of Chernobyl-released radionuclides through trophic chains, considering the composition of ichthyofauna and the high level of contamination of ecosystem components. Studies were conducted by the Institute of Hydrobiology of the National Academy of Sciences of Ukraine.

Coefficients of ^{137}Cs transfer were calculated for radionuclide concentration in muscles of fish. Silver carp was selected to represent vegetarian fish for the secondary link in the food chain of the ecosystem. Phytoplankton, filamentous algae, plankton, and vegetarian detritus form the diet of the adults. Cladocera organisms were also often found in the bowels content, and sometimes Copepoda and Oligochaeta.

The basic food of redeye is aquatic plants; insect larvae are a secondary source. Filamentous algae and young sprouts of the highest aquatic plants were also found in their diet, mostly in spring, as were roe of mollusks in summer. Sometimes residual vegetarian detritus and soil are found in the bowels.

Among the fish feeding on zoo-plankton and zoo-bentos (the third link of the food chain), bream and silver bream were examined. The basic food of adult fish is small mussel mollusks (*Dreissena polymorpha*), gasteropod mollusks of the class Gastropoda and Chyronomides. The secondary food is plankton. Residual vegetation detritus and soil are also found in the bowels.

The diet of predatory fish (pike, sheat-fish), depending on size, ranges from small insects to fish. The basic ration of young fish is small zoo-plankton and changes to large zoo-plankton as the fish grows, then to insect larvae and at last to fish. The secondary food of full-grown predators is large insects, larvae of chyronomides, and amphipoda. Sometimes large crayfish and even large gasteropod mollusks are found.

The average values of ^{137}Cs transfer coefficients (coefficient of travel) calculated for fish from the Chernobyl cooling pond are given in Table 3.16. This table shows the highest C_t in predators, followed by carnivorous and vegetarian fish in decreasing order.

Similar results were obtained for the ^{137}Cs transfer coefficient in silver bream (the basic food of mollusks) and zander (basic food of silver bream) in Kanev Reservoir during 1986–1996 (Zarubin 1997). According to these data, C_t of carnivore silver bream rose from 0.12 to 1.46 by 1992. By the end of 1986 C_t of predator-zander exceeded 1 and since 1988 varies from 4 to 6.

Table 3.16. Coefficients of travel for various fish living in Chernobyl cooling pond

Group of fish	Coefficients of travel
Vegetarian	3.7
Carnivore	14.8
Predators	24.2

These data agree with the common understanding of radionuclide transfer in the trophic chain. The dynamics of transfer coefficients reflects the changes in the radioecological situation in the water body, possibly allowing its use to predict radionuclide transfer in an aquatic food chain. Nevertheless, large scatters of C_t values for the same fish in the different water bodies may pose a difficulty in its practical use. Using C_t with confidence to predict the radionuclide transfer in the fish food chain is possible only after creation of large database that includes C_t values for commercial fish of specific water bodies and their dependence on the specific ecological conditions of habitat.

3.2.2 Role of Radionuclides in Contamination, and Distribution in Fish

In the summer of 1986, almost all Chernobyl-released radionuclides were found in fish caught in water bodies contaminated by the Chernobyl accident. Among gamma-emitting nuclides, the highest concentrations of ^{131}I and ^{137}Cs were observed in roach, silver bream, and bream, which accumulated the widest spectrum of radionuclides.

Since 1987, ^{137}Cs is the prevailing radionuclide in fish of the Dnieper reservoirs. The ^{137}Cs concentrations in tissues of 10 kinds of fish exceeded 1986 values by 1.7 times and constituted 56 to 83 percent of the total radioactivity. Similar values of ^{137}Cs contribution to total radioactivity in nine kinds of fish were obtained in 1988 (69 to 76 percent) and eight in 1989 (70 to 81 percent). The slight reduction of ^{137}Cs percentage in the total contamination of fish of the Dnieper cascade is shown in Table 3.17, moving away from the region of the accident.

Table 3.17. Relative contribution of ^{137}Cs to total radionuclide content of fish in the Dnieper cascade of reservoirs (%)

Reservoir	1986	1987	1988	1989
Kiev	54.0	91.5	90.5	94.5
Kanev	15.5	83.3	-	68.8
Kremenchug	-	83.4	72.0	61.0
Kakhovka	-	64.5	-	-
Dnieper-Bug Estuary	-	51.0	-	53.0

As evidenced by these data, from the beginning of 1987, ^{137}Cs is the dominant radionuclide in fish in the Dnieper River, especially in Kiev Reservoir. The ^{90}Sr contribution is significantly less; however, its relative importance increases with downstream distance, as indicated in Table 3.18.

For radioactive contamination of fish, it is important to evaluate the characteristics of radionuclide distributions by organs and tissues. Prior to the Chernobyl disaster, several studies were performed by examining radionuclide behavior in global releases (Ilchenko 1968; Kulikov and Molchanova 1975; Shehanova 1983). It was found that radionuclide distributions in fish organs and tissues (1) follow general patterns of the element analogs, (2) do not depend on the temperature of the environment, and (3) are determined by tissue metabolism. As an osteotrophic element, ^{90}Sr selectively accumulates in the bony tissue of skeleton, scales, and fins. Up to 90 percent of ^{90}Sr entering the organism is concentrated. Cesium-137 accumulates mostly in soft tissues. The highest content is observed in muscles, liver, and kidneys and least in blood and bones.

Biota study results at the Chernobyl cooling pond showed these expected features of radionuclide distributions in the organs and tissues of fish in the radioecological conditions of the CEZ, as shown in Tables 3.19 through 3.21.

Represented by a study of bream from Kiev Reservoir (Table 3.19), the muscles have the highest ^{137}Cs concentration. The muscle tissue of immature bony fish is 48.5 percent of the body mass, and it is a repository of ^{137}Cs in an

Table 3.18. Relative contribution of ^{90}Sr to the total radionuclide content of fish in the Dnieper reservoirs and the Dnieper - Bug Estuary (%)

Reservoir	**1987**	**1988**	**1989**
Kiev	3.6	8.6	4.9
Kanev	11.0	-	32.0
Kremenchug	16.4	28.0	39.0
Kakhovka	35.2	-	-
Dnieper-Bug Estuary	8.3	-	47.0

Table 3.19. Distributions of ^{137}Cs and ^{90}Sr in organs and tissues of bream relative to concentrations in muscles in Kiev Reservoir

Organs and tissues	^{137}**Cs**	^{90}**Sr**
Bones	0.35	5.4
Scales	0.22	3.7
Fins	0.04	13.3
Head	0.70	4.4
Inner organs	0.65	1.2

Table 3.20. Distributions of ^{137}Cs in organs and tissues of fish relative to concentrations in muscles in Chernobyl cooling pond

Species	Hard roe	Soft roe	Liver	Alimentary canal	Bone
Carp	0.41	2.3	0.71±0.37	1.3 ±4.8	0.36±0.24
Silver carp		0.89±0.39	0.32±0.02	0.47±0.11	0.32±0.05
Bream	1.47±0.46		0.4±0.11	0.76	0.44
Catfish	0.47±0.14	0.05	0.28±0.03	0.47±0.11	0.34±0.06
Pike perch		0.1	0.86±0.09	0.91±0.31	0.58±0.2

Table 3.21. Distributions of ^{90}Sr in organs and tissues of fish in Chernobyl cooling pond relative to concentrations in muscles

Species	Hard roe	Soft roe	Liver	Alimentary canal	Bone
Silver bream	0.4	-	0.06		86
Crucian	-	0.14	3.4	0.8	106
Catfish	3.1	1.44	0.32±0.09	7.3±2.8	92±43
Carp	3.2±3.0	-	2.2±2	9.3±8 9	152±93
Silver Carp	0.6	0.96±0.98	0.29±0. 16	7.38±9.26	91±26

organism. Strontium-90 is concentrated mostly in fins, bones, and scales. The latter are 15.3 and 6.9 percent, respectively, of the mass of the fish fully covered with scales (Shehanova 1983). The Chernobyl cooling pond data do not show major variations of ^{137}Cs and ^{90}Sr distributions in the soft tissue.

3.2.3 Dynamics of ^{137}Cs Content in Kiev Reservoir Commercial Fish

Kiev Reservoir is the uppermost reservoir of the Dnieper cascade and the most contaminated from the 1986 accident. As a result, fish in Kiev Reservoir have the greatest contribution to the individual effective exposure dose of the population. Tables 3.22 and 3.23 present the data obtained by the Institute of Fish Management of the National Academy of Sciences of Ukraine. They do not include the whole population of fish, only the commercial fish. Thus, they have a partial volubility for ichthyologic analysis of the radioecological situation in the reservoir. However, these data can be useful in calculating exposure doses through consumption of fish products.

3.2.4 Radionuclide Accumulation in Commercial Fish of Dnieper Reservoirs

Various hydrobionts can concentrate radionuclides in their bodies, and modern radioecology commonly accepts the concentration factor (F_c) or the coefficient of accumulation, C_a, a ratio of concentration of an isotope in an organism to its concentration in water. To be valid, the dynamic equilibrium in the ecosystem is assumed in the intercomponent transfer of radionuclides from water to fish and relative steadiness of their concentration in water.

Table 3.22. Content of ^{137}Cs in commercial fish in Kiev Reservoir in 1986 (Bq kg^{-1} of wet weight)

Season	Bream			Roach			Silver bream			Pike			Pike perch		
	min	avg	max	min	avg	max	min	avg	max	min	avg	max	min	avg	max
Spring	210	420	940	59	130	510	62	140	530	120	410	930	80	350	1000
Summer	290	570	1007	39	360	194	80	380	1630	370	540	1870	330	600	1920
Autumn	230	710	1200	160	530	1630	120	550	1700	520	1110	1870	460	1000	1650

Table 3.23. Dynamics of ^{137}Cs concentrations in fish from commercial fishing in Kiev Reservoir (Bq kg^{-1} of wet weight)[(a)]

Year	Bream		Roach		Silver bream		Pike		Pike perch	
	average	(±)	average	(±)	average	(±)	average	(±)	average	(±)
1987	930	170	1100	530	1160	570	1420	360	1160	180
1988	718	73	700	210	720	260	1080	140	940	170
1989	800	280	880	520	890	460	1210	350	850	380
1990	270	130	323	75	280	130	520	200	470	170
1991	258	78	310	140	250	130	370	220	300	160
1992	106	12	190	150	150	150	270	110	208	90
1993	178	78	140	100	142	42	166	61	129	49
1994	131	49	188	32	140	72	212	50	205	34
1995	37	13	28.3	6,2	39	11	48	16	69	17

(a) ± refers to the accuracy of measurement.

A considerable number of laboratory experiments and field data are available to characterize the F_c values of radionuclides in hydrobionts of different trophic levels. The values vary for different hydrobionts depending on their ecological features, metabolism, stage of ontogenesis, and the physical and chemical parameters of their environments, as is commonly accepted by researchers (e.g., Shehanova 1983). The established regularities are analogous to those that take place under exchange of mineral substances between hydrobionts and the environment.

The Chernobyl work presented here is for contaminated areas where the scale and features of radionuclide contamination and physical and chemical properties of fallouts have no analogues.

Fish begin to accumulate radionuclides when they enter the water. However, the limit of accumulation of radioactive elements in the organism is a function of time. In experiments, the final ^{90}Sr F_c was found on the 90th day and the ^{137}Cs F_c in four months. However, the real-world processes are not in equilibrium, considering time-varying activity of the transformation and inter-component redistributions of radionuclides in the ecosystem, in washoff from the adjacent areas, in the water bodies, the processes of accumulation and excretion of radionuclides by fish in the first months after the accident, etc. Consequently, the F_c calculated with these data reflects these varying conditions. Nevertheless, calculations of this coefficient may be useful to determine the extent of contamination of the water ecosystem components.

In June 1986, "fish-water" F_c values were 25 to 29,400 in the upper Kiev Reservoir. The lower limit of values was for ^{137}Cs in perch. Among short-lived isotopes, the F_c for ^{131}I varies from 5,400 in perch to 29,400 in roach; for ^{140}La, from 700 in silver bream to 900 in roach; for ^{140}Ba, from 310 in pike to 2,400 in roach. Roach has the highest F_c values because of feeding habits.

The dynamics of average F_c for ^{137}Cs in various Kiev Reservoir areas for fish is presented in Figure 3.10 for 1987–1995. As indicated in this figure, the F_c of ^{137}Cs for all fish increased until 1989, showing a more rapid decrease in water than in fish; then it decreased in 1990 due to reduced concentrations in fish. F_c values then went up and down, perhaps from changes in ecological conditions, especially the ^{137}Cs concentration in river water.

The variation of F_c values for fish in the lower Dnieper River exhibits a trend similar to that shown in Figure 3.10. The differences are in specific values. The maximum values of F_c of ^{137}Cs are 4,000 to 6,500 in those fish in Kremenchug Reservoir. From 1990 to 1995, ^{137}Cs in Kremenchug Reservoir reached steady-state values of 400 to 1,100.

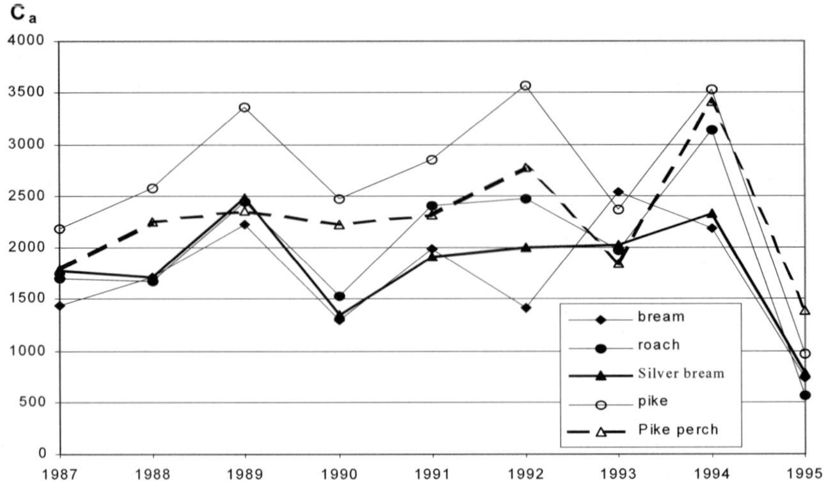

Figure 3.10. F_c of ^{137}Cs (L kg^{-1}) in commercial fish from Kiev Reservoir

The coefficient of accumulation, as a factor characterizing the capability of hydrobionts (fish in this case) to concentrate radionuclides in their bodies, is not an inner characteristic of the organism but depends on the intensity of radionuclide exchange between the fish and water through a trophic chain, the fish's metabolism, and the ecological conditions of their habitat. In the radioecology of the Dnieper River, F_c reflects not only physiological features of hydrobionts but radionuclide distributions in inner- and interecosystem components. The oscillation of ^{137}Cs concentrations in fish in Figure 3.10 can be explained with variations of these factors which influence ^{137}Cs concentrations in both water and fish.

The observed sharp (several times) decrease of ^{137}Cs concentrations and F_c values in commercial fish in Kiev Reservoir in 1995 has no clear explanation, but the same results were obtained in radioecological investigations of fish conducted in the upper Kiev Reservoir (Pripyat and Teteriv Bays) (Ryabov 1998). Figure 3.11, taken from the TRANSAQUA database (2003), illustrates the dynamics of ^{137}Cs concentrations in predatory and nonpredatory fish in the upper Kiev reservoir and demonstrates comparatively low ^{137}Cs concentrations in fish in 1995.

The average ^{137}Cs concentration factors of predatory fish exceed those of nonpredatory fish by two to four times in the Dnieper reservoirs, and the ratio tends to increase with downstream distance in the cascade (see Figure 3.12).

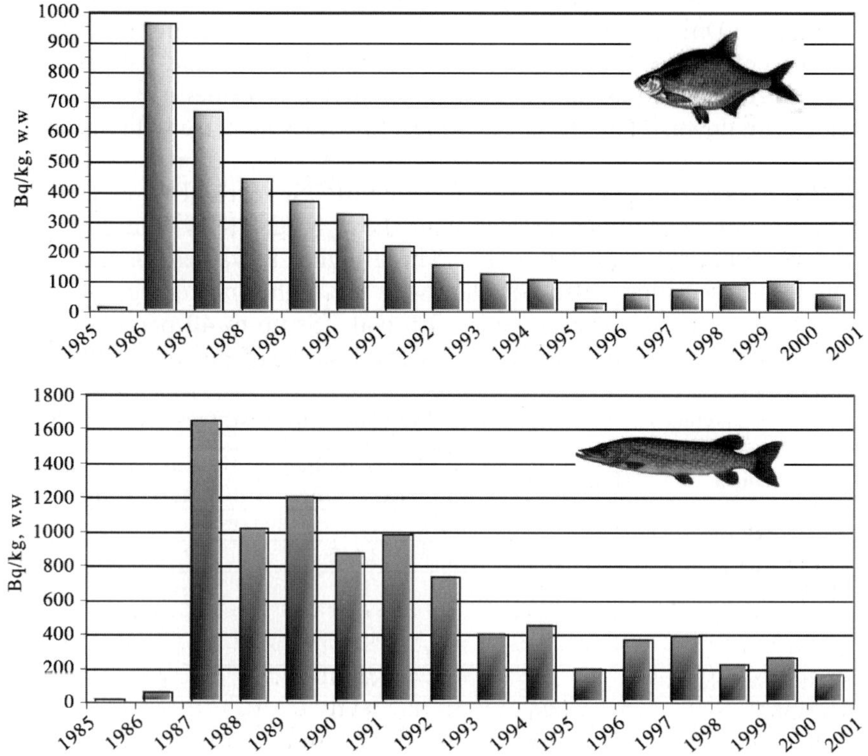

Figure 3.11. Average amount of ^{137}Cs in nonpredatory (Bream) and predatory (Pike) fish from upper part of Kiev reservoir (from TRANSAQUA 2003)

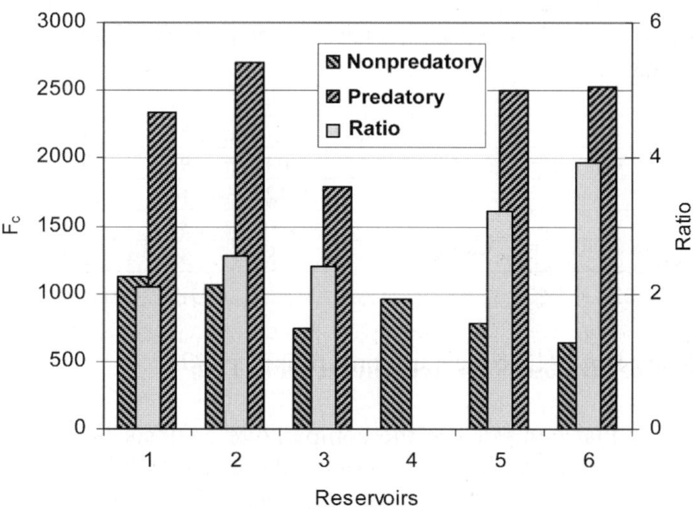

Figure 3.12. ^{137}Cs concentration factors for fish in the Dnieper reservoirs (2003): 1–Kiev, 2–Kanev, 3–Kremenchug, 4–Dnieprodzerzhinsk, 5–Zaporozhje, 6–Kakhovka

Table 3.24 shows F_c values of ^{137}Cs and ^{90}Sr for aquatic plants, mollusks, and fish in the Dnieper reservoirs. These results may shed light on the effects to fish as they move away from the contaminated source and habitat within the overall ecological system. These data show that F_c values of ^{137}Cs for fish decrease with downstream distance, while ^{90}Sr values increase with distance.

In the Dnieper-Bug Estuary, the F_c value of ^{90}Sr reached 54 in 1992 and was seven and three times greater than those in the Kiev and Kakhovka reservoirs, respectively (Table 3.25). These higher F_c values for the fish in the Dnieper-Bug Estuary also correlate with the higher percentage of ^{90}Sr in the total radioactivity. The estuarine fish had ^{90}Sr up to 40 percent of total and the fresh water fish in Kiev Reservoir had less than 5 percent in 1990.

Table 3.24. Average F_c of ^{137}Cs and ^{90}Sr for hydrobionts in different reservoirs in 1992

Biota	Kiev Reservoir	Kremenchug Reservoir	Kakhovka Reservoir
	^{137}Cs		
High water plants	3508	1447	427
Mollusks	294	190	560
Fish	1037	343	340
	^{90}Sr		
High water plants	422	61.9	67.5
Mollusks	28.3	35.0	87.0
Fish	7.8	1.7	17.0

Table 3.25. ^{90}Sr F_c estimated for fish in the Dnieper-Bug Estuary (lower Dnieper) in 1990

Species	1987	1988	1989	1990	1991	1992	1993
Silver bream					100	52	111
Bream	34	39		74	61	51	88
Perch		28			61	47	85
Roach	42	55		64	94	76	134
Sazan				32	14	23	
Pike perch		43			33	38	
Sabre carp		130		38	52	57	
Pike	46	93				92	

3.2.5 Effect of Size on ^{137}Cs Accumulation in Fish

One factor that complicates the comparative analysis of radioecology in fish as well as dynamics for water bodies is the so-called "dimensional effect" in accumulation of ^{137}Cs in fish; that is, the dependence of ^{137}Cs concentration in a fish on its mass. During sampling, fish of various sizes are usually selected randomly. Accounting for the dimensional effect when comparing two or more samplings of various sizes helps to obtain correct results.

The positive size effect, increasing ^{137}Cs concentration with increasing mass, was described for similar fish in different habitats (Andersson 1989; Elliot et al. 1992a, b; Haddering 1989; Linder et al. 1989). However, other size effects can be found in the same kinds of fish living in other lakes and rivers. Various dimensional effects can be found in different kinds of fish. Distinctions in different fish in the same water body may be governed by metabolism and diet, but the distinctions between the same kinds living in different water bodies may be governed by limnological features such as the nature of bottom sediments and soils of the catchments (Eliot et al. 1992a, b).

The analysis of manifestations of the dimensional (size) effect was provided for fish obtained at the following four sites:

- The northern part of Kiev Reservoir near Strakholesye village. In this area the sampling was provided at three stations at distances of 0.3, 0.7, and 1.2 km from the bank. This area is about 1 m deep and covered with dense thickets of higher aquatic plants merged with floating leaves and air-aquatic forms. In the spring this area is used by several kinds of fish, especially carp, for spawning.
- The old channel of the Pripyat River, upstream from the inflow into Kiev Reservoir. The maximum depth is about 9 m. The coastal shallow waters are full of dense thickets of merged and air-aquatic plants. Samples were taken in the old channel and also in the shallow waters about 500 m from the old channel.
- The cooling pond of Chernobyl nuclear power plant. Samples were taken in so-called warm and cold areas.
- Kozhanovskoye Lake in the Bryansk region of Russia. Samples were taken at three places in the open part of the lake by branchiate net and near the eastern bank in the thickets of aquatic plants by a sweep net.

A regression analysis was used to determine the dependency of ^{137}Cs concentration on fish size (Sansone and Voitsekhovich 1996). The following parameters were determined:

a = point of intersection of regression line with ordinate
b = slope of regression line with confidence interval $\pm 2\ SD$
r^2 = coefficient of determination.

The coefficient of determination is a measure of ^{137}Cs concentrations in isolated samples (individual or uniform) determined by mass variations. The dimensional effect was considered positive (+) if the slope $b \pm 2SD$ had a positive value, negative (–) if the slope $b \pm 2SD$ had a negative value, and neutral (0) if the interval $[b-2SD, b+2SD]$ had the point $b = 0$.

The ^{137}Cs concentration was studied in 12 species: pike, roach, redeye, sabre, tench, silver bream, bream, crucian, asp, sheat-fish, perch, and zander. The 49 data sets collected in 1992–1994 were analyzed. Data from different seasons were analyzed separately to exclude possible seasonal changes. The results are presented in Table 3.26, where fish are listed by decreasing dimensional effect. To compare ^{137}Cs concentrations in fish from the study sites, concentrations for a standard 0.5-kg fish were also calculated with regression equations. The standard pike was 2.5 kg because it is a large fish. Concentrations for standard fish are presented in the last column of the table (Sansone and Voitsekhovich 1996). In these tables, a is the point of intersection of the regression line with the ordinate, a.a. and ww are the mass of sample at natural humidity, d.a. and dw are dry mass, t is whole body, and m is muscles. The + and – signs show positive and negative dimensional effects, while 0 indicates no dimensional effect. As stated before, b is the slope of regression line with a two standard deviation confidence interval, and r^2 is the coefficient of determination.

Table 3.27 summarizes results published by Sansone and Voitsekhovich (1996). A positive size effect was found in perch, zander, pike, saber, and asp but not in bream and tench. In most cases, the ^{137}Cs concentration did not change with mass (dimensional effect = 0). For others (crucian, roach, redeye, silver bream), the dependence was neutral or negative (0 or -). In the Chernobyl cooling pond, the fish exhibiting a clear size effect have the highest ^{137}Cs concentrations. The largest size effect was found in perch, zander, sabre, asp, bream, and crucian in the cooling pond. For these, the most common fish in the cooling pond, the slope of the regression line (coefficient b) was greater than 0.44.

The difference in ^{137}Cs concentrations in fish from the four sites is due to the level of water pollution and the size of fish sampled. The variation in F_c, however, the ratio of radionuclide concentration in fish to that of water, is less than 2 and much less than variations in ^{137}Cs in fish from these sites (Table 3.28). Perch, zander, and pike are predators (in higher trophic levels) with F_c from 5,440 to 13,000 and an average of 7,800±2,500 (n=8). Vegetarian fish ranged from 1,200 to 3,200, averaging 1,900±600 (n=13). The results indicate that accumulation of ^{137}Cs in fish expressed by F_c is similar in the four sites and ^{137}Cs concentration itself does not significantly affect F_c.

Table 3.26. Data on fish from four habitats, showing sample amounts, regressional parameters, estimates of dimensional effect, and ^{137}Cs concentration (Bq kg^{-1} a.a.) in muscles of reference fish (mass 0.5 kg) and pike (mass 2.5 kg) (from Sansone and Voitsekhovich 1996)

Fish Species	Parameters of regression line log(a)	r^2	b	Location	Date	Number of fish	Size effect (d.a.)	^{137}Cs Bq kg^{-1} (a.a.)
Perch dw/t	2.63	0.81	0.32 ± 0.1	Stracholesye	05/92	12	+	1868
Perch dw/t	2.72	0.34	0.21 ± 0.19	Stracholesye	11/92	12	+	1117
Perch dw/t	2.62	0.035	0.18 ± 0.44	Stracholesye	04/93	20	0	765
perch dw/t	2.83	0.10	0.08 ± 0.06	Stracholesye	09/93	66	+	642
Perch dw/t	1.74	0.78	0.57 ± 0.17	Pripyat	05/92	14	+	1138
Perch dw/t	2.01	0.61	0.37 ± 0.33	Pripyat	11/92	5	+	634
Perch dw/t	4.90	0.97	0.25 ± 0.04	Kozhanovskoe	03/93	7	+	90148
Perch dw/m	4.95	0.55	0.15 ± 0.06	Kozhanovskoe	09/94	20	+	43013
Perch ww/m	3.21	0.53	0.46 ± 0.19	Cooling Pond	1994	24	+	28283
Zander ww/m	3.08	0.42	0.49 ± 0.39	Cooling Pond	1994	11	+	25264
Pike ww/m	1.84	0.78	0.32 ± 0.14	Stracholesye	05/92	8	+	846
Pike ww/m	0.78	0.64	0.57 ± 0.39	Pripyat	05/92	7	+	521
Pike dw/m	4.29	0.99	0.31 ± 0.04	Kozhanovskoe	04/93	3	+	44096
Pike dw/m	4.30	0.78	0.27 ± 0.09	Kozhanovskoe	10/93	11	+	37947
Pike dw/m	4.22	0.98	0.28 ± 0.04	Kozhanovskoe	09/94	8	+	32647
Sheat-fish ww/m	3.07	0.60	0.31 ± 0.07	Cooling pond	1993	54	+	8066
Sheat-fish ww/m	3.52	0.14	0.13 ± 0.07	Cooling pond	1994	90	+	7428
Sabre ww/m	2.79	0.76	0.52 ± 0.10	Cooling pond	1993	36	+	15612
Sabre ww/m	2.85	0.70	0.45 ± 0.09	Cooling pond	1994	43	+	11602
Asp ww/m	2.69	0.47	0.51 ± 0.28	Cooling pond	1994	17	+	11654
Tench dw/t	2.38	0.20	0.19 ± 0.35	Stracholesye	05/92	7	0	327
Tench dw/t	2.07	0.29	0.28 ± 0.30	Stracholesye	04/93	10	0	268
Tench dw/t	3.27	0.054	-0.23 ±0.96	Stracholesye	09/93	6	0	187
Tench dw/t	3.34	0.33	-0.21 ±0.59	Pripyat	05/92	3	0	270

Table 3.26 (contd)

Fish Species	Parameters of regression line log(a)	r^2	b	Location	Date	Number of fish	Size effect (d.a.)	^{137}Cs Bq kg^{-1} (a.a.)
Tench dw/t	1.98	0.23	0.18 ± 0.30	Pripyat	11/92	7	0	128
Bream dw/t	2.56	< 0.01	0.01 ± 0.37	Stracholesye	05/92	7	0	185
Bream dw/t	2.53	< 0.01	0.01 ± 0.38	Stracholesye	04/93	13	0	154
Bream dw/t	2.21	0.69	0.10 ± 0.07	Pripyat	05/92	6	+	143
Bream ww/m	3.48	0.30	0.14 ± 0.08	Cooling Pond	1993	31	+	7209
Bream ww/m	2.57	0.27	0.44 ± 0.51	Cooling Pond	1994	10	0	5722
Goldfish dw/m	5.60	0.49	-0.33 ±0.17	Kozhanovskoe	10/93	17	-	10242
Goldfish dw/m	5.25	0.08	-0.17 ±0.24	Kozhanovskoe	09/94	26	0	12366
Goldfish ww/m	2.56	0.46	0.51 ± 0.42	Cooling Pond	1993	9	+	8639
Goldfish ww/m	3.56	0.037	0.13 ± 0.65	Cooling Pond	1994	6	0	8145
Roach dw/t	2.70	0.18	-0.09 ±0.12	Stracholesye	04/93	11	0	155
Roach dw/t	2.89	0.35	-0.19 ±0.08	Stracholesye	09/93	41	-	113
Roach dw/m	4.89	0.041	-0.08 ±0.15	Kozhanovskoe	10/93	27	0	14295
Roach dw/m	4.81	0.062	-0.09 ±0.15	Kozhanovskoe	09/94	24	0	13432
Roach ww/m	3.28	0.39	0.17 ± 0.18	Cooling Pond	1994	8	0	5481
Redeye dw/t	2.80	0.027	-0.06 ±0.42	Stracholesye	05/92	5	0	222
Redeye dw/t	3.63	0.17	-0.49 ±1.51	Stracholesye	11/92	4	0	111
Redeye dw/t	2.79	0.11	-0.15 ±0.28	Stracholesye	04/93	12	0	115
Redeye dw/t	3.17	0.73	-0.38 ±0.13	Stracholesye	09/93	15	-	66
Silver bream dw/t	2.77	0.40	-0.10 ±0.10	Stracholesye	05/92	8	-	173
Silver bream dw/t	2.88	0.397	-0.17 ±0.12	Stracholesye	04/93	13	-	127
Silver bream dw/t	2.33	0.77	0.10 ± 0.07	Pripyat	05/92	4	+	184
Silver bream dw/t	3.06	0.77	-0.34 ±0.25	Pripyat	11/92	4	-	78
Silver bream ww/m	3.48	0.058	0.12 ± 0.16	Cooling Pond	1993	35	0	6366
Silver bream ww/m	3.18	0.066	0.18 ± 0.25	Cooling Pond	1994	32	0	4632

Table 3.27. Size effects for fish from 49 data sets

Fish species	Stracholesye (Kiev Res.)	Pripyat River inflow to Kiev Res.-Ir	Cooling pond	Kozhanovskoe Lake (Russia)	Total		
					+	0	-
Perch	+ + + 0*	+ +	+	+ +	8	1	0
Zander			+		1	0	0
Pike	+	+		+ + +	5	0	0
Sheat fish			+ +		2	0	0
Sabre fish			+ +		2	0	0
Asp			+		1	0	0
Tench	0 0 0	0 0			0	5	0
Bream	0 0	+	+ 0		2	3	0
Crucian			+ 0	0 -	1	2	1
Roach	0 -		0	0 0	0	4	1
Redeye	0 0 0 -				0	3	1
Silver bream	- -	+ -	0		1	2	3
* quantity of signs equals number of cases resulting in one or another effect.							

Table 3.28. Relationship of ^{137}Cs concentrations in fish muscles (Bq kg^{-1} a.a.) and water (Bq/L) in four habitats during 1992–1994

Fish Species	Pripyat R. (Kiev Res.) 1992	Stracholesye (Kiev Res.) 1993	Chernobyl Cooling Pond 1994	Kozhanovskoe Lake 1994
Perch	13400	8280	7070	7170
Zander	-	-	6316	-
Pike	6130	8460	-	5440
Tench	3180	2680	-	-
Bream	1680	1540	1430	-
Crucian	-	-	2040	2060
Roach	-	1550	1370	2240
Silver bream	2170	1270	1160	-

The potassium concentration of less than 2 mg/L in water may lead to an exponential rise in ^{137}Cs in fresh water fish (Fleishman 1973). The measured concentrations of potassium in the water of all four sites are above this critical value; at Stracholesye the level was 4.1 mg/L in 1993, at Pripyat 3.4 mg/L in 1992, in the cooling pond 4 mg/L and in Kozhanovskoe Lake 1.5 to 2.6 mg/L in 1993. Almost all F_c values obtained are within 200–8,000, mentioned by Blaylock (1982) for different kinds of fish inhabiting water with a potassium concentration of about 2 mg/L.

3.3 Conclusions

Many research institutes have conducted radioecological monitoring of up to 1.6 million hectares of irrigated land in southern Ukrainian since the 1986 Chernobyl accident. Despite insufficient high-quality data, these studies led to the following conclusions.

Irrigation significantly affects radionuclide distribution in soil and irrigated crops, especially in rice paddies. In the initial period of contamination (1987–1988), radionuclide concentrations in the 0.5-cm topsoil layer increased in one growing season from contaminated irrigation water. However, over the next three to five years, radionuclide concentrations in the Dnieper reservoirs decreased by an order of magnitude, and transport of radionuclides by irrigation decreased proportionally. Washout of radioactivity from the top 1 cm of soil to the lower layers was observed in irrigated lands. At present, the low radionuclide influx into the soil of irrigated lands is balanced by washout into drainage water and radionuclide decay.

Irrigation of the main agricultural plants with Dnieper River water increased contamination by five to ten times in 1986–1990. Relative contamination of food from irrigated land in southern Ukraine remained 10–20 times lower than ^{137}Cs concentrations in food products from the dryland areas of northern Ukraine, where soil contamination was up to 1 Ci km^{-2}. During the initial period after the accident (1987–1988), ^{137}Cs contamination of crops was five to ten times higher than ^{90}Sr contamination, but by 1996 the concentrations in crops were basically the same because of greater ^{90}Sr root uptake and reduced ^{137}Cs in irrigation water by an order of magnitude. The ^{137}Cs uptake by the irrigated rice crop stabilized at 1 Bq kg^{-1} for grain and 2 Bq kg^{-1} for straw, while ^{90}Sr concentrations continued to increase in both grain and straw.

The transfer coefficients of ^{137}Cs and ^{90}Sr into crops from the irrigated lands of Ukraine were determined from a large number of field data. They can be used to predict radionuclide transport into the various crops.

During the first years after the accident, ^{137}Cs and ^{90}Sr transferred mainly from leaves to the plants. For ^{137}Cs that is the main pathway of contamination of plants. For ^{90}Sr, root transfer of radionuclides is initially the main pathway for the majority of agriculture plants. The time to reach final equilibrium between root and non-root ^{90}Sr transport has been determined as 14 to 16 years for corn and grain plants and at least two to six years for alfalfa.

The method of irrigation and quality of water (hydrochemical composition and mineralization) also affect contamination. Irrigation by artificial rain increases the ^{137}Cs and ^{90}Sr contamination of crops by a factor of 2 compared with drip irrigation. The least accumulation (two to five times less) is achieved in crops irrigated with highly mineralized water (590 mg/L).

These conclusions should be used carefully because they are based on poor-quality data. Thus, further research should be undertaken, including developing a scientifically defensible methodology and computer simulation method for predicting radioactive contamination of agricultural products and a better formulation for radiation doses of populations through irrigated agricultural food chains.

After the 1986 Chernobyl accident, a large number of experimental and field data on fish contamination were taken and analyzed. However, they have not yet been comprehensively and methodically assessed. Evaluating time-varying, ^{137}Cs accumulations in fish would enable quantification of the important mechanisms of radionuclide accumulation in the fish. The coefficient of accumulation (or bioaccumulation factor), F_c, for fish was relatively constant from 1987 to 1994. In 1995, this trend was broken. For example, F_c values in Kiev Reservoir were 1,500 to 3,000 in 1995. Since then, a significant change in F_c has occurred in Kiev Reservoir, with values ranging from 600 to 1,200 for no obvious reason. This change in F_c also occurred in other Dnieper reservoirs. The results of these studies can provide an important contribution to the current understanding of radioactive accumulation in fish.

One significant result from these studies is clearer "fish-size effect" on ^{137}Cs accumulation in and release from fish. This clarification allows a more accurate estimation of ^{137}Cs accumulation during fish growth and feeding regime changes. With regressional analysis, the fish size dependency can be incorporated to improve radioecological models.

Further studies of radionuclide accumulations in fish and irrigated crops would improve estimates of population exposure doses via the aquatic pathways affected by the 1986 accident at the Chernobyl nuclear power plant.

3.4 References

Aleksahin RM, OI Bufatin, VG Malikov, GP Perepelyatnikov, et al. 1985. "Radioecology of Irrigated Agriculture." *Energoatomizdat*, p. 224 (in Russian).

Andersson E. 1989. "Incorporation of ^{137}Cs into Fishes and other Organisms." *The Radioecology of Natural and Artificial Radionuclides, Proceedings of XVth Regional Congress of IRPA*. Visby, Gotland, Sweden, pp. 312-317.

Blaylock BG. 1982. "Radionuclide Data Bases Available for Bioaccumulation Factors for Freshwater Biota." *Nuclear Safety*, 23:427.

Doroguntsov SI, MA Hvesik, and MN Palamarchuk. March 1992. "Problems of Ecological Optimisation of Water Use in the Dnieper Basin and Their Solutions." Proc. of Int. Scientific Conference, *Problems of Ecological Optimisation of Land Use and Hydroeconomic Construction in the Dnieper Basin.* Kiev, pp. 8-13 (in Russian).

Elliot JM, J Hilton, E Rigg, PA Tullet, DJ Swift, and DRP Leonard. 1992a. "Sources of Variation in Post-Chernobyl Radiocesium in Fish from Two Cumbrian Lakes (Northwest England)." *Journal of Applied Ecology*, 29:108-119.

Elliot JM, JA Elliot, and J Hilton. 1992b. "Sources of Variation in Post-Chernobyl Radiocesium in Brown Trout, *Salmo Trutta L.*, and Arctic Charr, *Salvelinus alpinus (L.)*, from Six Cumbrian Lakes (Northwest England)." *Annls. Limnol.* 29:79-98.

Fleishman DG. 1973. "Accumulation of Artificial Radionuclides in Freshwater Fish." *Radioecology*. John Wiley & Sons, New York, pp. 347-370.

Haddering RH. September 1989. "Distribution of ^{137}Cs in the Aquatic Food Chain of the Ijsselmeer after the Chernobyl Accident." *The Radioecology of Natural and Artificial Radionuclides, Proceedings of the XVth Regional Congress of IRPA.* Visby, Gotland, Sweden, pp. 325-330.

Ilchenko AI. 1968. "Radioecology of Fresh Waters." *Issues of Ichthyology,* 9:2(55):324-337 (in Russian).

Koulikov AO and IN Ryabov. 1992. "Specific Caesium Activity in Freshwater Fish and Size Effect." *The Science of the Total Environment*, 112:125-142.

Kulikov NV and IV Molchanova. 1975. "Continental radioecology (soil and fresh water ecosystems)." *M. Nauka*, p. 182 (in Russian).

Linder G, W Pfeiffer, JA Robbins, and E Recknagel. September 1989. "Long-Lived Chernobyl Radionuclides in Lake Constance: Speciation, Sedimentation and Biological Transfer." *The Radioecology of Natural and Artificial Radionuclides, Proceedings of the XVth Regional Congress of IRPA.* Visby, Gotland, Sweden, pp. 295-300.

Palamarchuk MN. March 1992. "Features of Territorial Arrangement of Water Use in the Dnieper Basin." *Proc. of Int. Scientific Conference, Problems of Ecological Optimisation of Land Use and Hydroeconomic Construction in the Dnieper Basin.* Kiev, pp. 31-36 (in Russian).

Perepelyatnikov GP. 1983. "Accumulation of ^{90}Sr and ^{137}Cs by Main Crops under the Conditions of Irrigation During Contamination of the Territory of Northern Caucasia by Products of Nuclear Explosions." Ph.D. dissertation, Krasnodar (in Russian).

Perepelyatnikov GP. 1993. "To the Issue of Radiation Situation on the Irrigated Lands." *Proc. of Ukrainian SRI of Agricultural Radiology: Problems of Agricultural Radiology.* Kiev, No. 3, pp. 111-115 (in Russian).

Perepelyatnikov GP, BS Prister, and AS Sobolev. 1993. "Migration of Radionuclides in the System Water-Soil-Plant at Lands Irrigated by the Dnieper after the Accident at ChNPP." *Proc. Ukrainian SRI of Agricultural Radiology: Problems of Agricultural Radiology*, Issue 1. Kiev, pp. 106-112 (in Russian).

Ryabov IN. 1998. "Peculiarities of fish ecology in water bodies contaminated with radionuclides as a result of the accident on Chernobyl NPP". Abstract of Doctor of biological sciences dissertation. Moscow, p. 50 (in Russian).

Sansone U and OV Voitsekhovich (Eds.). 1996. *Modeling and Study of the Mechanisms of the Transfer of Radionuclides from the Terrestrial Ecosystem to and in Water Bodies around Chernobyl.* Final report, ECP-3 Project, EUR 16529 EN. European Commission, Luxemburg.

Sanzharova NI. 1997. "Radioecological Monitoring of Agroecosystems and Farming in the Zone Affected by Nuclear Power Plants." Ph.D. dissertation. Obninsk, p. 32 (in Russian).

Shehanova IA. 1983. "Radioecology of Fishes." *M. Light and Food Industry*, p. 208 (in Russian).

TRANSAQUA. 2003. "Transfer in Aquatic Ecosystems." Final report of the French-German Initiative for Chernobyl, CD database, IRSN, Cadarach.

Voitsekhovich OV (Ed.). 1997. "Radioecology of Water Bodies of the Zone of Chernobyl NPP Accident Effect." *Chernobylinterinform*, Vol. 1, p. 308 (in Russian).

Zarubin OL. January 1997. "^{137}Cs in Components of Trophic Chain of Kanev Reservoir after the Accident at ChNPP." *Proceedings of Annual Scientific Conference of Institute of Nuclear Research.* NASU, Kiev, pp. 349-353 (in Russian).

Chapter 4
Population Dose Estimate Due to Aquatic Pathways

Vladimir Berkovskiy
Radiation Protection Institute, Kiev, Ukraine

Oleg V. Voitsekhovich
Ukrainian Hydrometeorological Institute, Kiev, Ukraine

This chapter presents dose estimates via contaminated water pathways for the population living near the Chernobyl nuclear power plant and in the Dnieper River Basin. Water is the main pathway for transferring radionuclides from the heavily contaminated CEZ and catchments in the Dnieper River basin to less contaminated areas.

The Dnieper River is the most important source of water in Ukraine. Its water is used by about 8 million people for drinking. More than 20 million people living in the Dnieper River Basin are exposed to its contaminated water through consumption of irrigated farm products and fish. The commercial fishing catch averages about 25,000 tons annually.

Current annual releases of ^{137}Cs and ^{90}Sr from the contaminated floodplains and catchments to the Dnieper River reach 2 to 4 trillion and 10 to 20 trillion Bq, respectively. These values indicate the availability of radionuclides from heavily contaminated catchments to the cleaner Dnieper River downstream areas, as the water travels through the Dnieper cascade of reservoirs. Cesium-137 in the river has mostly accumulated in the six reservoirs, whereas a soluble form (a majority) of ^{90}Sr has actually reached the Black Sea. Population dose estimates for the middle to lower Dnieper Basin are based on monitoring studies performed in the years following the Chernobyl accident.

4.1 Dose Exposure Models for Water Pathways

This section describes the methodology used to determine radioactive contamination for various water pathways. Estimated radionuclide transfer through agricultural ecosystems is based on International Atomic Energy Agency (IAEA) recommendations (IAEA 1982) and official data from Ukraine Ministry of Health and Statistics. Dose exposure models were developed to estimate contaminated foodstuffs via water pathways. The

Y. Onishi, O.V. Voitsekhovich and M.J. Zheleznyak (eds.), Chernobyl – What Have We Learned?, 87–107.

ECOSYS-87 model (Muller and Prohl 1993) was used for irrigation effects. The aquatic pathways considered for internal exposure dose are

- drinking water from the Dnieper River
- fish from the Dnieper River
- irrigated food products.

4.1.1 Drinking Water

Drinking water is one of the critical pathways of dose exposure. A water purification station as a rule cannot significantly reduce most soluble radionuclides in contaminated water. It can reduce those radionuclides that have been transferred from water to suspended particulate sediments. The Chernobyl experience demonstrates that water purification can reduce cesium and strontium concentrations in the catchments by 1.5 to 4 times. In this evaluation, the water purification was assumed to be 2.

Age-dependent, daily uptake of radionuclides through drinking water during period *t* may be estimated as

$$A_w(t,\tau) = C_w(t) K_c v_w(\tau) \tag{4.1}$$

where

τ = age (or exposure period)
$A_W(t, \tau)$= daily uptake of radionuclides through drinking water (Bq/day)
$C_W(t)$ = radionuclide concentration (Bq/L) in water
K_C = water purification factor (0.5)
$v_W(\tau)$ = daily water consumption (L/day).

Age-dependent daily water consumption for five age groups is 0.5, 0.7, 1, 1.25, and 1.5 liters per day.

4.1.2 Fishing

Age-dependent daily radionuclide uptake from eating contaminated fish may be calculated as

$$A_f(t,\tau) = C_f(t) v_f(\tau) K_{f,r} \tag{4.2}$$

where

$C_f(t)$ = radionuclide concentration in edible fish parts (Bq/kg)
$v_f(\tau)$ = daily fish consumption (kg/day)
$K_{f,r}$ = fraction of Dnieper fish in total consumed fish.

Estimating $v_f(\tau)$ used in equation 4.2 is discussed in Section 4.2.6. The radionuclide concentration in edible fish parts is calculated as

$$C_f(t) = AwK_fTW \tag{4.3}$$

where

Aw = radionuclide concentration in water, Bq/L
K_f = radionuclide bio-accumulation factor for certain species
T = time factor for radionuclide uptake
W = portion of radionuclide accumulated in muscle tissue.

The estimated multiplication factors (K_f, T W) for ^{137}Cs and ^{90}Sr were 1,000 and 30, respectively (IAEA 1982). These factors can vary over a wide range, but the values selected for various fish species may reduce uncertainty. Some experimentally determined values are presented in this chapter.

4.1.3 Irrigation

Accumulation of radionuclides in plants on irrigated land occurs as a result of root uptake of radionuclides (1) deposited in soil after the first fallout from Chernobyl and from pre-Chernobyl origin and (2) introduced to the soil through irrigation with contaminated water and incorporated directly from the atmosphere through the leaves of the plants. Experimental data on the effects of using contaminated water as secondary contamination of irrigation crops are presented in previous Section 3.1. Section 4.2 describes the method used to calculate human exposure due to consumption of irrigated products.

4.2 Modeling Radionuclide Transport from Irrigated Foods

Total concentration absorbed by the green part of irrigated plants is

$$A_i = f_{w,i}\, A_w, \tag{4.4}$$

where

A_i = total radionuclide concentration (Bq/m^2) in plant species i
$f_{w,I}$ = radionuclide portion kept in plant species i
A_w = surface radioactivity introduced through irrigated water (Bq/m^2).

The amount of radionuclides in plant i species, $f_{w,i}$, may be determined as

$$f_{w,i} = \frac{LAI_i \cdot S_i}{R} \cdot (1 - \exp(\frac{-\ln 2}{3 \cdot S_i} \cdot R)) \tag{4.5}$$

where

S_i = effective water thickness (mm) for plant species *i*

LAI_i = leaf index (leaf surface in relation to total plant area)

R = irrigation water layer per leaf square given in mm.

When the radionuclide fraction exceeds 1, the value of $f_{w,i}$ is 1. Table 4.1 presents irrigation values for some plants. Typical model parameters, S_i, for grass, herbs, and maize are estimated as 0.2 to 0.4 and for other plants from 0.3 to 0.6. *LAI* values depend on the season. Table 3.12 in Chapter 3 presents *LAI* values for every plant species considered.

Table 4.1. Average irrigation rates used in Ukraine (10^3 m^3/ha)

Plant	Rate
Grass	0.6 – 1.5
Winter wheat	1.0 – 2.0
Spring wheat	1.0 – 2.0
Winter barley	1.0 – 2.0
Spring barley	1.0 – 2.0
Oats	1.0 – 2.0
Rye	1.0 – 2.0
Maize	1.0 – 1.5
Beet	0.6 – 1.0
Potatoes	0.6 – 1.0
Green vegetables	2.0 – 3.0
Vegetables	1.0 – 1.5
Root vegetables	0.6 – 2.0

The *LAI* for grass is best expressed by the yield. The following function presents *LAI* dependence on the yield:

$$\mathrm{LAI_g = LAI_{g,max}(1\text{-}exp(\text{-}k\ Y_g))} \tag{4.6}$$

where

LAI_g = leaves index for grass surface

$LAI_{g,max}$ = maximum *LAI* value for grass (m^2/m^2)

k = normalization factor (m^2/kg)

Y_g = grass yield in irrigation time (kg/m^2).

Table 3.12 in Chapter 3 shows plant yield as a function of time. Radionuclide concentrations in vegetables due to leaf absorption and root uptake are

$$\mathrm{C_i(t) = C_{i,l}(t) + C_{i,r}(t)} \tag{4.7}$$

where

$C_i(t)$ = total radionuclide concentration in plant species *i* (Bq/kg)
$C_{i,l}(t)$ = concentration absorbed through leaves (Bq/kg)
$C_{i,r}(t)$ = concentration absorbed through roots (Bq/kg).

4.2.1 Radionuclide Penetration Through Leaves

Radionuclide concentrations in plants are calculated by distinguishing between wholly consumed (e.g., green vegetables, grass) and partially consumed plants (e.g., herbs, potatoes). Radionuclide concentration $C_{i,l}(t)$ at time *t* after irrigation is determined by initial activity, loss of activity due to weather conditions (rain, wind), radioactive decay, and dilution effects due to growth of plant biomass. For wholly consumed plants except pasture grass, biomass growth is implicitly considered because the leaf-absorbed concentration depends on the yield. Thus, concentration can be expressed as

$$C_{i,l}(\Delta t) = \frac{A_i}{Y_i}\exp(-(\lambda_w + \lambda_r)\Delta t) \tag{4.8}$$

where

$C_{i,l}(\Delta t)$ = radionuclide concentration in species *i* at harvest (Bq/kg)
A_i = total specific activity (Bq/m^2) by plant species *i* depending on *LAI* at irrigation time
Y_i = yield (kg/m^2) of plant species *i*
λ_w = concentration loss rate (day^{-1}) due to weather conditions
λ_t = radioactive decay factor (day^{-1})
Δt = time interval from irrigation to harvest (days).

Tables 3.12 and 3.13 in Chapter 3 show plant yields and harvest periods. A longer period is used for pasture grass because it is used to feed animals and its concentration is reduced due to plant biomass dilution. Otherwise, the concentration of a radionuclide that is highly mobile in organic soils and transported through roots and its subsequent redistribution are expressed by

$$C_{g,1}(t) = \frac{A_g}{Y_g}[(1-a)\exp(-(\lambda_b + \lambda_w + \lambda_r)t + a\exp(-(\lambda_t + \lambda_r)t)] \tag{4.9}$$

where

$C_{g,l}(t)$ = concentration (Bq/kg) in grass in time *t* after irrigation
A_g = total specific activity of grass (Bq/m^2)
Y_g = grass yield (kg/m^2)
a = part transferred to root area
λ_b = dilution rate due to biomass growth (day^{-1})

λ_t = concentration decrease rate due to transfer to root area (day^{-1})
t = time period after irrigation (day).

The value of a 25-day half-purification period is λ_w; the value of λ_b is seasonally dependent (Table 4.2). Values λ_w and λ_b give the effective period of half-purification as 10 to 16 days; λ_t is 1.16×10^{-2}/d (biological half–purification period 60 days) at $a = 0.05$, based on Chernobyl data.

Table 4.2. Radionuclide dilution rate due to biomass growth and half-purification rate for grasses

Month	Dilution rate (day^{-1})	Half purification rate (day^{-1})
January-March	0.0	
April	1.65×10^{-2}	42
May	3.85×10^{-2}	18
June	$3.47\times10\text{-}^{2}$	20
July	$3.65\times10\text{-}^{2}$	19
August	$2.89\times10\text{-}^{2}$	24
September	$2.57\times10\text{-}^{2}$	27
October	$1.65\times10\text{-}^{2}$	42
November-December	0.0	--

Radionuclide concentrations in grass and straw silage are proportional to levels in grass gathered from May 15 to September 15. In the first part of this period the specific silage activity equals 70 percent of the specific activity in grass and in the second 30 percent, reflecting growth in these months.

For plants only partially consumed as food, radionuclide transfer from leaves to edible parts is considered. This process depends on the physiological properties of the radionuclides. It is important for cesium but not for strontium. Only direct incorporation of ^{90}Sr into edible parts was considered. The degree of transferred activity also depends on time, Δt, between irrigation and harvest.

Transient factor $T_i(\Delta t)$ is an activity fraction from leaves to edible plant parts. The transfer factor depends on the elements, plant species, and time from irrigation to harvest. The radionuclide concentration in plant species i, gathered after irrigation in Δt days, is defined with the following equation:

$$C_{i,l}(\Delta t) = \frac{A_i}{Y_i} T_i(\Delta t)\exp(-\lambda_r \Delta t) \qquad (4.10)$$

where

$T_i(\Delta t)$ = transfer factor for plant species i

Y_i = edible plant parts yield for species *i*.

4.2.2 Radionuclide Root Uptake

The radionuclide concentration in a plant from root uptake is calculated using the concentration in soil and accumulation factor TF_i, which is the ratio of the activity in plants (wet weight) to that in soil (dry weight) (Table 4.3).

$$C_{i,r}(t) = TF_i\, C_S(t) \tag{4.11}$$

where

$C_{i,r}(t)$ = radionuclide concentration in plant species *i* due to root uptake during time *t* since irrigation (Bq/kg)

TF_i = soil-plant accumulation factor for species *i*

$C_s(t)$ = radionuclide concentration in root area of soil at time *t* (Bq/kg).

Table 4.3. Soil plant accumulation factors, TF_i (Bq/kg)/(Bq/kg)[(a)]

Plants	^{137}Cs	^{90}Sr
Grass	0.05	0.5
Potatoes	0.01	0.02
Herbs	0.02	0.03
Green vegetables	0.02	0.03
Root vegetables	0.01	0.02
Vegetables	0.01	0.02
Distribution factor	0.001	0.01

(a) Values may differ from accumulation factors in ECOSYS model and some recommended distribution coefficients, K_d, for a soil-water system but were selected based on experimental data presented in Section 3.1.

Root uptake that takes place during plant growth is less and is corrected by a factor related to time from irrigation to harvest and vegetation. The radionuclide concentration in the soil of the root layer is calculated by

$$C_s(t) = \frac{A_s}{L\delta}\exp(-(\lambda_s + \lambda_f + \lambda_r)t) \tag{4.12}$$

where

A_S = total activity in soil (Bq/m^2)

L = root layer depth (m)

δ = soil density (kg/m^3)

λ_s = decreasing activity rate due to its transfer from root layer (day^{-1})

λ_f = radionuclide fixation rate in soil (day^{-1}).

$$\lambda_s = \frac{v_a}{L(1 + \frac{K_d \delta}{\Theta})} \qquad (4.13)$$

where

v_a = water infiltration rate through the soil layer (m/yr)

K_d = radionuclide distribution coefficient (cm^3/g)

Θ = amount of water in soil (g/g).

Root layer depths for arable land and pastures are 0.25 and 0.1 m, respectively. Annual water penetration averages 2 m/yr, soil density is $1.4x10^3$ kg/m^3, and average water content in soil is 20 percent. Distribution coefficients, $K_{d,}$ are 100 cm^3/g for ^{90}Sr and 1,000 cm^3/g for ^{137}Cs. The soil fixation rate, λ_f, was selected as $2.2x10^{-4}$/d for ^{137}Cs and $9x10^5$/d for ^{90}Sr.

Tables 4.4 through 4.6 present additional information for radionuclide transfer from soils to plants to agricultural products.

Table 4.4. Different yields per Dnieper Basin region (10^{-1} tons per ha)

Region	Cereal	Winter/ spring wheat	Maize	Winter/ spring barley	Potato	Veg.	White beet	Fruit and berry
Chernigov	20.3	23.2	33.4	19.9	137	122	233	20.1
Kiev	26.6	30.5	39.5	23.3	110	132	269	12.9
Cherkassy	28.1	32.9	35.0	25.7	113	120	258	10.6
Poltava	28.2	31.3	36.7	28.2	93	111	214	30.9
Kirovograd	26.8	29.3	30.3	28.3	56	87	213	10.4
Dnipropetrovsk	28.4	32.9	24.7	28.1	76	130	229	23.8
Zaporoziye	25.8	29.2	26.4	22.1	77	119	153	27.5
Kherson	25.4	28.6	34.5	20.4	69	149	-	25.8
Nikolajev	27.0	29.1	26.3	28.7	67	128	226	9.2
Crimea	30.7	31.0	36.9	30.0	105	169	-	36.4

Table 4.5. Average values of radionuclide T_f (transfer factor) from irrigation water to irrigated plants 10^{-3} ($Bq/kg_{\text{wet weight plants}}$)/($Bq/m^2_{\text{soil}}$)

Radio-nuclide	Irrig. type	Winter wheat	Alfalfa	Maize	Beet	Tomato	Cucumber	Cabbage
^{137}Cs	Furrow	1.0	2.5	0.4	0.6	0.3	0.4	0.5
	Sprinklers	2.0	6.0	0.6	0.8	0.6	0.6	0.8
^{90}Sr	Furrow	3.0	7.0	0.06	0.8	0.3	0.4	0.8
	Sprinklers	4.0	7.0	0.13	0.8	1.0	0.4	1.0
Note: Variation coefficient (C_v) is 50–100 percent).								

Table 4.6. Average transfer factor T_f from soil to plants 10^{-3} (Bq/kg)/(Bq/m^2)

Product	Sodi-podzolic sandy soil		Sodi-podzolic loamy soil		Grey forest loamy soil		Ordinary black soil		Heavy black soil	
	^{90}Sr	^{137}Cs	^{90}Sr	^{137}Cs	^{90}Sr	^{137}Cs	^{90}Sr	^{137}Cs	^{90}Sr	^{137}Cs
Cereals	3.0	0.81	0.72	0.25	0.16	0.12	0.13	0.062	0.056	0.042
Edible root		0.30	0.65	0.8	0.30	0.04	0.12	0.02	0.055	0.01
Potatoes	2.0	0.30	0.48	0.10	0.11	0.05	0.09	0.03	0.04	0.02
Veg. parts			2.2	0.35	0.50	0.16	0.40	0.10	0.18	0.07
Nat. grass hay	420	13	95	4.4	21	1.9	17	1.1	7.5	0.68
Note: Variation coefficient (C_v) is 50–100 percent.										

4.2.3 Animal Food Products

Animals that ingest contaminated feed and water produce contaminated food products. The activity concentration in food product by animal origin *m* and forage *k* is determined by equation 4.14 with the equilibrium accumulation factor TF_m (Table 4.7). Table 4.8 presents daily ration values, $I_{k,m}$.

$$C_m(t) = TF_m C_k(t) Kw_k I_{k,m}(t) \qquad (4.14)$$

where

$C_m(t)$ = radionuclide concentration (Bq/kg) in food product *m* at time *t*
TF_m = accumulation factor (day/kg) for food product by type *m*
$C_k(t)$ = radionuclide concentration (Bq/kg) in forage *k*
$I_{k,m}(t)$ = daily ration (kg/day) of forage *k* for animal *m*
Kw_k = part of forage *k* cultivated in irrigated field in total feed by type *k*.

Table 4.7. Transfer (accumulation) factor "forage-animal product" TF_m (IAEA 1982)

Animal product	^{137}Cs	^{90}Sr
Milk, day/L	1	0.1
Beef, day/kg	4	0.02
Chicken, day/kg	0.4	0.03
Eggs, day/kg	0.03	0.02

Table 4.8. Daily forage allowance (ration)

Animal	Forage	Intake (kg/day of fresh feed)
Cow	Grass	50[(a)]
Pig	Winter barley	3
Chicken	Winter wheat	0.09
(a) Green grass; in winter it would be hay and silage.		

Due to insufficient data, the Kw_k value in equation 4.14 is considered to be proportional to an irrigated common sown area where forage *k* grows:

$$Kw_k = \frac{SwK_{ik}}{S_{ik}} \tag{4.15}$$

where

Sw = total irrigated area

K_{ik} = part of area for plant *i*, from which the forage of *k* type is produced on an irrigated area

S_{ik} = total sown area *i* for plant species in a specific region.

4.2.4 Animal Food Product Contamination due to Contaminated Water

A considerable fraction of radionuclides enter an animal's body through drinking water. This pathway is especially important for radionuclides in the milk and meat of cows, an essential part of the human diet. Contamination levels of animal products due to water consumption may be estimated by equation 4.15. The radionuclide uptake by animals drinking contaminated water is expressed as

$$A_m(t) = C_w(t) I_{w,m}(t) \tag{4.16}$$

where

$A_m(t$ = radionuclide uptake into animal's body *m*, (Bq/day)

$C_w(t)$ = radionuclide concentration (Bq/L) in drinking water

$I_{w,m}(t)$ = daily (L/day) water consumption for animal food product *m*, as shown in Table 4.9 (Golosov et al. 1967).

Table 4.9. Daily average water consumption rate in liters per animal

Animal	On Farm		On Pasture	
	With water-supply	Without water supply	Ordinary district	Desolate district
Cows hand-milked	90	70	60-70	50-60
Cows milked by machine	115	95	-	-
Cows not milked	60	50	50	45
Bulls	60	50	50	45
Calf (4–6 months)	20	18	-	-
Calf (up to 2 years)	35	30	35	30

Animal water consumption depends on species, age, productivity, operational conditions, feeding, meteorological conditions, water temperature, and individual features. In Ukraine's steppe and forest-steppe regions cattle consume less water, but productivity is at the standard level. Calves consume twice as much water by weight as adult animals. Milking cows that yield 12 L of milk drink 35 to 40 L of water a day, and those yielding 40 to 45 L of milk drink up to 110 L of water a day (Golosov et al. 1967).

4.2.5 Food Processing Influence on Product Contamination

Food can also be contaminated by processing and storage. The radionuclide concentration in food k is expressed as

$$C_k(t) = C_{k0}(t - t_{pk})P_k \tag{4.17}$$

where

$C_k(t)$ = concentration (Bq/kg) in prepared food k at time t
$C_{ko}(t)$ = concentration (Bq/kg) in fresh product at time t
P_k = food processing factor k.

Table 4.10 lists food processing factors. Concentration changes in prepared products are not considered.

Table 4.10. Food processing factors

Foodstuff	Processing factor (radionuclide content in fresh product = 1)		
	^{90}Sr	^{137}Cs	Others
Wheat flour	0.5	0.5	0.5
Wheat bran	3	3	3
Rye flour	0.5	0.6	0.5
Rye bran	3.5	2.7	3
Peeled potatoes	0.8	0.8	0.8
Vegetables	0.8	0.8	0.8
Butter	0.2	0.2	1
Cream (30% rich)	0.4	0.7	1
Cheese	6	0.6	1
Cottage cheese	0.8	0.6	1
Condensed milk	2.7	2.7	2.7

4.2.6 Radionuclide Uptake into the Human Body

Radionuclide uptake into the human body is calculated with time-dependent food concentrations and the average consumption by various population groups:

$$A_h(t,\tau) = \sum_k C_k(t) v_k(\tau) K w_k \tag{4.18}$$

where

$A_h(t)$ = radionuclide uptake with food (Bq/day)
$C_k(t)$ = radionuclide concentration (Bq/kg) in product k
$v_k(\tau)$ = daily ration (kg/day)
τ = age of person at time t
Kw_k = ratio of the diet containing the plant k species grown on irrigated lands to total consumption of food product of the same species k. (For animal products this parameter is 1, due to the condition applied for equation 4.15.

Time-dependent daily rations $v_k(\tau)$ are calculated by

$$v_k(\tau) = K_k(\tau)\frac{S_k \sum_{\tau} N(\tau)}{\sum_{\tau} K_k(\tau)N(\tau)} \tag{4.19}$$

where

$K_k(\tau)$ = ratio of consumed product k by age group τ to consumption k by adults (value for adults is 1)
S_k = total produce consumption k by all age population groups
$N(\tau)$ = number of people in age group τ.

4.3 Reconstruction and Prediction of Exposure Dose for Population Living in Dnieper Basin

One of the serious consequences of the Chernobyl accident is the transport and redistribution of radionuclides from highly contaminated to relatively clean territories by the Dnieper River system. This process enlarges the scale of the accident and requires evaluation of radiation impacts on the population living along the Dnieper reservoirs downstream of the CEZ. The exposure dose estimation due to water pathways is very difficult because of the large territory and because it is difficult to separate the contamination pathways of foodstuffs based solely on concentration of activity per ration.

Long-term dose forecasts were performed using actual monitoring data of radionuclide concentrations in water, long-term probabilistic hydrological scenarios simulated for Pripyat and Dnieper river runoff, and radionuclide transfer models of the food chain. These models use all accessible information on the properties of agricultural production, structure of irrigated fields, fishing, and drinking water (ICRP-56 1989; Berkovskiy et al. 1994,

1995). The dose assessment data were the basis for creating the concept of water protection measures in the CEZ (Voitsekhovich et al. 1996a, b).

4.3.1 Internal Dose Components

Internal exposure doses of the population in the Dnieper River Basin were caused by the initial radioactive contamination of the territories and the consumption of contaminated drinking water, fish, animals, and agricultural products from irrigated fields. The main radionuclides for determining exposure doses in the early phase of the Chernobyl accident were ^{131}I and ^{137}Cs; ^{90}Sr came in a later phase.

Estimating thyroid exposure doses of radioactive iodine is interesting mainly for epidemiological investigations because iodine is a problem only in the initial period after the accident. This dose cannot be reduced by countermeasures because the impact has already been realized and cannot be remediated. Doses due to transuranic isotopes can be neglected due to their low concentrations and low solubilities (Berkovskiy et al. 1995).

External exposure mainly depends on the length of time spent in contaminated places. For exposure doses due to water pathways, external doses are negligible compared with internal doses. Thus, the total exposure dose was estimated from the following three aquatic pathways:

- drinking water consumption from the Dnieper reservoirs
- consumption of fish from the Dnieper reservoirs
- consumption of agricultural products grown on lands irrigated by water from the Dnieper reservoirs.

4.3.2 Dose Calculation Principles

Some explanations of exposure dose for a population are in order. As a rule, the log-normal distribution of individual dose takes place in any large population. There are some with very high doses and many with relatively small doses. A group of these persons is a critical group. Second, the principal ratio between annual dose rate and expected exposure dose: or ^{137}Cs the internal exposure dose is realized mostly in the first year, so the dynamics of average annual dose rate and expected annual dose are basically the same. For ^{90}Sr, dose accumulation continues over a lifetime. Even after termination of ^{90}Sr intake, irradiation of bone tissues continues from ^{90}Sr in the skeleton.

Third, it is not difficult to estimate the exposed population during the period following the accident. However, predicting future dose is very complicated because the most probable events and conditions must be predicted accurately. The proper selection of a population group for collective dose estimation is also an important factor.

A computer model (Berkovskiy et al. 1994) was developed to calculate internal exposure doses for populations in regions of Ukraine adjacent to the Dnieper cascade and in Crimea, where people consume water from Kakhovka Reservoir. The dose estimate used region-specific data (e.g., age, consumption of agricultural products and fish, irrigation). Simulated metabolic processes and dose levels in the human body are based on ICRP-56 (1989), ICRP-60 (1990), and ICRP-67 (1993). Potential population exposure was provided in terms of annual effective dose (AED). In addition, age-dependent collective dose commitment (70-year period) (CDC70) was calculated.

4.3.3 Water Use

Table 4.11 provides water uses from the Dnieper Basin.

4.3.4 Age and Gender of Population

Table 4.12 presents age and gender data of the populations from various regions of the Dnieper Basin.

4.3.5 Long-Term Human Exposure due to Dnieper River Contamination

Long-term hydrological contamination scenarios were developed by the computer model WATOX (Zheleznyak et al. 1992), with discharges from the Dnieper River and its tributaries averaged over three months beginning in 1995. A high-flow period from 1970 through 1992 and a low-flow period from 1912 through 1950 provided the forecasting template. From 1986 to 2000, actual monitoring data were used. Estimations from 1986 to 2000 assumed that water-soluble ^{90}Sr would be constant due to competing ^{90}Sr decay processes and leaching from fuel particles in the watershed's contaminated soils. Calculations beyond 2001 used an exponential factor to correct the radionuclide concentration with a half-life twice that of ^{90}Sr. ^{90}Sr concentrations averaged over three months would be 0.1 to 0.2 Bq/L in 2056. The ^{137}Cs and ^{90}Sr concentrations in the Dnieper reservoirs were used to determine population exposure doses (Berkovskiy et al. 1996a, b).

The dose assessment for populations living along the Dnieper River have also been carried out by others using different methods (e.g., Voitsekhovich et al. 1998; Los' 1993). The results were not significantly different from the estimates by Berkovskiy et al. (1996a, b).

Table 4.11. Water use in the Dnieper Cascade

Region	Rivers and reservoirs	Population (x 10^6)	Population drinking water from Dnieper (x 10^6)	Irrigated area (10^3 ha)	Total agricultural land (10^3 ha)
Chernigov	Dnieper and Desna Rivers	1.4	-	6	2034
Kiev	Kiev and Kanev Reservoirs	4.5	0.75	116	1670
Cherkassy	Kremenchug	1.5	0.2	66	1420
Kirovograd	Kremenchug	1.2	0.4	56	2015
Poltava	Kremenchug	1.7	0.3	57	2086
Dnepropetrovsk	Dneprodzerzhynsk Zaporozhiye, Kakhovka	3.8	2.0	254	2387
Zaporozhie	Kakhovka	2	1.0	272	2225
Nikolayev	Kakhovka	1.3	0.4	190	1953
Kharkov	Dneprodzerzhynsk	3.2	0.4	-	-
Lugansk	Dneprodzerzhynsk	2.9	0.1	-	-
Donetsk	Dneprodzerzhynsk	5.3	2.2	-	-
Kherson	Kakhovka	1.2	-	464	1932
Crimea	Kakhovka	2.5	0.5	390	1748
Total		32.5	8.1	1871	19470

Table 4.12. Population in the Dnieper Basin on January 1, 1992

Ukraine Regions	1 to 2 yr	2 to 7 yr	7 to 12 yr	12 to 17 yr	Adults
Chernigov	31851	91010	90023	90584	1077197
Kiev	117506	331162	326605	31854	3407819
Cherkassy	38912	105647	101303	104901	1158351
Poltava	42952	120248	114013	117667	1343154
Kirovograd	321095	88948	83893	84970	930652
Dnepropetrovsk	98363	287685	273048	270244	2932633
Zaporozhiye	53754	154723	148573	148110	1571070
Kherson	36469	100793	97190	96046	919193
Nikolaev	37109	106867	100649	99560	986422
Crimea	68379	198054	190058	184347	1895634

4.3.5.1 Drinking water

The Dnieper River provides drinking water for more than 8 million people in the Dnepropetrovsk and Donetsk regions. Around Kiev, the Dnieper provides water for more than 750,000 people. Because the Kiev water station is so close to the initially contaminated area, Kiev was considered a critical area for drinking water. The age-dependent parameters are 0.20 L/d for a three-month-old child, 0.50 L/d for a one-year old, 0.70 L/d for a five-year old, 1.00 L/d for a 10-year old, 1.25 L/d for a 15-year old, and 1.50 L/d for adults.

The water cleaning factors for ^{137}Cs and ^{90}Sr are equal to 2. Figures 4.1 through 4.3 show annual collective effective doses (ACED) for Kiev, the Poltava region, and the Crimea regions without water protection measures. A contribution of ^{131}I in Kiev water to ACED was estimated with measurements of water samples from the Kiev reservoir (Vakulovskiy et al. 1990). The estimated dose in 1986 is about 100 man-Sv for 750,000 people.

4.3.5.2 Commercial fishing

The Dnieper reservoirs are used extensively for commercial fishing, which produces an average of 24,000 tons annually. There was no decrease in fishing after the Chernobyl accident. For this dose estimate, actual values of fishing and contamination levels of fish were used (Napier et al. 1992). Fish contamination levels were predicted using simple bio-accumulation models.

Figures 4.1 through 4.3 show the results of fish consumption compared with total collective dose for Kiev, Poltava, and Crimea with no remediation countermeasures carried out beyond 2000.

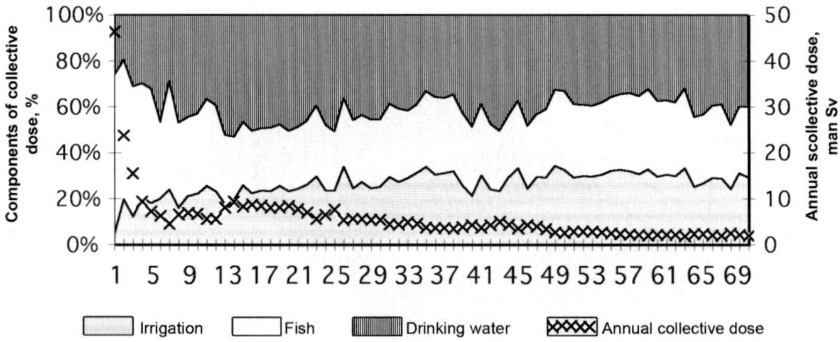

Figure 4.1. Annual collective effective dose for population of Kiev region from water pathway

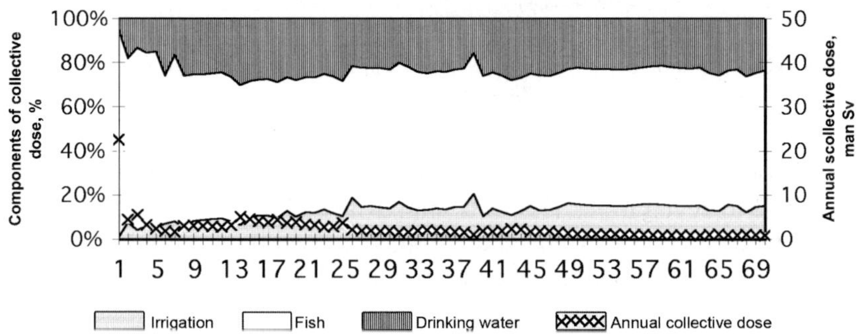

Figure 4.2. Annual collective effective dose forming for population of Poltava region due to water pathways (water from Kremenchug Reservoir)

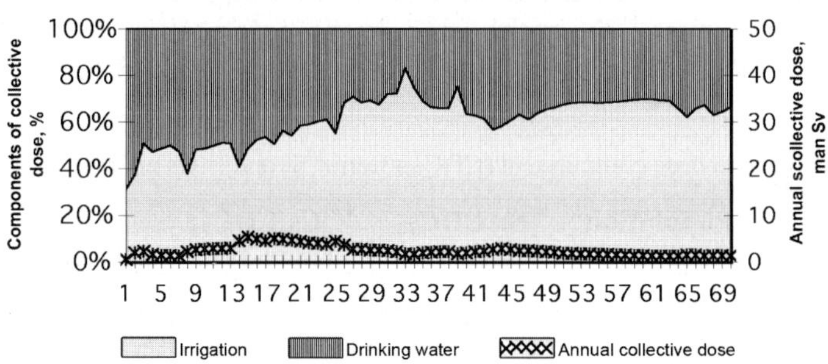

Figure 4.3. Annual collective effective dose for the Crimean population from water consumption (water from Kakhovka Reservoir)

4.3.5.3 Agricultural water

In the Dnieper basin, more than 1.8 million hectares are irrigated for agriculture. Fifty percent of this area is planted in forage and 10 percent in vegetables. The Kakhovka Reservoir is the primary irrigation source for 72 percent of this region (Table 4.11). Figure 4.4 presents ^{90}Sr content in irrigated crops grown in the Kiev region. The highest radionuclide concentrations are found in vegetables, bread products, milk, and potatoes.

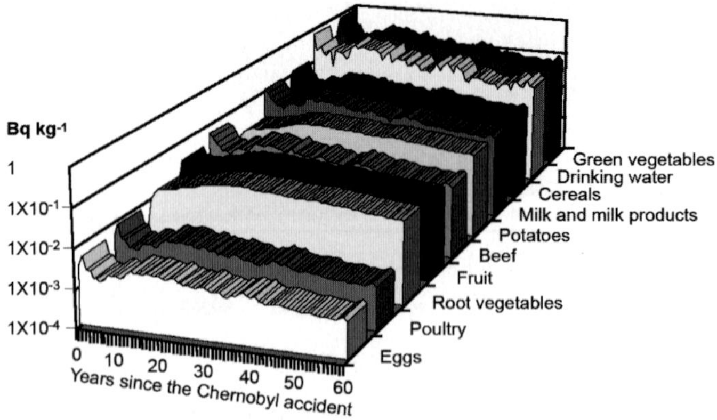

Figure 4.4. Concentration of ^{90}Sr in products grown in Kiev region using water from Kiev and Kanev reservoirs

The contribution of water pathways to the collective dose for the Kiev region is five to seven times less now than in 1986, as shown in Figure 4.1. A reverse effect is observed for the Crimea due to low initial fallout then increased contamination due to downstream transport by the Dnieper River (see Figure 4.3).

Table 4.13 lists doses caused by ^{90}Sr and ^{137}Cs in the Dnieper River Basin. These calculations are based on the absence of water remediation measures. The results show that ^{90}Sr is the primary dose formation factor for water pathways, and its dose can exceed the cesium dose by 2 to 35 times.

Table 4.13. Collective dose commitment (CDC_{70}) from ^{90}Sr and ^{137}Cs in Dnieper River Basin

Region	Population (millions)	^{90}Sr CDC_{70} (man-sv)	^{137}Cs CDC_{70} (man-sv)	Ratio ^{90}Sr $CDC_{70}/^{137}Cs$ CDC_{70}
Chernigov	1.4	4	2	2
Kiev	4.5	290	190	1.5
Cherkassy	1.5	115	50	2.3
Kirovograd	1.2	140	40	3.5
Poltava	1.7	130	60	2.2
Dnepropetrovsk	3.8	610	75	8
Zaporojie	2	320	35	9
Nikolaev	1.3	150	20	8
Kharkov	3.2	60	4	15
Lugansk	2.9	15	1	15
Donetsk	5.3	330	20	17
Kherson	1.2	100	20	5
Crimea	2.5	175	5	35
Total	32.5	2439	522	5

4.4 Conclusions

This chapter describes the methodology applied to dose calculations from the Chernobyl accident and the results of its application. This methodology can be applied to other industrial sources in ordinary and emergency situations. Its individual elements have been used to develop standards for permissible discharge levels in nuclear power plant operations.

Water pathways are a small part of the total dose received by the population from all dose-forming factors. However, the Dnieper River system transfers radionuclides from the CEZ and other highly contaminated areas to the cleaner areas, increasing the number of contaminated areas and the exposed population. Radionuclide transfer and exposure from drinking water, fish, and irrigated crops affect millions of people in the downstream region. Thus, radionuclide mobility in the river system deserves special attention.

Exposure estimates from water pathways measure radiation risk to populations in water use regions and are crucial in developing optimal water protection measures in areas affected by the Chernobyl accident. The estimates must consider the aquatic system as a whole, incorporating radionuclide migration, water use, and population dose. For example, the expense of hydraulic structures and water decontamination schemes in the CEZ solved some local problems of a specific reservoir or stream but did not significantly reduce the exposure doses for the population. Remedial countermeasures are generally more effective when protective measures are installed in a localized contamination area or areas to prevent future water contamination.

Doses can reach 3,000 man-Sv due mainly due to ^{137}Cs and ^{90}Sr in water if all current protective activity implemented before 1991 is terminated. The contribution of ^{137}Cs to the exposure dose is mostly restricted to Ukraine's northern regions. But due to the relatively uniform ^{90}Sr distribution in the Dnieper cascade of reservoirs, the ^{90}Sr contribution to the expected collective dose of the population is relatively uniform.

Protective measures like dike construction that were implemented in 1992 and 1993 along the Pripyat River's left (northeast) bank floodplain decreased exposure doses by about 700 man-Sv. Similar measures on the right (southwest) bank would decrease the dose by another 200 to 300 man-Sv.

The modeling results presented in this chapter are based on the best available information on radionuclide concentrations in rivers and reservoirs, radiological information on soil, agricultural products from irrigated lands, and fish contamination. However, the reliability of dose estimation is determined by the accuracy of the required information, such as long-term changes in radionuclide contents of the rivers and reliability of forecasting radiological contamination levels of soil, agricultural products of irrigated lands and fish, and future population data. The studies in this field should be continued to improve the predictive methodology and required databases.

4.5 References

Berkovskiy V, VS Repin, G Ratia, et al. 1994. *Estimation of Dnieper Water Contribution to Dose Formation of Ukrainian Population. Final Report.* Research Contract No. 35-93. Ministry on Chernobyl Affairs, Kiev, p. 152.

Berkovskiy V, V Repin, and G Ratia. 1995. *Estimation of potential risk for population of Ukraine from transuranium isotopes in and out of Chernobyl Exclusion Zone.* Ministry Chernobyl Affairs of Ukraine, Kiev, p. 179 (in Russian).

Berkovskiy V, O Voitsekhovich, O Nasvit, M Zheleznyak, and U Sansone. 1996a. "Exposures from Aquatic Pathways." *First International Conference of European Commission, Belarus, Russian Federation, and Ukraine on Radiological Consequences of Chernobyl Accident.* Minsk, pp. 283-294.

Berkovskiy V, G Ratia, and O Nasvit. 1996b. "Internal Doses to Ukrainian Population Using Dnieper River Water." *Health Phys.* 71(1):37-44.

Golosov IM, AN Boltushkin, and PF Pribytkov. 1967. "Sanitary Estimation of Water in Stock Raising." *Kolos*, Leningrad (in Russian).

IAEA. 1982. *Generic Models and Parameters for Assessing Environmental Transfer of Radionuclides from Routine Releases. Exposure of Critical Groups.* IAEA Safety Series No. 57.

ICRP-56. 1989. *Age-dependent doses to members of the public from intake of radionuclides.* ICRP Publication 56 Part 1, Pergamon Press, Oxford, U.K., p. 122.

ICRP-60. 1990. *Recommendation of the International Commission on Radiological Protection.* ICRP Publication 60, Pergamon Press, Oxford, UK.

ICRP-67. 1993. "Age-Dependent Doses to Public from Intake of Radionuclides: Part 2, Ingestion Dose Coefficients." ICRP Publication 67, *Annals of the ICRP* 23(3/4):167.

Los' IP. 1993. "Hygienic Assessment of Dose-Forming Sources of Ionizing Radiation of Natural and Man-Induced Origin and Exposure Doses of Population of Ukraine." Abstract of Ph.D. dissertation, Kiev (in Russian).

Muller H and G Prohl. 1993. "ECOSYS-87: A Dynamic Model for Assessment of Radiological Consequences of Nuclear Accidents." *Health Physics,* Vol. 64, pp. 232-252.

Napier BA, WL Templeton, IN Ryabov, II Kryshev, and TG Sazykina. 1992. *Long-Term Radiation Dose and Effects from Contamination of Dnieper Aquatic Ecosystem.* Pacific Northwest National Laboratory, Richland, Washington.

Vakulovskij SM, OV Voitsekhovich, IY Katrich, VI Medinets, AI Nikitin, and VB Chumichev. 1990. "Radioactive Contamination of Water Systems in the Area Affected by Releases from the Chernobyl Nuclear Power Plant Accident." *Environmental Contamination Following a Major Nuclear Accident*, Vol. 1, pp. 231-246. IAEA, Vienna.

Voitsekhovich OV, V Kanivets, et al. 1992. "Hydrological Processes and Their Influence on Radionuclide Behaviour and Transport by Surface Water Pathways as Applied to Water Protection after Chernobyl Accident." *Proc. UNECSO, Hydrological Impact of NPP.* Paris, pp. 85–105.

Voitsekhovich O, U Sansone, M Zheleznyak, and D Bugai. March 1996a. "Water Quality Management of Contaminated Areas and its Effects on Doses from Aquatic Pathways." *First International Conference of the European Commission, Belarus, Russian Federation and Ukraine on the Radiological Consequences of the Chernobyl Accident.* Minsk, Belarus, pp. 401-410.

Voitsekhovich O, B Prister, O Nasvit, I Los,' and V Berkovskiy. 1996b. "Present Concept on Water Protection and Remediation Activities for Areas Contaminated by the 1986 Chernobyl Accident." *Health Phys.* 71(1):19-28.

Voitsekhovich OV and EL Panasevich. 1998. "Dose and Social and Economical Expediency of Modern Water Protective Activity in ChNPP Exclusion Zone." *Bulletin of Ecological State of Exclusion Zone* №12, pp. 3-8 (in Ukrainian).

Zheleznyak MJ et al. 1992. "Mathematical Modeling of Radionuclide Dispersion in the Pripyat-Dnieper Aquatic System after the Chernobyl Accident." *Science of the Total Environment*, 112:89-114.

Chapter 5
Radiation Risk Assessment and Countermeasure Justification

I. P. Los'y, I. Segeda, and K. Shepelevich
Research Center for Radiation Medicine, Kiev, Ukraine

O. V. Voitsekhovich
Ukrainian Hydrometeorological Institute, Kiev, Ukraine

Yasuo Onishi
Washington State University, Richland, WA, USA
Formerly of Pacific Northwest National Laboratory, Richland, WA, USA

This chapter presents conceptual approaches to determining what remedial intervention would protect the population of Ukraine from the harm caused by radiation from the 1986 Chernobyl accident. Also discussed is the feasibility of developing remediation criteria and intervention levels for water protection. Specific recommended countermeasures are based on total population dose data obtained after the accident. Estimates are given for radiation-induced cancer mortality from accidental and nonaccidental exposure sources. Health, economic, social, and psychological criteria are presented to support the interventions as are methods used to justify and optimize specific countermeasures.

5.1 Conceptual Approaches to Population Protection Countermeasures

After the 1986 Chernobyl accident, many countermeasures were implemented to reduce the adverse health impacts to the local population and environmental impacts from potential secondary radiation exposures. Many were successful; however, some turned out to be unjustifiable because they required too many resources, were wrong, or were not implemented timely enough to produce benefits.

There are many examples of inadequate planning and implementation of the countermeasures to prevent or reduce radiation exposure via water pathways. The most important outcome from these experiences is a guideline that can be used to select and implement effective countermeasures. This will

Y. Onishi, O.V. Voitsekhovich and M.J. Zheleznyak (eds.), Chernobyl – What Have We Learned?, 109–140.

be accomplished by assessing successful and failed remediation activities after the Chernobyl accident with commonly accepted approaches.

To achieve this goal, the general principles of countermeasure planning were studied to protect the public in case of a radiation accident. Without understanding the general principles of radiation protection, water protection measures would not be correct. Remediation measures taken after the 1986 Chernobyl accident would have been easier to implement and would have produced more benefits had there been more advance planning, better understanding of the time-varying factors and conditions, and a more accurate interpretation of actual and predicted exposure doses.

Since 1986 much has been learned about effective radiation countermeasures. The following sections cover planning for countermeasures to protect the population and water based on general radiation protection principles.

5.1.1 General Radiological Protection Principles

The International Commission on Radiological Protection's (ICRP) recommended radiation protection principles, especially for accident situations, are based on past scientific investigations as well as experiments of ionizing radiation on individuals. These studies include epidemiological observation of survivors of the Hiroshima and Nagasaki nuclear bombs, clinical records of people who became ill after exposure to ionizing radiation and sought medical treatment, data on uranium miners, and laboratory experiments on radiation effects in living organisms.

These original recommendations (ICRP 1977) were updated with a large amount of Chernobyl information, and the revised recommendations were released as Publication 60 (ICRP 1991). Publication 60 contained several changes from the 1977 recommendations, including new terms for some dose values, modified values of radiation and tissue-suspending coefficients of irradiation, readjusted risks of stochastic consequences of exposure, new limits of the deterministic effects of exposure, and modified principles of radiological protection. New approaches were also proposed for implementing practical radiation protection activities and for intervention in various situations. In addition, more rigid dose limits were established for both population and individual exposures.

The modern radiological protection methodology divides radiation protection activities into *practical activity* and *intervention*, as shown in Figure 5.1.

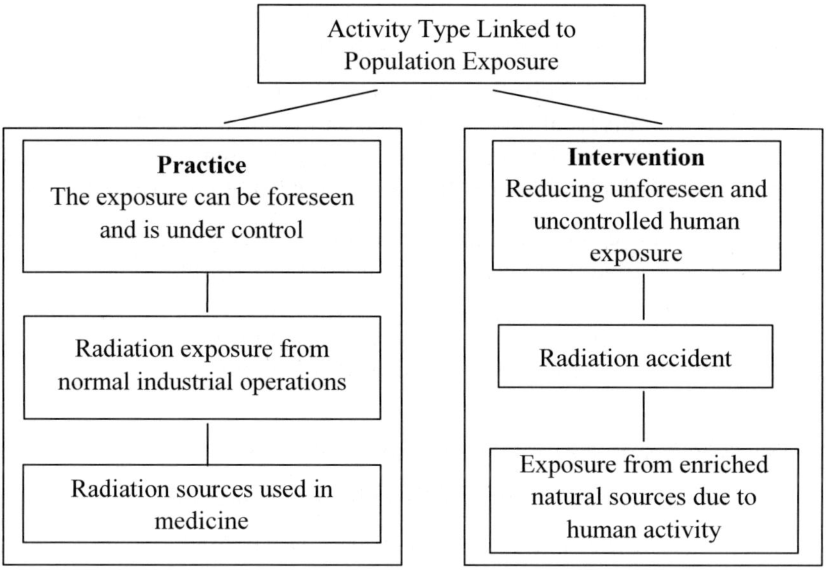

Figure 5.1. International recommendations for human radiological protection

Practical activities ("Practice" in Figure 5.1) use radiation sources to achieve material or other benefits and may lead to:

- increased exposure doses
- additional exposure paths
- more people exposed
- changed structure of irradiation paths from all sources related to the present activity.

Practical activities that may increase the exposure probability or number of people exposed to radiation include:

- radiation source production
- radiation source and radioactive substance uses in medicine, research, industry, agriculture, education, and others
- nuclear energy production, including all elements of fuel-energy cycle
- conservation and transport of ionizing radiation sources
- radioactive waste management.

Radiation protection from these practical activities follows the following principles (ICRP 1991; IAEA 1994):

- The practical activity must provide sufficient benefit to the people exposed to justify any harm caused (principle of justification).
- Optimization to limit the number of exposed population and an amount of individual dose should be as low as reasonably achievable to achieve the maximum benefit compared with the sacrifice for economic or societal factors (principle of protection optimization).
- Individual doses must be limited so that all activities combined do not exceed individual limits (principle of limitation of individual dose).

Interventions (see Figure 5.1) decrease or prevent uncontrolled and unforeseen exposure or reduce the probability of exposure in the following situations:

- emergency exposure (acute or chronic)
- chronic exposure from artificially enhanced sources of natural origin
- other temporary exposure situations as defined by a radiation regulatory body.

Radiation safety protection under the intervention category follows these general principles:

- All possible efforts should be taken to prevent grave deterministic effects (principle of inadmissibility of grave deterministic effects).
- Intervention must produce more benefit than harm (principle of intervention justification).
- The level and duration of the intervention must be optimized to achieve the maximum benefit (principle of optimization of form, scale, and duration of intervention).

5.1.2 Evolution of International Recommendations for Intervention

In the late 1970s, international organizations began developing intervention principles and criteria to apply to emergency situations caused by ionizing radiation. The Commission of European Communities (CEC) was the first international organization to establish exposure dose levels (CEC 1982). In 1991, ICRP developed "Principles of Intervention for Population Protection in Case of an Radiation Emergency Situation" by including more adaptable approaches that justify intervention and optimize the duration and scale of an intervention effort (ICRP 1991).

The International Atomic Energy Agency (IAEA) also revised and expanded their intervention criteria in Publication 109, *Intervention Levels for*

Nuclear or Radiation Emergency Situations (IAEA 1994). This document superseded the previously published IAEA document, *Principles of Intervention Levels Established for Population Protection in Case of Nuclear Accident or Radiation Emergency Situation* (IAEA 1985). Publication 109 is the basis for intervention standards (regulations) and their numeric values involving international basic safety standards (IAEA 1997).

Previously, the intervention levels for countermeasures were based on expected exposure doses without countermeasures (IAEA 1985; ICRP 1987). The current approach to determining intervention levels uses "potentially averted dose." Figure 5.2 illustrates that, after a long post-accident period, countermeasures are not very effective because the primary dose is already formed. Expected equivalent doses are given for the first year only. Interventions are proposed when levels range between upper dose limits, which require countermeasures, and lower dose limits, where intervention may not be justified (Table 5.1). A range between the upper and lower levels is selected as an intervention level by decision makers. As indicated in Table 5.1, the recommendations of international organizations for radiological protection before 1985 referred to "extraordinary" and "intermediate" phases of the accident and did not involve countermeasures that were implemented later in the post-accident period.

The newer guides (ICRP 1991; IAEA 1994) recommend more rigid values for intervention levels for all stages of a radiation accident. These guides also recommend dividing countermeasures into emergency and long-term, as shown in Table 5.2. Emergency is for the early post-accident phase, and long-term applies to intermediate or late phases. These recommendations

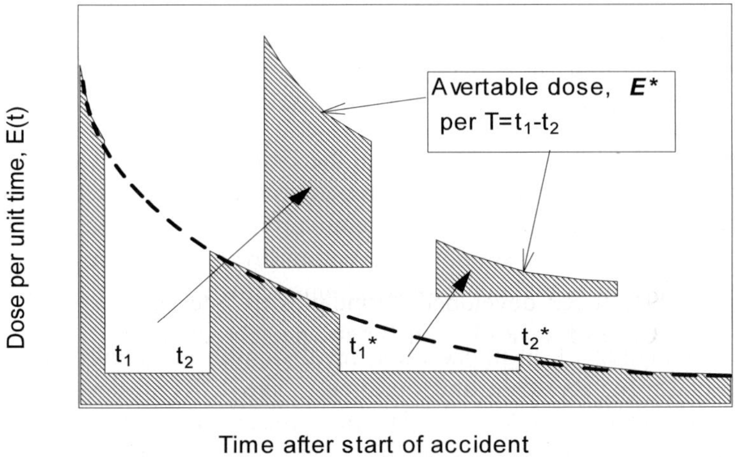

Figure 5.2. Dose averted from applied countermeasure during period Δt

Table 5.1. Intervention levels for post-accident countermeasures recommended by IAEA and ICRP prior to 1985

Post-Accident Period	**Countermeasure**	**Predicted Equivalent (absorbed) dose for the whole body**[a]	
		Lower limit	Upper limit
Early[b]	Sheltering	5 mSv	50 mSv
	Iodine prophylaxis (dose only for thyroid)	50 mGy	500 mGy
	Evacuation	50 mSv	500 mSv
Intermediate[b]	Food product control	5 mSv	50 mSv
	Temporary resettlement	50 mSv	500 mSv

(a) Internal dose for thyroid glands.
(b) For the early post-accident period, this value is not accurately determined.

Table 5.2. Post-accident intervention levels recommended by ICRP (1991) and IAEA (1994)

Countermeasure	**Countermeasure**	**Prevented effective (absorbed) dose**[a]	
		IAEA optimal intervention level[b]	ICRP optimal range values
Emergency	Sheltering (less than 2 days)	10 mSv	5-50 mSv
	Iodine prophylaxis (thyroid dose)	100 mGy	50-500 mGy
	Evacuation (week)	50 mSv	50-500 mSv
Long-Term	Temporary resettlement	Begin at 30 mSv per month, up to 10 mSv/month	Almost justified at 1 Sv, optimized at 5-15 mSv/mo
	Permanent resettlement	Living dose exceeds 1 Sv	

(a) Internal dose for thyroid gland.
(b) Optimal intervention level refers to average dose for a properly selected group, not for critical individuals with the most exposure, whose predicted dose should be maintained lower than the threshold of the occurrence of deterministic effects.

on countermeasures are not restricted to the first post-accident year. Furthermore, IAEA suggests that "general intervention levels" be used for only one intervention level for every concrete countermeasure in a basic emergency situation. In an accident, similar intervention levels should be used and then readjusted when the actual conditions become known.

Implementation of an intervention level should be concurrently justified and optimized for its scale and duration. The ICRP follows a two-level concept for intervention levels. An intervention is always justified if the upper limit is exceeded but not justified below the lower limit, as shown in Table 5.2. The concept of "action levels" was derived (back-calculated) from dose values to be used for radiation accident monitoring and forms a quantitative basis of the go or no-go decision making for countermeasure planning, especially during the rehabilitation phase of a radiation accident.

The action level can be selected for dose rate inhaled; radiation activity found in air, water, and food products; fallout activity on soil; etc. Action levels can be used to regulate contaminated food products (IAEA 1994) and can overcome the difficulties of radiation monitoring. Action levels thus determine the maximum admissible contamination level without countermeasures.

The most important element of the international recommendations is to differentiate radiation protection approaches for a properly functioning nuclear technology case and an intervention situation for radionuclide accidents. The task to limit exposure doses is a quality control issue for radiation-nuclear operation, while the reduction of exposure dose is in the intervention situation. Therefore, regulations for functioning radiation-nuclear technologies are not a part of these international recommendations.

5.1.3 Criteria Development and Intervention Levels in the Former Soviet Union

In 1970, the former Soviet Union approved the first decision-making procedure, "Temporal Method Directives of Measures to Be Developed for Population Protection in Case of Accidents at Nuclear Reactors." The objective was to reduce the number of exposed people and doses from radioactive contamination. Based on this document, each reactor site developed a detailed plan to reduce population exposure after a nuclear accident.

Table 5.3 shows these decision-making criteria and relative protective measures. Interventions are shown as a two-level system based on predicted dose values, conforming to international practices of that time period.

Table 5.3. FSU intervention criteria for nuclear accident initial phase, adapted in 1970

Intervention level	Protective measure	Total external exposure dose, mGy
I	Evacuation	≥750
I–II	Temporary shelter on premises, restriction on being in open, leather and clothing decontamination, limitation on contaminated food consumption	250–750
II	No need for these measures	<250

In 1981, *Sanitary Regulations of Design and Operation of Nuclear Power Plants* (USSR 1981) set engineering safety requirements for nuclear power plants. These regulations set the expected individual dose for a child's thyroid at a level not to exceed 300 mSv iodine isotopes at and outside of the boundary of the sanitary protective zone under the maximum designed accident in the worst weather conditions. It also limits the expected dose of external radiation on the whole body and any other organs of exposure (excluding thyroid) to a level not to exceed 100 mSv.

In 1983, *The Criteria of Decisions to Be Taken for Measures of People Protection in Case of a Reactor Accident* (Il'in and Avetisov 1983) adopted two radiation effect levels. Table 5.4 shows the criteria for exposure or contamination. If the radiation contamination does not exceed Level A, no emergency measures are necessary. When exposure or contamination exceeds Level A but not level B, decisions are made considering the specific situation and local conditions; if exposure exceeds Level B, emergency measures are necessary to protect the population. The most important change from the previous criteria of SR NPP-79 (1981) is that the lower intervention level is more rigid for all radiation effects and intervention levels.

Table 5.4. FSU radiation protection criteria for nuclear accidents adapted in 1983

Nature of Effect	**Intervention level**	
	A	B
External gamma exposure, mGy	250	750
Thyroid exposure from radioactive iodine, mGy	250	2500

In 1990, the Ministry of Public Health of the former Soviet Union approved *The Criteria of Decisions to Be Taken Concerning Population Protection in Case of Accidents at Nuclear Reactors* (Ministry of Health 1990). This regulation set the criteria for major actions to be taken to protect the population from radioactivity after a reactor accident. It required making decisions on countermeasures during the initial and intermediate accident phases by comparing estimated doses (predicted during progress of the accident) with the dose criteria in Tables 5.5 and 5.6 for possible radiation effects at the upper and lower levels.

The criteria and intervention levels in the former Soviet Union before the 1986 accident are similar to those in effect throughout the world at that time. They were nevertheless stamped "Confidential" and "Internal Use Only," and their availability was restricted. Thus, the basic working document in Ukraine was the "Radiation Safety Standard" of 1987, where only proper operation was considered.

Table 5.5. Intervention criteria during initial phase of a nuclear accident (adapted in 1990)

Protective Measures	**Dose criteria (predicted first 10 days) mSv**			
	Whole body		**Individual organs**	
	Lower Level	Upper Level	Lower Level	Upper Level
Sheltering, inhalation protection, organs and skin covering	5	50	50	500
Iodine prophylaxis: adult, children, pregnant women	- -	- -	50[(a)] 50[(a)]	500[(a)] 250[(a)]
Evacuation: adult, children, pregnant women	50 10	500 50	500 200[(a)]	5000 500[(a)]
(a) Only for thyroid.				

Table 5.6. Radiation protection criteria during mid-phase of a nuclear accident in progress (adapted in 1990)

Protective measures	**Dose criteria predicted for first year after accident, mSv**			
	Whole body		Individual organs	
	Lower level	Upper level	Lower level	Upper level
Limit on contaminated food and water consumption	5	50	50	500
Resettlement or evacuation	50	500	Not established	

5.1.4 Chernobyl Intervention Analysis

Ukrainian intervention policies after the Chernobyl accident are divided into two periods: the beginning of the accident until 1991 (when Ukraine was part of the Soviet Union) and after independence from the Soviet Union in 1991. Up to 1991, decisions and directives were dictated by Moscow. The basis of radiation protection measures was set up by the ICRP (1990).

Because of the Chernobyl accident, the former Soviet Union, with the participation of Ukraine, suggested a dose for life as 350 mSv for 70 years from all external and internal sources. This suggestion was rejected due to negative public reaction. Subsequently, a two-level concept of dose for life was adopted (70 mSv and 150 mSv for 70 years) during 1989–1991.

The concept of dose for a year adopted in 1991 consisted of a total annual external gamma radiation dose and an internal dose from consumption of local contaminated products (with ^{137}Cs and ^{90}Sr) and inhaled plutonium. Intervention levels were 1 and 5 mSv/yr at that time because 1994 levels were set at 0.5 man-sv/yr. These values were not aligned with international recommendations. An intervention level of 1 mSv/yr for a residual dose was considered, and countermeasures below this level were not considered justified. This was erroneously confused with the annual population dose limit for practical activity (ICRP 1987).

After the Chernobyl accident, the intervention criteria in Ukraine had activity levels expressed in annual doses above which various protective measures should be implemented. The intervention levels with the corresponding prevented doses are lower than those recommended by international organizations. Thus, the new *Radiation Safety Standards of Ukraine* (RSSU 1998) became effective on January 1, 1998 based on modern radiological protection principles (ICRP 1991; IAEA 1994, 1997).

Table 5.7 presents this two-level system that is similar to the ICRP system (1991). Comparing Tables 5.2 and 5.7, intervention levels in RSSU-97 agree with international recommendations based on predicted dose. As Table 5.8 indicates, water contamination was always below the assigned temporary admissible levels (TPLs). However, the maximum radioactivity levels in Kiev Reservoir for drinking water in the first days after the accident equaled $1x10^{-7}$ of gross beta activity. During May and June of 1986, this level was determined by iodine-131 levels (^{131}I), which reached 80 to 90 percent of the total radioactivity in May and 30 percent in June. Such levels demanded decisive intervention. While some foods can be banned for a time, water cannot be banned completely.

Table 5.7. Ukrainian minimum justification levels and undoubted justification levels for emergency countermeasures according to RSSU-97 (1998)

Countermeasures	**Prevented dose for first two weeks after accident**					
	Justification levels			Undoubted justification levels		
	mSv	mGy		mSv	mGy	
	whole body	thyroid	tissue	whole body	thyroid	tissue
Sheltering	5	50	100	50	300	500
Evacuation	50	300	500	500	1000	3000
Iodine prophylaxis						
for children	-	50	-	-	200	-
for adults	-	200	-	-	500	-
Limitations in open air						
for children	1	20	50	10	100	300
for adults	2	100	200	20	300	1000

Table 5.8. Temporary admissible level (TPL) change in radioactive content in drinking water after the Chernobyl accident

Radionuclides	Established Standard, Bq l^{-1} (Ci l^{-1})	Actual Content, Bq l^{-1} (Ci l^{-1})	TPL Approved Date	Document
^{131}I	3,700 ($1x10^{-7}$)	148 ($4x10^{-9}$) (Apr. 5, '86)	May 6, 1986	TPL № 4404-86. Radioactive ^{131}I in drinking water and food products for elimination period after accident. Approved by P. Byrgasov, USSR Chief Sanitary Doctor.
Total beta activity	370 ($1x10^{-8}$)	111 ($3x10^{-9}$) (May 6, '86)	May 30, 1986	TPL №129-252. ^{131}I in food products, drinking water, medicinal herbs. Approved by P.N. Byrgasov, USSR Chief Sanitary Doctor.
Total $^{134,137}Cs$	18.5 ($5x10^{-10}$)	0.9 ($2.5x10^{-11}$) (Apr. 12, '87)	Dec. 15, 1987	TPL №129-252-1. ^{134}Cs and ^{137}Cs in food products and drinking water (TPL-88). Approved by A.I. Kondrysev, USSR Chief Sanitary Doctor.
Total $^{134,137}Cs$	18.5 ($5x10^{-10}$)	(a)	Oct. 6, 1988	TPL ^{134}Cs and ^{137}Cs in food products and drinking water TAL-88. Approved by A. I. Kondrysev, USSR Chief Sanitary Doctor
Total $^{134,137}Cs$ ^{90}Sr	18.5 ($5x10^{-10}$) 3.7 ($1x10^{-10}$)	(a)	Jan. 22, 1991	TPL ^{137}Cs and ^{90}Sr in food products and drinking water (TAL-91). Approved by A. I. Kondrysev, USSR Chief Sanitary Doctor.
Total $^{134,137}Cs$ ^{90}Sr	1.1 ($3x10^{-11}$) 0.3 ($7x10^{-12}$)	(a)	Not approved	Protocol № 1, Jan. 29, 1993. "Permissible levels of radionuclides $^{137+134}Cs$ and ^{90}Sr for food products and drinking water." Adopted by Ukraine National Commission on Radiation Protection.
^{137}Cs ^{90}Sr	2 ($5.4x10^{-11}$) 2 ($5.4x10^{-11}$)	(a)	Jun. 25, 1997	Permissible ^{137}Cs and ^{90}Sr levels in food products and drinking water (AL-97). Approved by L. S. Nekrasova, Chief Sanitary Doctor of Ukraine, enforced since Jan. 1, 1998 by Order № 255 on Aug. 19, 1997

(a) There are no data available on the radionuclide content mentioned for the corresponding period.

In addition, uncertain conditions and reactor behavior (e.g., whether radionuclide releases would continue or stop, type of radiation, location of radionuclides, and the increased probability of an accident in Unit 3, etc.) after the accident reaffirmed the need for protective measures. Furthermore, protective measures might be justified by the exposure dose or other factors of a social or political nature.

5.2 Total Accidental Doses and Water Contribution to Population Exposure

5.2.1 Total Ukrainian Population Dose

Countermeasure implementation depends on the values of accident exposure doses and their dynamics. The contribution of the water component to the total dose depends on the period after the accident. The estimate of accidental exposure doses is shown in Figure 5.3. As illustrated, almost 50 percent of the equivalent expected dose for 70 years, from 1986 to 2056, occurred in the first post-accident year. Approximately 30 percent of the dose was from ^{131}I and was realized within the first two months after the accident. This supports the notion that in a reactor accident the most crucial period for dose formation from the water pathway is the early post-accident period.

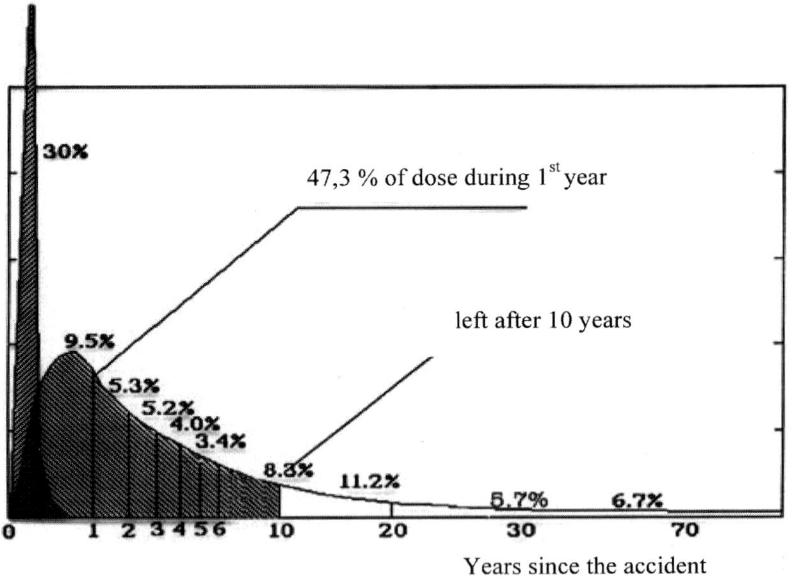

Figure 5.3. Estimated exposure doses for children born in 1986, from the beginning of the accident up to 70 years (2056) (from Los'y et al. 2001)

Chapter 4 analyzed the Dnieper River system exposure doses due to the Chernobyl accident. According to the data, Kiev's reservoir reached maximum levels (3,700 Bq/L) during the first days after the accident (Likhtariov et al. 1988). Iodine-137 accounts for 80 to 90 percent of the total. At such levels the internal exposure of a Kiev citizen rapidly increased from drinking water, and countermeasures were critically needed.

Although it is difficult to determine the fraction of exposure dose from drinking Dnieper River water to that from all water used in Ukraine, the Dnieper River does provides the major fraction. In open water the maximum contamination occurred when the radioactive cloud passed. Runoff from adjacent areas had relatively moderate contamination compared with that in the CEZ. Therefore, it is reasonable to assume that the exposure doses from the water component come mainly from drinking Dnieper River water.

Expected collective exposure dose for 70 years after the accident is predicted to be 2,500 to 3,000 man-sv from ^{90}Sr and 500 man-sv from^{137}Cs (see Section 4.3). Estimated total population exposure will be 55,000 to 70,000 man-Sv, with 4 to 5 percent coming from water. Figure 5.4 shows individual doses in μSv/yr, and Figure 5.5 shows collective doses. The water contribution to the total accident dose is given in Figure 5.6. Some studies also estimated dosage due to contamination in the Dnieper reservoirs (Los'y et al. 2001; Zelensky et al. 1993).

According to Figures 5.4 and 5.6, water contamination is highest in regions with low individual doses (excluding Kiev, which has relatively high doses). Because the radioactive cloud moved away from the Chernobyl plant,

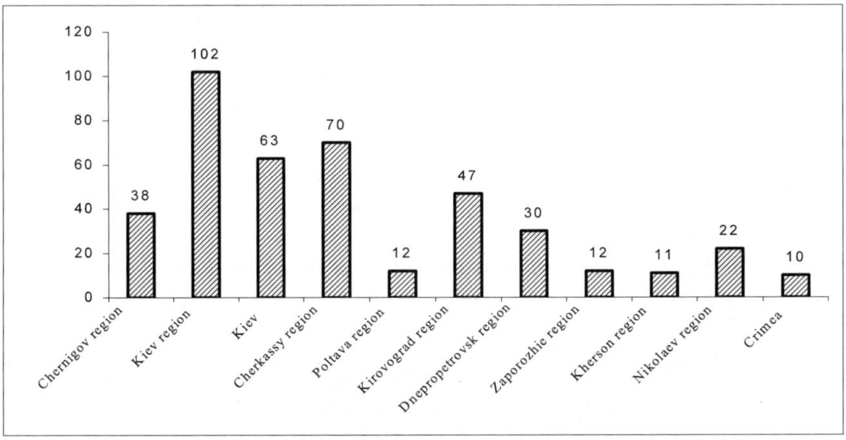

Figure 5.4. Average individual accident effective dose for population using Dnieper River water in Ukraine in 1993

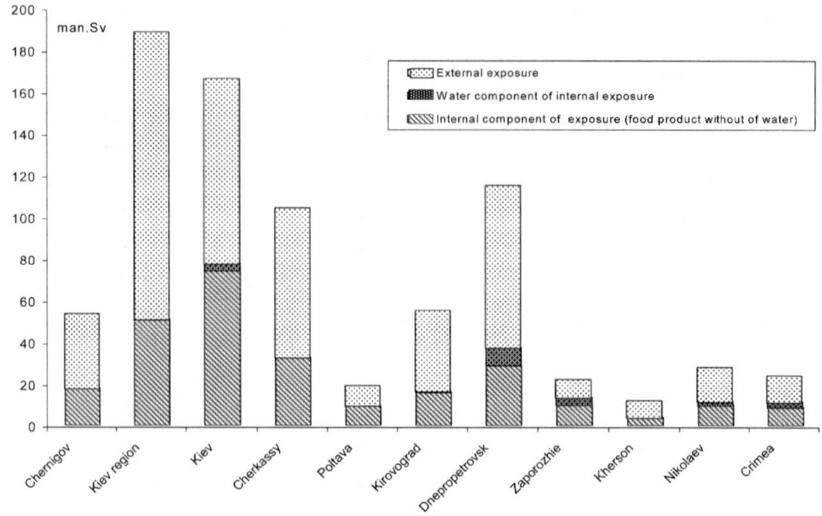

Figure 5.5. Expected collective doses of population using the Dnieper water in the various Ukrainian regions in 1993

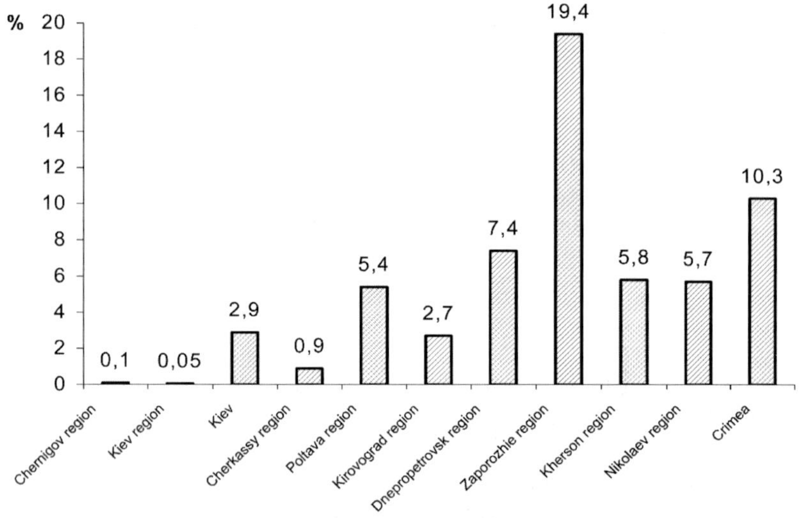

Figure 5.6. Dnieper water contribution to total collective population dose in various regions of Ukraine in 1993

exposure was reduced. The relative water contribution, however, increased as the contaminated water moved downstream. Overall total individual effective doses are low, 10 μSv/yr in Crimea and 102 μSv/yr in the Kiev region. With values less than 1 mSv/yr, countermeasures are ineffective.

It is difficult to estimate exposure doses from the water component in other regions of Ukraine, particularly in the Zhytomir, Rivno, and Volyn' regions, and the Ivankov and Polessje districts of Kiev region. The water contribution to the accidental internal exposure dose was estimated for the most contaminated regions, Kiev and Chernigov, as shown in Figure 5.7.

A more significant contribution from water to the total accidental dose is found in the urban population because contamination of food products is significantly less. The data presented above confirm that, in the first days after the Chernobyl accident, drinking water contamination reached levels that justified countermeasures.

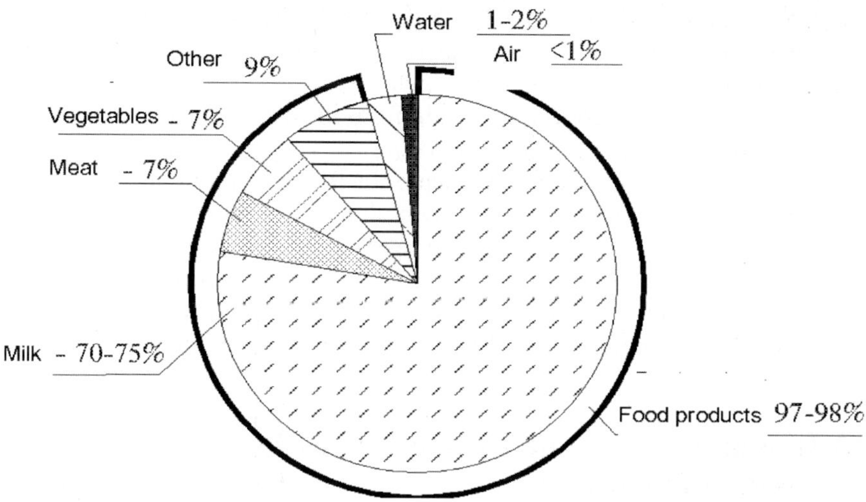

Figure 5.7. Accidental internal dose for rural Ukrainian population in 1993 (average estimates for most contaminated areas)

5.2.2 Population Exposure from Chernobyl Accident Versus Exposure from Natural Origin

The population of Ukraine is mostly exposed to naturally occurring radionuclides in open water bodies (^{3}H, ^{40}K) and groundwater sources (^{222}Rn, ^{226}Ra, ^{238}U). Exposure from a nuclear accident may be compared to that from natural radionuclides.

The source of Ukraine's drinking water is 67 percent rivers and other surface water, 22 percent artesian wells, and 21 percent shallow wells. The average suspended annual collective dose exposure to a resident of Ukraine is approximately 3,800 man-Sv derived from the average age-dependent volume consumed per person (RSSU-97 1998), the person's age, the average natural origin radionuclide content from surface water bodies and groundwater, and dose factors entering the digestive canal (UNSCEAR 1993; IAEA 1997). The anticipated collective exposure dose will be approximately 270,000 man-Sv for 70 years.

In Ukraine, 8.1 million people drink Dnieper River water. The collective annual dose from ^{40}K and ^{3}H in drinking water is 11 man-Sv for these people, and the collective anticipated dose from these radionuclides is 760 man-Sv for 70 years. Collective exposure doses from natural origin exceed doses from the accident by 1.5 times in the regions using the Dnieper for drinking water. The Cherkassy and Poltava regions are highest (4.3 times), and the Dniepropetrovsk and Zaporozhiye regions are lowest (1.2 times). Drinking water dose is one-half the accident dose. For the 70 years between 1986 and 2056, the anticipated collective dose from natural origins will be double the Chernobyl accident dose (about 385 man-Sv) for the given population (8.1 million people) due to reduced exposure from the accident over time.

Another source of radiation exposure for the population of Ukraine living downstream of Dnieprodzerzhinsk Town along the Dnieper River is a former uranium facility (tailings and waste storage). It is west of the town and causes a relatively low but still significant exposure for the local population (UNDP 2003). This factor was also considered in developing the water remedial action plan for the regions affected by Chernobyl radionuclides.

5.3 Radiation Doses and Loss of Life Risks from Contaminated Water

5.3.1 Mortality Cancer Risk Calculated from Accidental and Natural Exposure to Radionuclides in Water

Everyday decisions are made based on risk assessments related to natural phenomena and personal activities. Risk assessment is important for many reasons, especially to set up regulations for priority tasks and make decisions on policy and public health. The population should be informed about risk factors related to environmental conditions. In some regions of the world this right is legislated; for example, in the United States, California adopted legislation related to the public's "right to know."

Risk assessment has qualitative and quantitative characteristics such as unfavorable health consequences from ecological factors. The present study assumed that ionizing radiation is a result of the Chernobyl accident rather than from other dangerous or unfavorable ecological factors.

This radiation risk calculation followed ICRP Publication 60 (1990), which used sex and age dependence of coefficients of relative risk to people in Japan, the United States, Puerto-Rico, Great Britain, and China. To extrapolate a risk estimate for Japan from atomic bombs in 1945 to Ukraine, three models (additive, multiplicative, and mixed) were used; the study indicated that the multiplicative model best fits the epidemiological data. To estimate cancer risk, the linear-quadratic dependence dose effect was accepted with a mean-weighted risk coefficient (by age). Each radiation exposure of 0.1 Sv^{-1} (in small doses, where somatic effects of exposure appeared) results in a loss of life from cancer. Adopting half of this risk coefficient value, 100 percent fatal cancer risk would occur at 0.05 Sv^{-1} (0.04 Sv^{-1} for adult workers) when the dose is less than 0.2 Sv, according to UNSCEAR (1993).

Figure 5.8 gives the results from the cancer death assessment due to accidental and natural radionuclides present in water. This figure shows that mortal cancer risk from the Chernobyl accident is two orders of magnitude less than the risk from nonaccidental radionuclides averaged for all ground- and surface drinking water sources.

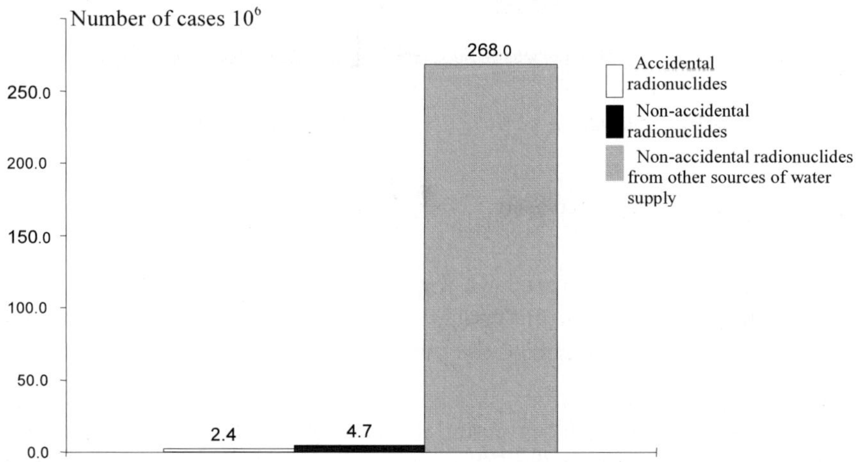

Figure 5.8. Cancer mortality numbers in Ukraine due to radiation doses received over 70-year post-accident period (1986 to 2056) from accidental and natural origin

5.3.2 Comparative Risk Analysis from Accidental and Nonaccidental Exposure

Comparative estimates of risk generally do not determine the actual risk levels for several reasons. First, each source has its own uncertainty factors. Second, quantitative parameters to determine the risk are different for each danger category. Third, different parts of the population are exposed to different risk factors. Most importantly, a simple risk comparison does not consider qualitative aspects. For example, most people are afraid of being killed by lightning, though this risk is very small. Some risks are considered necessary, like driving automobiles. Some people take unnecessary risks by choosing to smoke even though they are aware of the health risks.

Figure 5.9 shows the cancer death probability from drinking water from the Dnieper River that is contaminated with radionuclides from the Chernobyl accident, compared with other dangers (railways, home accidents, and fires), including spontaneous cancer growth (Ukraine 1993).

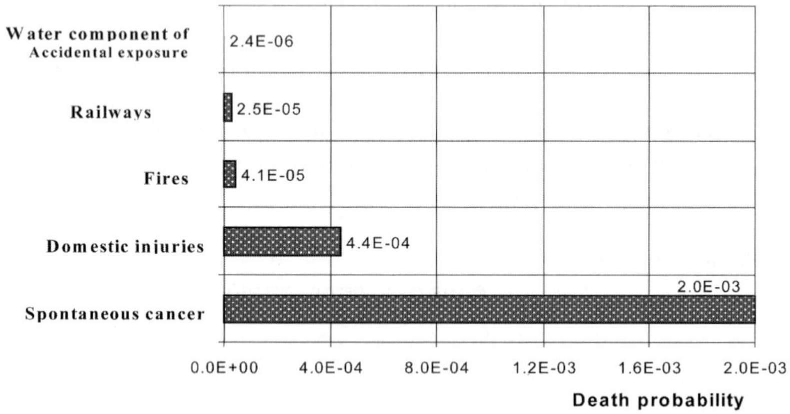

Figure 5.9. Death probability comparing cancer from radiation exposure to other dangers

Fish consumption and water use for irrigation during the same reference period (70 years) can double the cancer risk but cannot significantly increase the integrated water component of the potentially expected radiation risk.

Results of the comparative assessment given in Figure 5.9 demonstrate that actual risk (cancer death) of radiation exposure due to water consumption from the Dnieper River over the 70-year post-accident period (1986 to 2056) is expected to be very low. However, the psychological effects to the millions of people living in the regions where alternatives for the water use are very limited are significant. Therefore, some remedial actions to reduce radionuclide washout and annual secondary radionuclide releases from the Chernobyl site are considered reasonable (see also Chapters 6 and 7). To select a reasonable set of remedial actions, the optimization principles were applied for justification of water protection strategies.

The basic principles applied to the water protection optimization process at the Chernobyl site were first suggested and later developed in further studies (Voitsekhovich at al. 1998, 2001; Lepicar and Droz 1999; Los'y et al. 2001), as discussed below.

5.4 Principles and Methodology for Evaluating Effectiveness of Countermeasures

5.4.1 Justification for Health, Economic, Social, and Psychological Reasons

Former Soviet Union intervention levels were based on countermeasure efficiency with no consideration for international justification and optimization criteria. Radiological protection and intervention should be the same as international standards. Figure 5.10 shows the structural approach needed to assess justification and optimization of radiation protection. The following terms describe the stages of strategy justification:

- Variant of protection—special project or individual countermeasure
- Reasonableness—measure of effectiveness from applied countermeasures expressed in terms of averted individual and collective dose
- Cost—direct and indirect financial and resource costs of a protective measure (e.g., for design or justification)
- Factor—definite value of cost or effectiveness of a measure, e.g., financial cost, collective dose, maximum individual dose, requirements for countermeasure provided, discomfort from implementation.
- Criteria—quantitative and qualitative values for one or more factors such as individual dose limit or collective dose unit cost used as the basis for comparing effectiveness or version cost.

Decision-making sequences follow these steps:

- Accept that the problem exists and perform the needed studies, considering possible strategies.
- Investigate factors and classify those related to radiation protection.
- Analyze (quantitative and qualitative) potential effectiveness and apply each factor to preferred criteria.
- Recommend solutions.
- Make final decisions on actions to be taken by considering radiation protection and other aspects.

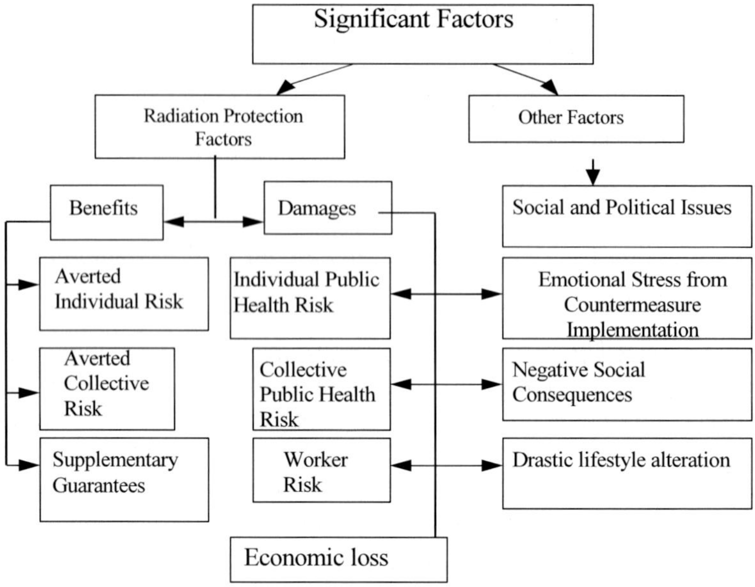

Figure 5.10. Main factors considered to optimize intervention

Political aspects may also be considered. The above simplified diagram does not include all capabilities; for example, reverse connection and ability to reflect incoming intermediate results on the assessment during analysis.

Radiological protection factors are related to the achieved level, like dose distribution, cost, and other applications. These factors fall into two categories: factors that are always included in the analytical procedure, especially cost and collective doses, and factors that are not always necessary. The latter factors include individual dose distribution or collective dose separation according to individual dose range, temporary dose distribution, population obtaining doses in different ranges at different times, probability of events, and reliability of the version being considered. Some factors are estimated only qualitatively, but they should be included for optimum results in recommending preferred countermeasures. Other nonradiation protection factors such as increased efficiency, aesthetic considerations, and public perception should also be considered but are not related to the level achieved.

Determining these factors initially is important. Some other factors are considered separately and, even if they influence the final decision, they are not included in the optimization of radiation protection. A multi-attribute benefit analysis for radiological protection, however, includes these other factors directly in the optimization structure.

The radiological and other factors given in Figure 5.10 are divided into those that reflect intervention benefits and those that reflect intervention damage. The selection consists of three parts: (1) postulate all scenarios, (2) conduct a preliminary assessment to exclude unfeasible scenarios, and (3) assess effectiveness of each selected scenario for each definite factor. The effectiveness of each scenario for each definite factor should be determined and expressed quantitatively as much as possible.

The effectiveness of scenario versions should be predicted using simulation methods. Many models have been developed, especially to predict radionuclide transfer and get a reasonably accurate prediction of exposure doses of the population. Because they are constantly improving, models may clearly predict the various specific situations. However, it is still necessary to include the criteria in the optimization procedure to compare the effectiveness of various versions for definite factors. These criteria are not determined by or in the procedure and should be set up in advance by competent people. Setting up in advance, criteria can be used to optimize countermeasures. One such criterion is dose limits, which can form limitations and monetary equivalents of collective exposure dose units. These monetary equivalents must be decided by the competent body or the highest level of a government.

The results of quantitative methods are known as analytical decisions. It is also possible to solve a reverse task, i.e., to find the monetary equivalent of a unit of collective exposure dose for the specific situation and then the government would make a decision concerning the availability of such value. The government also decides the monetary equivalent for collective exposure dose in situations where nonquantitative protection factors are considered. In this case, the analytical decision would not necessarily be optimal because qualitative and quantitative factors are combined. Qualitative factors should be combined with analytical decisions so a true optimum is achieved.

Sometimes factors unrelated to radiological protection complicate countermeasure decisions. One such factor is the attitude of the population to the danger of using contaminated Dnieper River water. To determine psychological and social reactions of the population on the radiation-hygienic situation, a questionnaire was provided (Los'y 1993). The questionnaire was distributed to 246 people in Kiev's Borschagovka, Syrets and Podol regions. It asked, "What contribution, in your opinion, do water, food products, air, or dwelling have on total exposure dose (insignificant, significant, very significant)?" and "What contributes the most to exposure dose from water: drinking water; watering vegetable gardens and fields with river water, consuming fish from the river, or swimming and resting on the river bank?"

The answers were transformed into numeric values. Public opinion placed the water contribution to total dose at 18 to 35 percent, an average of 27 percent (Table 5.9). Public opinion on partial contributions of water use is illustrated in Table 5.10. This shows that subjective assessment of water contribution to total accidental exposure exceeds the true value by 4.5 times on average (from 3 to 5.6 times).

Table 5.9. Public opinion about total exposure dose in Kiev

Radioactivity dose components	**Values of partial contribution, %**		
	Minimum	Maximum	Average
Food Products	22	39	30
Water	18	35	27
Air	17	34	26
Dwelling	8	26	17

Table 5.10. Public opinion about accident exposure from contaminated river water in Kiev

Dose Components	**Value of partial contributions, %**		
	Minimum	Maximum	Average
Drinking water	25	45	35
Products of irrigated lands	10	30	20
Fish product	15	35	25
Recreation	10	30	20

The results also show that people believe that water and food have the same exposure dose. Thirty-five percent believe drinking water prevails among all water components and products from irrigated lands, and river fish and recreation have approximately the same contribution level. These results exceed the true contamination factors from consuming fish (by 25 times) and recreation (by more than one million times).

Therefore, the government directed that countermeasures be imposed to reduce the contribution of water use to individual and collective doses of exposure, and alleviate the psychological and social reactions of the people of Kiev. Analytical decisions resulting from optimization in true post-accident situations are not always the final ones. In some situations other factors may prevail, as this example indicates.

Health safety standards are expressed by residual available doses of combined external and internal exposures. Intervention below these doses is unjustified, and intervention is necessary but not obligatory when they exceed a specific level. Similar levels are set up in the *Radiation Safety Standard of Ukraine* (1997) and corresponding international documents (IAEA 1997).

An economic criterion to justify countermeasures is expressed in the unit cost for collective exposure dose available to society (designated by α), i.e., total money that society or state spends to prevent 1 man-Sv dose. There are several methods to determine the quantitative value of this criterion. Cost-benefit analysis associated with dose reduction from Pripyat River dike construction relied on monetary value per man-Sv. The basic approach considered a constant value of 1 man-Sv for the entire referenced time period derived from the gross national product (GDP) of the considered country:

$$\alpha = GDP * LL * p \qquad (5.1)$$

where

α = monetary value of 1 man-Sv
LL = loss of life expected due to a radiation induced cancer
p = probability of radiation-induced cancer for general population (lifetime risk), associated with collective dose of 1 man-Sv.

Loss of life expectancy associated with radiation-induced cancer is estimated to be 16 years, which agrees with ICRP calculations. GDP is valued at $1,200 (U.S. dollars) in 1999. Based on these figures plus social attitudes, a corrected man-Sv is recommended at $8,000 to 10,000 U.S. or 5,000 1987 FSU rubles (FCO/ODA 1993). The value α was estimated as 2,850 1987 FSU rubles (IAEA 1992), considering medical costs, economic changes (growth of GDP, inflation, etc.), and uncertainties for different parameters. Value α is given for various countries in Table 5.11.

Great Britain's National Radiation Protection Board (NRPB) defines the monetary value α as a combination of monetary stochastic effects for health α_d with additional monetary value α_{nd} attributed to high individual doses (Webb et al. 1992). Most earlier attempts to combine these factors led to some intermediate values regardless of individual dose. These values were either too low to provide adequate protection of critical groups or the cost was too high to protect people from very low individual risks.

Table 5.11. Prevention costs per man-Sv collective dose unit; international recommendations (from IAEA 1992)

Country or International Organization	**α, U.S. dollars**
Nuclear Regulatory Commission of USA	100,000
National Radiation Protection Board of Great Britain	10,000
Centre on Study and Estimation of Protective Measures in Nuclear Area of France	1,800–5,400
The countries of northern Europe	20,000
IAEA	3,000

Table 5.12 recommends using α_d as a base value for collective doses consisting of individual doses at any level. For higher individual doses α_{nd} assumes additional costs to prevent doses close to the limits.

Table 5.12. Recommended unit collective dose monetary values of α by annual individual dose limits (from Webb et al. 1982)

Annual individual dose range limits, Sv	Percent of 5 mSv limits of annual effective dose for population	Value α for collective dose consisting of individual doses, pound sterling (£)/man-sv
$5x10^{-5}$	<1	2,000
$5x10^{-5}$ - $5x10^{-4}$	1-10	10,000
$5x10^{-4}$ - $5x10^{-3}$	10-100	50,000

5.4.2 Methods to Justify and Optimize Countermeasures

Justification and optimization of countermeasures must determine the optimal intervention parameter values like form, scale, and duration. Several quantitative methods are used: cost-efficiency analysis, cost-benefit analysis, extended cost-benefit analysis, and more complex multi-attribute utility analysis or multi-criteria outranking analysis (ICRP 1985, 1988).

5.4.2.1 Cost-efficiency analysis

A cost-efficiency analysis is the simplest way to express relationships by plotting the cost relative to the collective dose, as shown in Figure 5.11. In this example, the options are discrete without intermediate selections. In Figures 5.11 and 5.12, the options are shown as points connected with straight

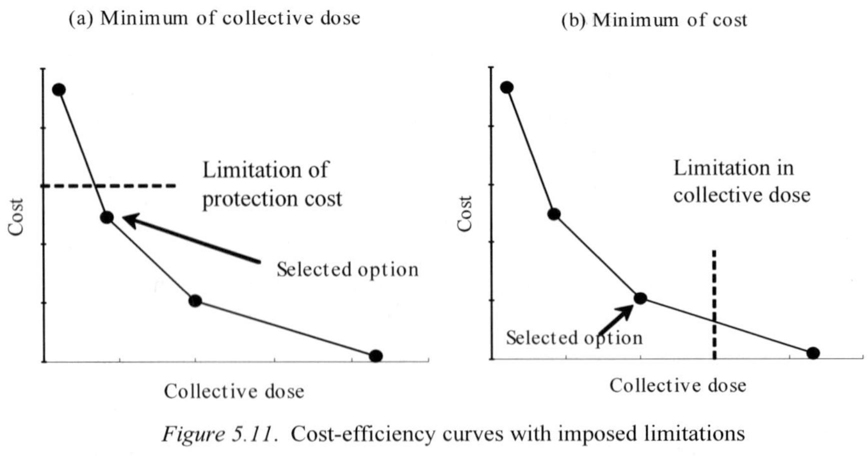

Figure 5.11. Cost-efficiency curves with imposed limitations

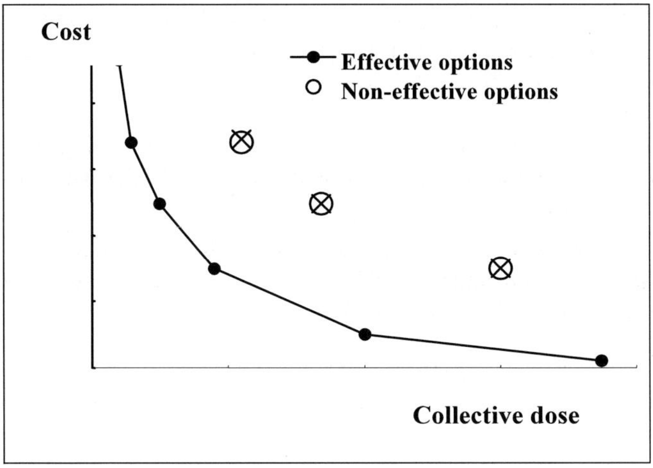

Figure 5.12. Typical cost-efficiency dependency

lines to improve readability. Including option variables helps smooth the curves. In Figure 5.12, a circle illustrates each protection option. Preliminary choice is obtained based on collective dose and cost. Some options correspond to the same collective doses with higher cost. A cross in the circle marks those excluded from consideration, meaning the other options are more cost-effective. This preliminary choice is useful but not enough to determine the optimal decision because the choice should be made from options that are different in cost. These data may be supplemented by imposing collective dose or cost limitations, as shown in Figure 5.11. Thus, in choosing an option, the collective dose is minimized by the fixed cost of protection or the cost is minimized by collective dose limitations.

5.4.2.2 Cost-benefit analysis

Cost-benefit analysis, the oldest method, was derived from an economic theory of welfare (Mishan 1977; Sudgen et al. 1978). It may be the most rectilinear quantitative method and relates benefits to costs incurred from actions taken. The characteristic feature of the cost-benefit analysis is that factors affecting the decision are expressed in monetary equivalent. Grouping monetary equivalents of cost and benefit determines the version with minimum costs to produce benefits. The assessment may be performed directly or with differential analysis. Optimization analysis considered only protective measure costs and appropriate level of prevented collective dose.

A simple cost-benefit analysis can transform collective dose into a monetary expression using the unit cost of collective dose α. Then the cost of protection, X, is added to the benefit cost, Y, to get the total cost (X+Y). This total cost is represented by an advantage curve and may be set up for each version considered. The analytical decision chooses the version with least total cost. However, a part of the total protection cost may include other factors to help make the final decision.

The collective dose is predicted relative to radiological protection measures and implementation costs (equation 5.2). Efficiency estimates provide economic criteria per unit cost of collective dose α saved (see Table 5.11). The costs of a unit of averted dose calculated for different countermeasures are compared. Countermeasures are considered effective when the following condition is satisfied:

$$\frac{X}{S} \leq \alpha \qquad (5.2)$$

where

S = collective dose prevented by countermeasure, man-Sv
X = cost of countermeasure in monetary equivalents
α = cost of a unit of collective dose, (monetary unit) (man-sv)$^{-1}$.

The main disadvantage of this simple cost-benefit analysis is that it is impossible to directly include individual dose distribution and other choice criteria except value α in weighing the benefits versus damages.

5.4.2.3 Extended cost-benefit analysis

An extended cost-benefit analysis includes individual dose distribution as a factor but not collective dose and protection costs. To obtain an analytical decision with this method, supplementary criteria are determined to have some admissible levels of individual dose.

Individual dose distributions are expressed as the difference between collective dose obtained from many low individual doses and collective doses obtained by a small population group with high exposure levels, especially if the latter doses are close to dose limits. To consider this effect, this method modifies the cost of a collective dose unit. The main cost α is supplemented with other factors estimating the cost of damage, as expressed in equations 5.3 and 5.4. Depending on individual dose levels, a collective dose unit may have a different cost. This new damage component would be expressed as a supplemental value, as described in ICRP Publication 37 (1985).

$$Y = \alpha S + \sum_j \beta_j S_j \qquad (5.3)$$

$$\sum S_j = S \qquad (5.4)$$

where

S_j = collective dose consisting of individual doses related to individual j
α_j = additional collective dose unit cost in j group (see Table 5.12).

5.4.2.4 Multi-attribute intervention evaluation

Interventions range from administrative actions on water use restrictions to fishing bans, chemical treatment, decontamination, or runoff control. Potential actions within interventions can be grouped into several categories in the Chernobyl case: social (administrative), physical (runoff and ground-water flux control) and chemical (change or control parameters of naturally occurring chemicals and radionuclides affecting migration in environment), physical stabilization of radionuclide migration, decontamination of sources or food products, and others. In some cases, combining actions may be the optimal strategy. Optimal strategy selection should be based on all relevant information, with the overall objective of minimizing contamination while accounting for other important environmental, social, and economic effects.

This approach is used in philosophy, engineering, and scientific management and is widely used for decision making. It helps to solve problems with factors that are difficult to express in monetary equivalent. It uses a scheme of magnitude (multi-attribute utility function) for such factors that if the magnitude (or utility) is the same for more than one option, no advantage exists for one over another. However, if the magnitude for option *i* exceeds the magnitude for option *m*, according to the definition, *i* is preferable to *m*.

First, factors of radiological protection are selected for an optimization study, and quantitative consequences are determined for each protection option in terms of these factors. Then an analysis is performed to determine their relative importance. This is performed by a utility function u_j, giving relative possible results desired for factor *j*. Generally, the best result or the least unfavorable consequences for each factor (for example, the least cost, minimum of collective dose) is determined by utility u_j equal to 1. For the worst consequence it is equal to 0. The main advantage of this method is that these utility functions can be nonlinear, allowing varying consequence values.

Figure 5.13 shows two utility functions, A and B, represented by the maximum individual dose. Function A is linear and reflects direct monotonic reduction of consequences. Function B is nonlinear and reduces faster in the worst cases (largest individual doses). This concentrates decisions on higher individual doses and is more sensitive to upper-limit consequences. Expressed in values of maximum dose in this example, similar considerations of dose distributions can be included in the extended cost-benefit analysis by assigning α values with greater weight on the high individual doses.

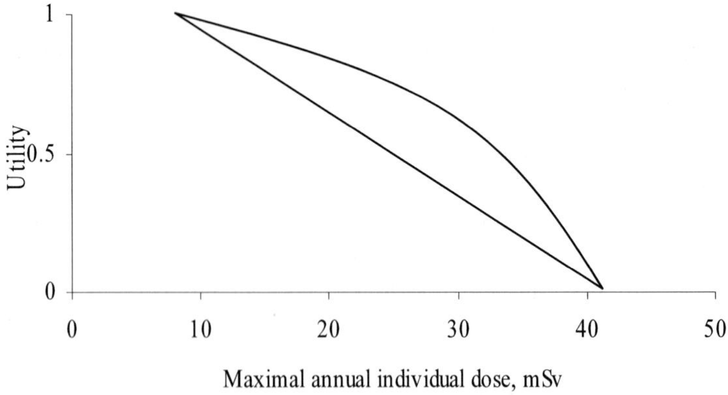

Figure 5.13. Multi-attribute utility functions for the factor of maximum individual dose

Flexible utility functions also allow consideration of other factors that do not translate quantitatively in monetary equivalent, like discomfort from countermeasures. The multi-attribute function, U_i, may be obtained from the single utility function, u_{ji}, expressing various utilities related to each version of protection, *i*. This function represents an advantage curve or full utility of each version *i*. When advantages for various factors are independent, multi-attribute utility function may be expressed by

$$U_i = \sum_{j=1} k_j u_{ji} \tag{5.5}$$

where k_j is a constant expressing relative importance or substantiality for each factor *j*. They are normalized as $\alpha k = 1$. The higher curve of U_i is ranked better, so the analytical decision will be the version that maximizes U_i.

The simple cost-benefit analysis may be considered as a special case of an additive multi-attribute benefit analysis. In this case all individual functions of benefit are linear, and the dimensional constants, k_j, become monetary equivalents for each unit of consequences.

5.4.2.5 Multi-criteria outranking analysis

The unitizing methods described combine all the attributes of the factors affecting the decisions into the unit curve of advantage. This is the total cost in a cost-benefit analysis or utility function in a multi-attribute analysis. However, all factors must cohere to the total value set for each protection option in monetary equivalent or utility expression to represent the consequences. A low quality factor may be fully compensated by the best characteristics of other factors such that all consequences may be interchanged.

Difficulties arise when factors are partially heterogeneous or estimated only qualitatively, e.g., public perceptions or attitudes of local authorities. If the options selected for protection are too incommensurable, for example, it may lead to the minimum protection cost and maximum damage cost, even if the total cost of the maximum protection cost and minimum damage cost are the same as the other. Under such circumstances the application of a multi-criteria outranking method may perform better than the unitizing methods.

Instead of expressing each option's effectiveness in terms of unit total curve of advantage, the outranking method compares each option *i* with option *m* to estimate whether *i* is preferable or superior to *m*. Comparison is based on two factors, the index of advantage and the exclusion criteria.

In the index of advantage, the expert expresses that option *i* is preferable to option m. This index, $Ad_{i,m}$, is equal to 1 if option *i* is preferable or equivalent to option *m* for all *j* factors; and is 0 if *i* is never preferable or equal to *m*. It varies from 0 to 1 if *i* is preferable or equal to *m* for some factors.

Exclusion criteria express the degree of disadvantage when the disadvantage of option *i* is more significant than option *m*. In the simplified option given here, this index, *Eci,m*, is equal to 1 when disadvantages of preferential option *i* over option *m* are too significant, and it is equal to 0 in the opposite case. If *Adi,m* is large enough and *Eci,m* is too small (in this simple option it is equal to 0), the option *i* prevails over option *m*.

Calculating the advantage index includes criteria related to factor k_j. In a simple case, weighted constants are determined as in multi-attribute utility analysis. Thus, the index of advantage will be equal to

$$Ad_{i,m} = \sum_j k_j a_j \tag{5.6}$$

where

a_j = index of advantage for factor *j*

a_j = 1 if the option *i* is better than or equal to the option *m* for this factor; in an opposite case it is equal to 0.

The main distinctive aspect of the multiple criteria ranking analysis is the criteria of exclusion. These are values to reject some options, not subject to fundamental interchange requirements of a unitizing method. For its application some qualitative or quantitative determination must be made for a point wherein disadvantages become too significant. This determination is known as the exclusion threshold and is an expression of decision makers. If a specific factor is not considered important enough to be excluded, the exclusion threshold for this factor is set up so that no pair comparison would give a criterion of exclusion of 1.

For each factor included in an optimization study, limitations (qualitatively or quantitatively) are determined, out of which the disadvantages of the option are determined as very significant. The option of protective measurements past all exclusion threshold checks is accepted as optimal.

The best-developed application for water protection optimization is MOIRA (Model-Based Computerized System for Management Support to Identify Optimal Remedial Strategies for Restoring Radionuclide Contaminated Aquatic Ecosystems and Drainage Areas). MOIRA uses a geographic information system (GIS) database and reliable validated models to predict temporal behavior of radionuclides in the fresh water environment and the ecological, social, and economic impacts. MOIRA also includes an evaluation module that facilitates decision making on intervention strategies, where the basic methodology is an analytical decision method.

MOIRA was implemented in a personal computer-based decision support system with all relevant information in the system. MOIRA can be used widely for decision-making procedures for water quality management at radioactively contaminated sites.

The methods described here were used to justify water remedial and restoration actions at the Chernobyl site. Chapter 7 presents some examples.

5.5 Conclusions

Actual cancer risk estimated from the expecting radiation dose exposure via aquatic pathways at the Dnieper reservoirs is very low, even compared with other nonradiation risk factors. However the aquatic pathways have been the largest factors which link the millions of people living along the Dnieper River to the radioactivity on the watershed area and in heavily contaminated lakes and wetlands at the CEZ. Therefore, it is natural that several water protection actions were implemented in this area.

Although the ALARA principles are usually recommended as a basic radiation protection and monetary value comparative approach to assess risk and cost benefit, they cannot be properly applied to justify the water remedial actions at the CEZ; social factors justified water protection measures.

5.6 References

CEC. 1982. *Radiological Protection Criteria for Controlling Doses to the Public in the Event of Accidental Releases of Radioactive Material.* General Employment, Social Affairs and Education, Health and Safety Directorate, Luxemburg, pp. 4-12.

FCO/ODA. 1993. "Chernobyl Cleanup Strategy, Post Chernobyl Remediation in Ukraine." Report to FCO/ODA Know How Fund.

IAEA. 1985. *Principles for Establishing Intervention Levels for Protection of Public in Event of a Nuclear Accident or Radiological Emergency.* IAEA Safety Series No. 72, p. 36. IAEA, Vienna.

IAEA. 1992. *International Chernobyl Project: Technical Report. Assessment of Radiological Consequences and Protective Measures: Report of International Consultative Committee.* International Atomic Energy Association, Vienna, p. 740.

IAEA. 1994. *Intervention Criteria in Nuclear or Radiation Emergency.* IAEA Safety Series No. 109, p. 119. International Atomic Energy Association, Vienna.

IAEA. 1997. *International Basic Safety Standards for Radiation Protection from Ionising Radiation and Safe Handling of Radiation Sources.* Safety Series No. 115, p. 383. International Atomic Energy Association, Vienna.

ICRP. 1977. *Recommendations of the International Commission on Radiological Protection.* ICRP Publication 26, Pergamon Press, New York, pp. 23-25.

ICRP. 1985. *Optimisation of Radiation Protection on the Basis of Cost-Benefit Analysis.* ICRP Publication 37, p. 95. Energoatomizdat, Moscow (in Russian).

ICRP. 1987. *Radiation Protection of Population.* ICRP Publication 40, p. 78. Energoatomizdat, Moscow (in Russian).

ICRP. 1988. *Optimisation and Decision-Making in Radiological Protection.* ICRP Publication 55, p. 60. Pergamon Press, New York.

ICRP. 1990. *Recommendations of the International Commission on Radiological Protection.* ICRP Publication 60, p. 197. Pergamon Press, New York.

ICRP. 1991. *Principles for Intervention for Protection of the Public in a Radiological Emergency.* ICRP Publication 63, p. 143. Pergamon Press, New York.

Il'in AL and GM Avetisov. 1983. "Criteria for Decision Making on Measures of Population Protection in Case of Reactor Accident." *Medical Radiology,* No. 5, p. 92 (in Russian).

Lepicar S and P Droz P (Eds.). 1999. "Chernobyl Exclusive Zone Protective Dike (UNDP expertise of justification)." *Collective dose reduction and cost-benefit analysis.* Geneva, Switzerland.

Likhtariov IA, RM Barhudarov, OA Bobyliova, et al. May 1988. "Radioactive Contamination of Water Ecosystems and Sources of Drinking Water Supply." *Medical Aspects of Chernobyl NPP Accident.* K. Zdorovie, p. 60 (in Russian).

Los'y IP. 1993. "Assessment of Contribution of Water Use in the Formation of Dose Loads of Population. Development of Criteria of Assessment of Water Protective Measure Efficiency in the ChNPP Accident Zone." Contract No. 35-93 with Minchernobyl, SPA ROSA, Kiev, p. 128 (in Russian).

Los'y IP, OV Voitsekhovich, and KI Shepelevich. 2001. *Radiation and Water. Experience Gained for Radiological Protection in Water Quality Management after Chernobyl Accident.* Kiev, p. 103 (in Russian).

Ministry of Health. 1990. "Criteria for Decision-Making on Measures of Population Protection in Case of Reactor Accident." Decree № 06-9/1549, Ministry of Health of the USSR (in Russian).

Mishan EJ. 1977. *Cost Benefit Analysis.* George Allen and Unwin, London.

RSSU-97. 1998. "Radiation Safety Standards of Ukraine; State Hygienic Norms." Ukrainian Centre of State Sanitary and Epidemiological Supervision of Ministry of Health of Ukraine, Kiev, p. 121 (in Ukrainian).

Sudgen R and A Williams. 1978. *Principles of Practical Cost Benefit Analysis.* Oxford University Press, UK.

Ukraine. 1993. *Economy of Ukraine in 1992: Statistical Annual.* Technika, Kiev, Ukraine, p. 464 (in Ukrainian).

UNDP. 2003. *Radioactive Contamination of the Dnieper River Basin: Contribution to the Ttransboundary Diagnostic Analysis and Recommendation for the Strategic Action Plan.* Summary report, Regional technical cooperation project RER/9/072, IAEA, Vienna, p. 24.

UNSCEAR. 1993. "Sources and Effects of Ionising Radiation: Report to the General Assembly." United Nations, New York, p. 922.

USSR. 1981. "Sanitary Rules of Design and Operation of Nuclear Power Plants." Institute of Biophysics of Ministry of Health of the USSR. SR-NPP-79, Energoatomizdat, Moscow, p. 36 (in Russian).

Voitsekhovich (Ed.). 1998. *Radioecology of the Water Bodies at the Chernobyl Affected Area.* Vol. 2, "Forecasts, Risk Assessment of Water Use and Effectiveness of Countermeasures for Water Bodies." Chernobylinterinform, Kiev.

Voitsekhovich OV. 2001. *Water quality management in the Chernobyl Exclusion Zone*. Kiev, p. 134.

Webb GAM, AB Fleishman, and MJ Clark. 1982. "Optimization of Protection of the Public: The Application of Cost-Benefit Analysis to the Dose Limitation System." *Proceedings of the Nuclear Fuel Cycle and Radiation Protection,* Madrid. International Atomic Energy Agency, Vienna, pp. 589-602.

Zelensky AV, MG Buzinny, and IP Los'y. 1993. "Measurements of ^{226}Ra, ^{222}Rn, and Uranium in Ukrainian Groundwater Using Ultra-Low-Level Liquid Scintillation Counting." *Proceedings of International Conference on Advances in Liquid Scintillation Spectrometry,* Vienna, pp. 405-411.

Chapter 6
Management of the Fresh Water Environment in the Chernobyl Affected Area

Oleg V. Voitsekhovich
Ukrainian Hydrometeorological Institute, Kiev, Ukraine

Yasuo Onishi
Washington State University, Richland, WA, USA
Formerly Pacific Northwest National Laboratory, Richland, WA, USA

This chapter evaluates the strategy and technologies applied to the CEZ. This evaluation is important to understanding the MOIRA (model-based computerized system for management support to identify optimal remedial strategies for restoring radionuclide-contaminated aquatic ecosystems and drainage areas) philosophy as a tool for the decision support system. We discuss actions taken for various water bodies and aquatic ecosystems in their historical context and evaluate examples for their effectiveness. The classification of countermeasures is a basic frame of the database containing the main findings of post-Chernobyl studies. It includes hydrological, physical, chemical, and social elements of the countermeasures as a potential set of measures in the environmental protection practice. The criteria for evaluation of the radiological impact of countermeasures are also discussed.

6.1 Introduction

In the history of Chernobyl, protection of water resources does not occupy the front page. And yet, 20 years after the Chernobyl accident, there is a need to ask why, not just because the water poses significant dangers to health, but first and foremost because it acts as a gauge of the emotional state of the society. It lets us see how the perspective of decision makers has changed since the days immediately after the Chernobyl accident.

From the earliest days after the accident one of the most serious problems to which the former Soviet Union government paid the greatest attention was how to prevent wholesale contamination of the Dnieper River water system with Chernobyl radionuclides, and how to provide people residing in the affected zone with clean drinking water and safe water use.

Y. Onishi, O.V. Voitsekhovich and M.J. Zheleznyak (eds.), Chernobyl – What Have We Learned?, 141–201.

The list of water protection measures developed due to the accident is large enough. It includes administrative measures to limit water use and technologies for regulating water flow, decontamination, and restoration of the water bodies affected by radioactive contamination. Not all of them were effective, and during the years that followed many of them were found to be harmful. A series of water protection projects has never been completed in the Chernobyl near zone, and these projects require not only additional financial resources but also justification to prove that they are reasonable.

The price was high for a society not prepared to solve the strategic and technologic problems of water quality management in the polluted regions. The assessment of radiological, economical, and social consequences of the countermeasures that were implemented in the affected zone was performed with the intention of finding the optimal strategy for water protection. Here we discuss a way to develop water remedial projects in light of the Chernobyl experience.

Today, issues designed to justify water protection strategies are well established; they are based on radiation protection activities commonly practiced around the world. It is useful to collect and analyze world experiences in selecting the best technologies for water protection.

6.2 Experience with Water Protection Activities Before the Chernobyl Accident

The operation of practically all nuclear power plants and production of nuclear weapons beginning in the 1940s have released radioactive substances into the environment. Routine and accidental radionuclide effluent releases to lakes, rivers, and seas were common at nuclear weapons facilities in the United States (e.g., Savannah River, Georgia and Oak Ridge, Tennessee) and the former Soviet Union (Chelyabinsk-40, Kyshtym), reprocessing plants (West Valley, New York) in the United States and Great Britain (Sellafield), nuclear laboratories in Canada (Chock River), and others (e.g., Rust et al. 1980; Pileev and Tischenko 1988). The radionuclide effluent releases to the Yenisey River (e.g., U.S. Navy 1996), the Danube River (Maringer et al. 1997), the Techa River (e.g., U.S. Navy 1996) and others are known. Since the late 1950s there has been leakage of liquid radioactive waste from waste tanks into the Columbia River (e.g., Beasly 1984) at the Hanford Site in the United States and into the Techa River in the South Urals, Russia.

Several decades before the accident at Chernobyl, many countries had faced the problem of accumulating radioactive substances in water bodies and had become aware of the danger of expanding the contamination zones due to contaminated water migration. Most nuclear power plants were located near rivers, lakes, or other water bodies; therefore, radioactive contamination from potential accidents was inevitable. This fact stimulated the development of water protection technologies long before the 1986 Chernobyl accident. Thus, prior to 1986, both in the West and in the former Soviet Union, the possible consequences of radioactive substances entering the water bodies were studied, and potential countermeasures were developed to reduce secondary contamination of water systems.

By the late 1970s, methods and engineering measures were developed to control or prevent the further spread of liquid radioactive substances in surface water and groundwater. The so-called static technology of contamination reduction was based on constructing artificial dams and barriers in the surface water and groundwater to change the direction and magnitude of radionuclide movement. The dams were built by injecting various solutions (silicate, bentonite soil, bentonite cement mixtures, lignins, organic polymers, etc.) into the soil through wells. Artificial waterproof dams and clayey shields were constructed where there were radioactive flows into the ground and surface water. Dynamic methods of regulating groundwater flow by pumping water into or out of the aquifers were widely used in Russia and the United States. In some cases it was proposed to build filtering walls with ion-exchange materials and natural sorbents in the path of polluted flows to decrease pollution levels of the groundwater entering surface water bodies.

In the 1960s and 1970s, the main method of purifying radioactively contaminated water was the use of ion exchange materials like zeolite, which was successfully used in the former Soviet Union (Kopeikin et al. 1988) and the United States (e.g., Nikiforov et al. 1985). For example, a zeolite exchange filter was used to remove radionuclides from water contaminated by the nuclear accident at Three Mile Island in Pennsylvania. However, even at that time, Oak Ridge National Laboratory showed that there was no universal method for chemical cosedimentation or adsorption of various radionuclides from water. In each case, specific recommendations were needed, depending on the content and the relationship between the radionuclides and the chemical composition of the water.

The experience of purifying radioactive flows was sufficiently well known to experts of the nuclear industry in the former Soviet Union, United States, Great Britain (Sellafield), Germany (Karlsruhe), and other countries (e.g., Nikiforov et al. 1985). Most of these methods were proposed for water purification under normal production conditions or for limited volumes of contaminated water. There was no large-scale decontamination of natural water bodies for emergency or industrial contamination before the Chernobyl accident because the known methods were too expensive. Thus, the main strategy was to dilute radioactive flow or localize radioactivity by engineering means. For example, the lakes of Kyshtym, Russia, where radioactive waste had accumulated, were filled with soil or covered by concrete. As a result, radioactivity entering the surface flows was reduced, but the problem of secondary contamination of the groundwater was not resolved. Problems of secondary contamination at significant distances by wind erosion and scattering of radioactivity by dust storms have appeared from partial draining of Lake Karachai at Kyshtym and others in the South Urals (Nikipelov et al. 1990; Ternovskiy et al. 1989).

Par Pond, South Carolina, U.S.A., was used as the cooling pond of a now-closed military nuclear reactor at the Savannah River Plant. Its bottom sediments under running water were contaminated with effluents of low-activity radioactive waste, including transuranic elements; more than 200 curies accumulated from 1954 to 1964 (Whicker et al. 1993). Until 1991 the pond was characterized as a relatively stable hydrologic regime with proper conditions for fish and other animal species. In 1991, after partial draining of the pond, the problem arose of retrieving the bottom sediments with their high content of radioactive substances and lead. A series of potential countermeasures was developed to reduce the ecological consequences. The main strategy in selecting the countermeasure was to limit or ban any economic activity in the area and to completely decontaminate the pond. However, for economic reasons, the final decision on decontamination of the reservoir bottom was not adopted. The expenditures were estimated from U.S. $10 million to $1 billion depending on the specific measure (Whicker et al. 1993). The high cost of retrieving the pond waste became the main factor for adopting an optimization methodology for selecting water protection measures despite significant local social-political pressures. The methodical experience of organizing surface and groundwater monitoring systems and the strategy of protecting the environment developed in the CEZ were used in the industrial park of the Savannah River Site.

A cooling pond for a nuclear reactor at Oak Ridge National Laboratory along the Clinch River (a tributary of the Tennessee River) accumulated several hundred curies of ^{137}Cs and other radionuclides from the flow of White Oak Brook during reactor operation from 1949 to 1990 (USGS 1991). Between 1959 and 1963 the pond was flushed out many times, contaminating the floodplains; this produced a threat of indirect human exposure from floodplains, cattle grazing, and fishing in polluted ponds. Some methods for minimizing radioactive contamination in the Tennessee River basin were available from studying the consequences of radioactive releases from the CEZ into the water system, water use, and restoration of Dnieper reservoirs.

The methods of mechanical, electrical, and chemical decontamination of soils in floodplains and the pond bottom were also well known. But the main limiting factors of the known mechanical or chemical technologies of decontamination in the CEZ were excessive power requirements, high cost, generation of a large quantity of radioactive wastes from decontamination, and the secondary negative ecological consequences of these technologies. Thus, despite of technological feasibilities, their use to retrieve the water system in the CEZ was problematic. Many of them could not be adapted to Chernobyl post-accident conditions due to incommensurably large scales, variability of natural conditions, and high levels of radioactive contamination compared with those for which they were developed.

Economic considerations were not the only factors to justify the proposed countermeasures during the first years after the Chernobyl accident. Only designs that appeared economically infeasible were rejected. For example, building a canal to transfer Pripyat River water to the Dnieper River to bypass the CEZ was not accepted. However, considering the economic prosperity of the former Soviet Union at that time, almost any design could potentially have been implemented. The country was just not prepared strategically or technologically to resolve the water protection problems, as evidenced by the lack of a reliable coordinated monitoring system and simulation methods to estimate the consequences of countermeasures and dose efficiency. A standard radiation protection and decision-making basis available before 1986 was adapted and optimized to justify proposed countermeasures.

It is significant that many of the experts involved in planning and monitoring measures to reduce radioactive contamination of surface and groundwater had no training in radioecology and radiation protection. Thus the self-sacrifice, good faith, and enthusiasm of many people involved in minimizing the consequences of the Chernobyl accident did not produce effective results. It is particularly true for proposed water protection measures. The results of implementing them could not be predicted objectively with regard to dose and economic and social justification.

6.3 Radionuclide Runoff as the Main Pathway of Secondary Radioactive Dispersion in Chernobyl Affected Areas

Water protection was one of the most significant post-Chernobyl accident countermeasure activities in Ukraine. The Dnieper River reservoir system, which is the main water supply of Ukraine, crosses the country from its border with the Russian Federation and the Republic of Belarus in the north to the Black Sea in the south. The Chernobyl nuclear power plant is near the bank of the Pripyat River, 30 km from its outflow to the Dnieper River at Kiev Reservoir (Figure 6.1). The floodplains near the power plant and the surrounding watersheds were heavily contaminated by ^{137}Cs, ^{90}Sr, and other long-lived radionuclides. Hot spots of radionuclide fallout from the accident are found in the upper Dnieper River watershed in Russia and Belarussia and on the entire Pripyat River watershed. Surface contamination led to permanent influxes of ^{137}Cs and ^{90}Sr into the Kiev Reservoir, the first of six on the Dnieper River. The river transports radionuclides through its reservoirs to the Black Sea, 900 km downstream. This is the main aquatic pathway dispersing radionuclides from the CEZ in the early accident phase.

Figure 6.1. The Dnieper River cascade; the main pathway of radionuclide transfer from the CEZ to the Black Sea

During the initial period, immediately after the April 26, 1986 Chernobyl accident, the surface water was directly contaminated by atmospheric fallout (Figures 2.2 and 6.2). The highest levels of contamination were observed during the first fortnight. A regular sampling program was established for the water bodies in the Chernobyl near zone, at all reservoirs on the Dnieper, and at the Belarussian part of the Pripyat River basin, as well as the lakes and rivers flowing from the catchments of the Bryansk (Russian-Belarussian) hot spots. Many institutions in Ukraine, Belarus, and Russia were involved in monitoring surface and groundwater contamination in and outside the CEZ.

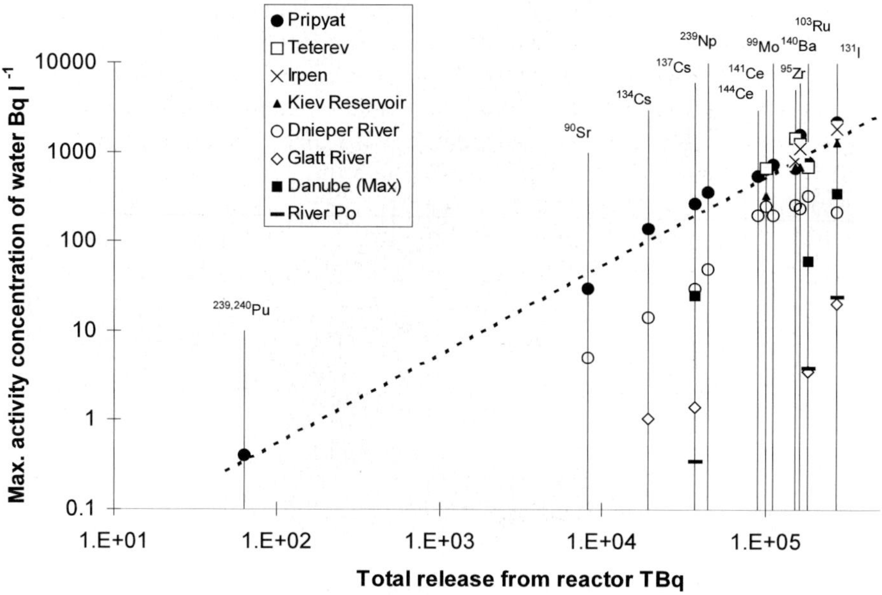

Figure 6.2. Maximum activity of radionuclides observed in the rivers of Ukraine and other European countries in the first week after the Chernobyl accident (from Smith et al. 2005)

During the initial period after the accident, surface water contamination was characterized with high levels of radiation over a wide spectrum of short-lived radionuclides. The content of gross total beta radionuclides in the open water bodies near the Chernobyl plant reached approximately $1x10^{-6}$ Ci/L (1 Ci = 37 GBq; 1 L = 10^{-3} m^3). The beta activity of the Pripyat River water downstream of the plant in early May 1986 exceeded $1x10^{-8}$ Ci/L. The range of radioactivity in the Dnieper River water near the main intake at Kiev (130 km downstream from the plant) was $1x10^{-10}$ and $1x10^{-8}$ Ci/L in May and June 1986, respectively. The largest contributor to radioactive contamination of water during the first months after the accident was ^{131}I. Since 1987, ^{137}Cs and ^{90}Sr have the largest influence on water contamination.

The program to sample water on a regular basis was organized to control radionuclide dispersion from the CEZ via the Pripyat River and into the whole Dnieper River basin. The detailed studies of watershed pollution demonstrate that the most contaminated areas that could be flooded were the left-bank (northeast) floodplain of the Pripyat River just upstream of the nuclear power plant (Figure 6.3). About 8,000 Ci of ^{90}Sr was estimated to have fallen on this rather small area of 10 km along the river. The parameters of radionuclide washout from the floodplain soil were studied in special laboratory experiments (Leptev and Voitskhovich 1993).

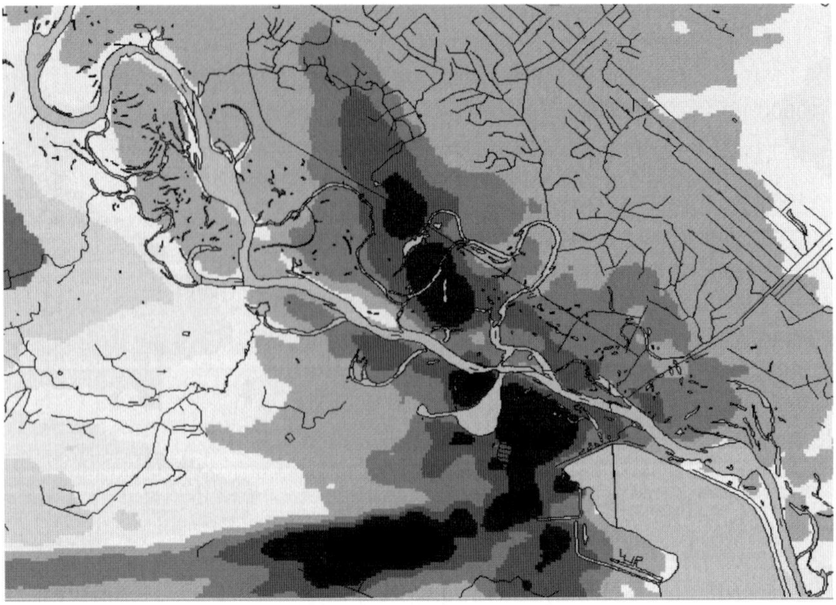

Figure 6.3. Radioactive ^{90}Sr spots covering the floodplain areas near the Chernobyl nuclear power plant (from Kashparov et al. 2003). Dark- and grey-colored spots in the floodplain area correspond to areas with contamination levels of 4,000-20,000 Bq/m^2

These studies showed that surface water runoff from contaminated watersheds and floodplains were and would remain major contributors to the Dnieper reservoirs' long-term contamination and sources for human exposure due to aquatic pathways. The groundwater is significantly polluted in the areas near the Chernobyl nuclear power plant, but the radioactive waste disposal sites contribute no more than 2 to 3 percent of the total transfer (washout) from the terrestrial environment. However, groundwater contamination in this area continues to be under long-term control.

The objectives of water remedial activities were to (1) prevent significant secondary contamination of the surface water bodies hydraulically linked with the heavy fallout areas and (2) mitigate expansion of groundwater contamination. The choice and design of the countermeasures were supported by radionuclide transport modeling of the aquatic system and by field and laboratory studies (Voitsekhovich et al. 1993, Voitsekhovich and Panasevich 1998).

6.3.1 Principal Events

Since the spring and summer of 1986, the most serious ^{137}Cs and ^{90}Sr contamination has been in the Dnieper River cascade downstream of Chernobyl and the Kiev Reservoir. After the spring and summer of 1986 (when the direct radioactive fallout onto the surface water took place), the most significant sources of surface water contamination of the Dnieper River were surface runoff from the initially contaminated floodplains and catchment areas and infiltration of heavily contaminated groundwater from the Chernobyl cooling pond and other water bodies to the river. The first flood after the accident occurred in 1987; it showed that the main sources of contamination of the Dnieper cascade were the catchments of the upper Pripyat River basin and a significant portion of the catchments in the upper Dnieper River (mainly the Sozh River in Russia). Since 1996, more than 70 percent of the annual radionuclide influx has been from sources within the CEZ (Voitsekhovich et al. 1997a). These monitored radionuclide data provided an important basis for the current strategy on water remediation.

There have been no long periods of high river flow or severe flooding in the contaminated areas since the 1986 accident. The Pripyat River spring flow was in the range of 800 to 2200 m^3/s, which did not exceed normal flood conditions. The maximum Pripyat River flow rate can exceed 5000 m^3/s, such as in the 1979 flood. Nonetheless, 1988 and 1991 flooding of the contaminated floodplains made it clear that unless mitigative actions were implemented, the floodplain would remain a potential hazardous source. Thus, the primary remedial strategy after 1992 was prevention of further significant transfer of radionuclides from the Pripyat floodplain in the Chernobyl near zone. Significant flooding events took place in 1994 and the spring of 1999, when the maximal water discharge reached about 3000 m^3/s. The 1999 flood lasted for about two weeks, and the heavily contaminated terraces of the river floodplain near the Chernobyl power plant were inundated significantly (see Figure 6.4). Due to inundation of the floodplain, the secondary radionuclide release to the river came mainly from the river's right (southwest) bank. These events of 1988, 1991, 1994, and 1999 show that a significant source of radionuclide runoff to the river was the CEZ.

Radioactive runoff from these sources was monitored to help determine management priorities. Unfortunately, in the first years after the accident, a lack of observation data resulted in an erroneous water protection strategy that focused on controlling runoff from small rivers crossing the CEZ. This mistaken assessment of the relative contribution of various sources led to many water engineering facilities that were of no use (Voitsekhovich 2001).

The lack of knowledge of the role of suspended particles in radionuclide transport with various hydraulic and sorption properties may be why sediment traps constructed in 1986 and 1987 did not catch radioactively contaminated alluvial sediment (Voitsekhovich et al. 1994; Voitsekhovich 2001). Many other erroneous management decisions failed to control surface runoff from the contaminated territories due to lack of data, understanding, and accurate

Figure 6.4. Dike on Pripyat Bay and inundated surrounding riverside at the CEZ during April 1999 flooding

determination of radionuclide contributions from various sources to the Dnieper River. These mistaken water management decisions could have been avoided had tasks been formulated correctly and monitoring locations and methods correctly chosen to measure the radioactive contamination of water bodies and water influxes from the accident zone to the main rivers.

In the initial period after the accident, ^{90}Sr runoff from catchments far beyond the CEZ was higher than it was near the Chernobyl plant. This was explained by the relatively low mobility of ^{90}Sr, which was part of the makeup of fuel particles during that period. As fuel particles were destructed, the ^{90}Sr mobility increased. Unlike ^{137}Cs, which attached quickly to soil minerals, ^{90}Sr retained the ability to migrate in the water and washed out faster to the surface and subsurface waters. Gradually, ^{90}Sr became a main radionuclide in river contamination and remains so. Figure 2.8 in Chapter 2 presents the dynamics of the ratio of soluble forms of the two radionuclides in the Pripyat River.

Figure 6.5 presents the ratio of ^{137}Cs concentrations attached to suspended sediment to the total ^{137}Cs concentration (sum of the soluble and sediment-suspended forms) in the Pripyat River. It shows that the suspended form of ^{137}Cs varies from 10 to 20 percent during the winter to 70 to 80 percent during spring inundation. The anti-erosion measures implemented in the CEZ reduced runoff of contaminated soils into the Pripyat River and thus beyond the borders of the near zone. The runoff of the ^{137}Cs soluble forms was also significantly reduced due to its fixation in soils and the deeper (downward) migration of the radionuclide into the lower layers of soil.

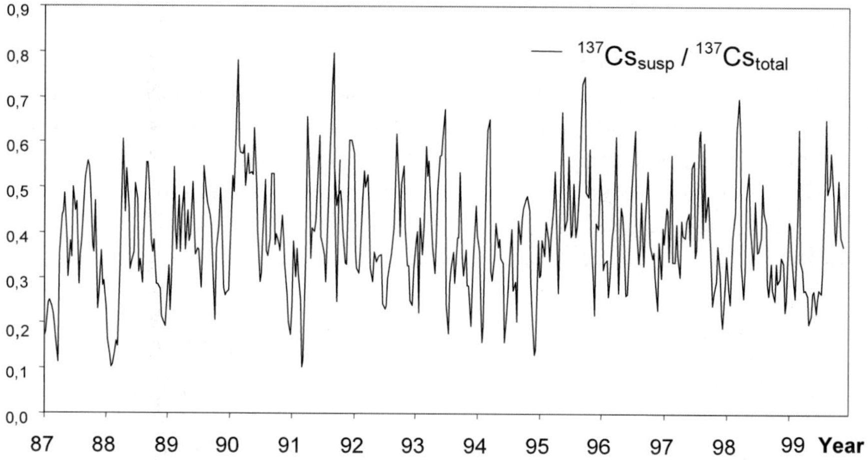

Figure 6.5. Dynamics of ratio of ^{137}Cs adsorbed by suspended particles to total ^{137}Cs content (soluble + sorbed forms) in Pripyat River at Chernobyl site, 1987–2000 (mean decade values)

The washout of ^{137}Cs beyond the borders of the CEZ was reduced by orders of magnitude from the level in the first years after the accident. Future ^{137}Cs removal from the CEZ will not significantly affect the radioecological state of the Dnieper reservoirs. On the contrary, ^{90}Sr concentrations will not change over time. The main sources of its removal from this zone will still be unprotected contaminated floodplains, groundwater filtering from the Chernobyl cooling pond, and runoff of polluted water from the Pripyat River's left-bank reclamation (polder) system (Voitsekhovich 2001).

A significantly reduced ^{90}Sr outflow from the CEZ may be expected after implementing a series of flood control (water protection) measures on the right (southwest) bank of the river, draining the cooling pond, and completing the runoff control action on the left (northeast)-bank polder. These actions are under way. During 2001, about 60 km of canals were dredged on the left bank of the polder (the reclamation system) and many hydraulic facilities (e.g., gates and drainages) were restored and repaired. These actions allow control of the runoff in the polder system in the heavily contaminated lands and prevent the possibility of secondary radioactive contamination of the river from these sources.

The radionuclide concentrations in the Dnieper River are not expected to pose a significant health risk. However, considering potential runoff and the risk of existing chemical pollution, some measures for preventing and mitigating the risk have been approved for economic and social reasons.

6.3.2 Natural Process of Water Self-Purification

If effective water protective measures cannot be implemented quickly enough or impose significant costs, it is sometimes worthwhile to assess natural processes that may purify the water. For the rivers in the accident zone, such processes may be assessed by observation (Figure 6.6) and generalized for the reservoirs of the Dnieper cascade (Voitsekhovich 2001).

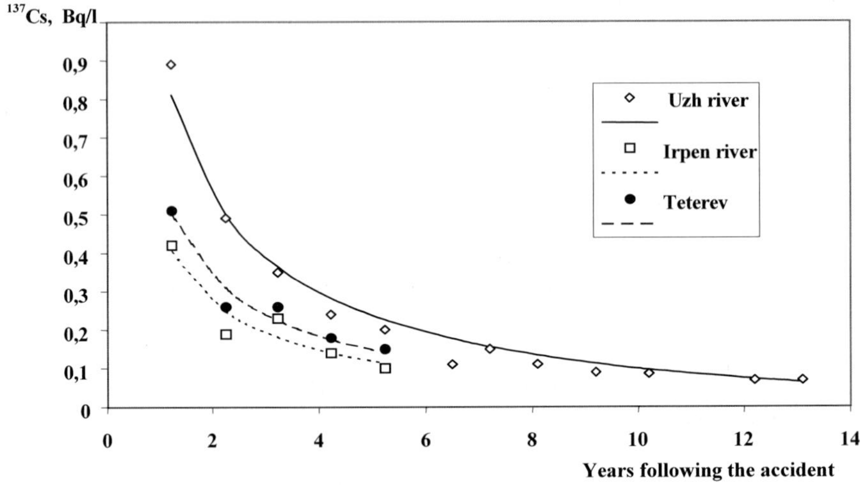

Figure 6.6. Comparison of observed and calculated (from Bulgakov et al. 2001) trends of ^{137}Cs activity in the rivers of the Ukrainian Polessye after Chernobyl accident

The ^{137}Cs contamination of the water was reduced faster than the ^{90}Sr contamination between 1987 and 2001 because these radionuclides have considerably different physical and chemical features and different adsorption by sediments and soils in the watersheds. The tendencies are similar in the Dnieper reservoirs (see also Figures 2.9 and 2.13). Sorption and sedimentation (deposition) processes play a major role in water purification for ^{137}Cs.

The most important processes promoting the natural reduction of radioactive contamination of the reservoir waters are the following:

- Hydrological mode of inflows and typical hydrodynamics
- Sedimentary processes (adsorption of radionuclides to the solid particles and their deposition on the river and reservoir bottom)
- Character and type of formation of hydrochemical composition
- Hydrobiological peculiarity of the water body
- Peculiarities of the water use and regulation of the water flow mode.

6.4 Water Protection Activities in the Chernobyl Zone

The main objective of the remedial activities was to prevent significant secondary contamination of surface water bodies hydraulically linked to the original contaminated area and to mitigate the expansion of groundwater contamination that was expected. The priority and available technologies for water remediation have changed over time. Most protective countermeasures were implemented in the CEZ; many other mitigative actions were applied at water intakes and irrigation channels. The remedial actions in Belarus and Russia were mainly focused on restricting water use, recreation, and fishing in the water affected by Chernobyl releases. The main thrust in Ukraine was to prevent subsequent radionuclide contamination in the Pripyat River and the downstream Dnieper reservoirs. The countermeasures required huge financial and human resources for implementation. Although some countermeasures and clean-up activities applied to radionuclide sources on catchments had positive effects, many others were ineffective or even useless. Thus, we must learn from our mistakes in post-Chernobyl radiation protection practices.

This section reviews past measures for preventing significant expansion of radioactive contamination beyond the CEZ, the specific methods applied, and new approaches based on risk assessment and cost-benefit analyses.

6.4.1 Classification of Water Remedial Actions

An analysis of the remedial actions taken to mitigate secondary water contamination after the Chernobyl accident can provide decision makers a unique opportunity to optimize their approaches to surface water and groundwater protection. The Chernobyl experience may teach lessons about controlling radionuclide redistribution via aquatic pathways (Voitsekhovich et al. 1993; Zheleznyak 1995; Kanivets and Voitsekhovich 2000; Voitsekhovich 2001). Because surface water and groundwater may act as secondary contamination sources, most engineering measures taken inside the 30-km CEZ were focused on protecting them.

Specifics of radionuclide transport processes defined strategies for aquatic countermeasures. Managing radionuclide transport is possible by controlling the hydrological flux and chemical properties of water-soil geochemical systems. Administrative resources and cleanup strategy can also be applied. Many remedial strategies that were proposed and implemented in the Chernobyl area may be found in a database developed by the Ukrainian Hydrometeorological Institute. The main framework suggests the following classification of water protection activities (Voitsekhovich 2001):

- Administrative countermeasures
- Hydrotechnical (engineering) countermeasures (flow control)

- Radionuclide flux control.
- Decontamination
- Radionuclide pathway regulation.

The database was developed under the framework of the SP-5 French–German initiative project "Radioecology" (1998–2001) and is part of the comprehensive database of countermeasures applied to protect agriculture and the environment (Deville-Cavelin et al. 2001).

Administrative countermeasures were mostly used during the initial stage of the accident. They focused on limitations, restrictions, or changes to normal water use in the area affected by radioactive fallout. Hydro-engineering measures were intended mainly to control runoff from the contaminated catchments or for water reclamation systems. These actions led to control of the radionuclide fluxes associated with runoff. By changing the geochemistry or other features of soil conditions or controlling the hydro-chemistry of the water bodies, it was also possible to control the radionuclide mass exchange between solid and liquid phases or between biotic and abiotic aspects of the water ecosystem. These actions that helped to control radio-nuclide fluxes (e.g., radionuclide transfer factors, retardation factors, or adsorption capacity of the particulate phases) are categorized in the database under "Radionuclide Flux Control." The direct radionuclide removal from contaminated environment is described in the database as "Decontamination." Control of radionuclides in food chains is described under the respective core in the database.

The database described above is based on the functional structure of the Chernobyl site. The actions can be divided into those applied to catchment (drainage) areas and those applied to the water bodies themselves. Actions in catchment (drainage) areas include:

- Removal of contaminated soil.
- Alterations of catchment area to minimize runoff from contaminated land to the water such as planting trees, digging channels and ditches, or adding chemicals to bind radioisotopes (lime, potash, dolomite).
- Prevention of flooding in the most contaminated areas next to a water body (e.g., construction of a dike in a floodplain).
- Construction to prevent radionuclide transport to surface water via groundwater flow (e.g., contra-seepage wall in soil).

Actions applied to the water bodies include:

- Hydraulic construction to increase sedimentation (particle deposition) of contaminated suspended materials in river (reservoir) bottom (e.g., quarry—a bottom trap for contaminated sediments), dams, ditches, and spurs.
- Hydraulic construction to separate the most contaminated parts of the water bodies from the main stream (e.g., dikes and dams dividing the water bodies).
- Dredging of contaminated deposits.
- Changing the operation mode of the Dnieper reservoir management to minimize radionuclide concentrations.
- Changing drinking water; e.g., recommending other water sources.

6.4.2 Runoff Control (water discharge)

As stated above, typical actions taken during the first years after the accident were to control water runoff into the rivers and reservoirs. The measures were applied to regulate rivers, channels, and reservoirs to reduce the radionuclide concentration mainly by diluting the water with clean water inflow into the system or retaining contaminated water in a reservoir having a clean water tributary. However, the flow rates also affect the resuspension rates of heavily contaminated sediments in the reservoirs and radionuclide desorption from contaminated bottom sediment.

The water discharge regulation would affect all these processes and thus was assessed carefully. For example, when a countermeasure to reduce radionuclide concentration by dilution was implemented in rivers, it increased the concentration due to greater desorption from the contaminated river bottom. Thus, the Dnieper reservoirs' operational mode was changed in the first year after the Chernobyl accident to diminish the maximum radionuclide concentration in the lower reservoirs and to prevent sharp resuspension of contaminated sediments in the Kiev Reservoir (Zheleznyak 1995).

Careful modeling to evaluate the efficiency of this remedial measure using an optimization procedure demonstrated that the maximum radionuclide concentration could be halved in the lower Dnieper reservoirs using a special operational mode in years of low water flow. During high-water periods this measure would not be effective due to the small capacity of the reservoirs compared with the volume of flood water. These measures could be applied to any regulated river; however, their efficiency would depend on the reservoir's regulating capacity, existing water use limitations, and contaminated tributaries in the aquatic system.

The efficiency of water discharge management for dilution and changes in desorption fluxes could be simulated using box (compartment) models (e.g., WATOX model applied to the Dnieper River) or one-dimensional models (Zheleznyak 1995). The evaluation of the reservoir discharge management to control resuspension of contaminated sediments could be analyzed using two- and three-dimensional models (Zheleznyak 1995). The costs of these measures would be calculated with known cost information on national water and energy pricing and technologies. The social effect of the measures to diminish maximum radionuclide concentration in the river and reservoir water was considered high even for a situation with a small averted collective dose.

6.4.3 Stages of Water Remedial Actions

Since the accident, engineering and administrative countermeasures have been performed to mitigate risk for the population residing along the Dnieper reservoir system. The chronology of government decisions on water protection activities from early May 1986 to 1989 are reported by Voitsekhovich (1997b). The following three phases of activities are apparent.

6.4.3.1 Emergency phase: (two to three months after the accident)

Countermeasures during this time were mainly based on administrative decisions and were intended to control the situation. Countermeasures include the following:

- Restricted water use and fishing, avoidance of contaminated surface water resources where possible.
- Supplemental purification of drinking water in municipal water treatment plants, including development of modern technology, sorbent materials, and better drinking water treatment methods.
- Attempts to regulate the flow of contaminated water through the Kiev Reservoir by a system of dams.
- Increased use of groundwater sources by municipalities and construction of supplementary groundwater supply wells.

Most of these measures were carried out without a cost-benefit analysis. Instead, consideration was given to the stress on the society and the availability of the resources of the former Soviet Union to reduce and eliminate the consequences of the accident. The main reasons for implementing only a limited number of the cost-effective actions during this early period were the lack of experience, time, and the required expertise to carry out the evaluation. Therefore, most measures to reduce radiation risk to the public from water use were expensive and limited in success (Voitsekhovich et al. 1995).

Decision makers also made many errors because they lacked adequate information on the spatial and temporal variations of water body and catchment contamination. For example, due to a lack of experimental data and controversy between knowledgeable scientists and inexperienced decision makers, the first assessment of the radionuclide adsorption/desorption parameters for the liquid-solid interaction was not correct, and radionuclide runoff from catchments to rivers (expressed as washoff coefficients) was greatly overestimated. As a result, many useless water protective actions were taken during the first months after the accident. For instance, ceolite (a natural material with high adsorption capacity) was washed away from Chernobyl and soil-clay barriers constructed along the Pripyat River to the river downstream.

Another failed example occurred in early May 1986. The surface gates of the Kiev Reservoir dam were opened and the bottom gates closed. It was believed that clean water was being let out of the reservoir from the surface gates and the highly contaminated near-bottom water including adsorbed radionuclides would be captured in the reservoir. In reality, during the first week the vertical mixing of water was slow, and the lower water layers of the reservoir were much less contaminated than the upper layers, which were contaminated directly by atmospheric fallout. A better approach to decreasing the activity level downstream of the Kiev Reservoir immediately after the accident would have been to open the bottom gates and close the surface gates at least for the first weeks in early May 1986. This action would have reduced the levels of radioactivity in downstream drinking water in the first weeks after the accident, when the main exposure was from drinking contaminated water.

The first data concerning the contamination of drinking water in the nearby Kiev catchment was up to 5×10^{-8} Ci/L (1 Ci = 3.7×10^{10} Bq) of total activity. This level of contamination exceeded that in the water before the accident by 1,000 to 5,000 times. All monitoring that had the technical ability to determine the content of radioactivity in surface water was involved in estimating the radiation situation. On April 29 to 30, 1986, pollution of the Pripyat River near Chernobyl Town reached 1.3×10^{-7} Ci/L. On May 2, 1986, the total activity levels of water near the Kiev watershed were 100 to 1,000 times greater than the natural background level.

It was clear that this contaminated water would be a real threat to the safety of the Kiev watershed. At the May 3, 1986 meeting of the Political Bureau of the Central Committee of the Communist Party of Ukraine, the first analytical reports on the scale of this tragedy were discussed. A. P. Lyashko, the chairman of the Soviet Ministers of Ukraine, declared at the meeting, "If a rather high water contamination is detected, we should use other water supply lines. If the water amount is not sufficient, we will regulate the rate of water consumption." As a result of this discussion at the highest administrative level, a number of ministries and agencies were commissioned to estimate the expected consequences of the accident on the water supply system for the population, to inspect all artesian wells, and to prepare proposals on establishing new drinking wells in Kiev and other Dnieper areas. On the same day, a confidential report of the Central Committee of the Communist Party of Ukraine (CPU) recommended that, "The emergency storage of water should be provided since the possibility that for some time the water from natural water bodies will be unfit for consumption could not be excluded" (Tolochko 1996).

The following are chronicles of major decisions on water protection while the Chernobyl accident was in progress. These were obtained from archived documents, published reports (e.g., Izrael 1990), and unpublished sources (departmental and personal).

- May 5, 1986. Temporary permissible levels were put into practice for ^{131}I activity in drinking water (1×10^{-7} Ci/L) and fish (1×10^{-7} Ci/kg).
- May 7, 1986. At the meeting of the Soviet of Ministers of Ukraine, the water supply for the population of territories exposed to radioactive contamination was discussed. Water supply problems were analyzed. After this, water protection questions were discussed almost daily at meetings of the newly created Working Commission on Problems of Water Supply, chaired by V. M. Shestopalov (corresponding member of the Academy of Sciences of Ukraine). The commission recommended implementing measures to reduce the water problems of the initial post-accident period. At the same time, studies showed that despite the large amount of work being done, countermeasures implemented in the initial post-accident period were inadequate due to "not-knowing," insufficient information, lack of analytical systems to select priorities, and inexperienced experts.
- May 7, 1986. A decision was made to build hundreds of water supply wells to increase groundwater use.
- May 8, 1986. A directive was issued to urgently increase production of activated charcoal for purifying water in the catchments.
- May 9, 1986. A decision was made not to use a sorbent to purify municipal sewage water due to high labor costs for filter construction.

- May 10, 1986. A directive was issued to provide the necessary quantity of natural sorbent materials for purifying natural water flow at the Dnieper catchment.
- May 11, 1986. A directive was established on the necessity of urgent contamination control for raw food materials, including water.
- May 12, 1986. A decision was made to build (clay barrier) walls in the ground to block groundwater flow moving from the zone of destroyed Unit No. 4 of the Chernobyl plant to the Pripyat River.
- May 13, 1986. A directive was issued to reorganize monitoring systems of surface water and groundwater at the Dnieper River basin in Ukraine.
- May 13, 1986. The government approved "primary measures on the prevention of contamination of the Kiev reservoir and the Dnieper River from radioactive substances." A new purification technology was recommended for the Dnieper at drinking water supply stations using more effective reagents (activated charcoal, clinoptillolite, and others). The experts estimated this technology would reduce the radioactivity of drinking water by half. The first strategy to localize radionuclides in the Chernobyl cooling pond was also prepared. A decision was approved by the government of Ukraine to build and strengthen earthen walls, waterproofing with a polyethylene film along the right bank of the Pripyat downstream of the CEZ.
- May 23, 1986. Directives were issued to design a drainage screen and a system of filtering water injection (reverse repumping) from the Chernobyl cooling pond to Pripyat River via the dam.
- May 24, 1986. The first official forecast was submitted to the government on the expected consequences of accidental radioactive pollution for the Dnieper River aquatic system.

The estimations were conservative and did not consider many factors and processes. However, such an approach was justifiable considering the lack of accurate data and the unprecedented large-scale radioactive contamination of the watershed of a large river such as the Dnieper. The worst scenarios showed the extremely high contamination levels of Kiev reservoir (up to $1x10^{-6}$ Ci/L). The predictions indicated that up to 15 percent of activity released within the Chernobyl industrial area would be washed out into the rivers. This water would be totally unfit for drinking—drinking just one liter of the water could result in a radiation exposure dose equal to 0.3 Sv. There were more realistic predictions; for example, according to data of the State Committee of Hydrometeorology of the former Soviet Union, predicted values of radionuclide washoff were considerably lower. They estimated that only a small part (less than 1 percent) of radionuclide fallout on the surface of the watershed would be transferred by runoff to the rivers during rains. The experts came to this conclusion based on the experience and knowledge obtained by secret studies on the consequences of the 1957 Ural nuclear accident. The analysis of the uniquely slow migration phenomena of "hot particle" fallout was revealed in the Chernobyl releases. Considering this slow migration, the worst scenario of radioactivity washoff outside the CEZ since the summer of 1986 would not result in contamination of the middle part of Kiev Reservoir beyond the total activity of $1x10^{-9}$ Ci/L.

In any case, there was no other reason to estimate the consequences to the safety of the water supply. The level of social tension from anticipated radioactive contamination of the Dnieper River was increasing as a result of the following statements on government decisions at that time. The Resolution of the Soviet of Ministers of Ukraine states that, "Urgent implementation of all engineering measures is vitally important to prevent washoff of radioactivity from the Chernobyl site and adjacent areas to save the Dnieper as a source of water supply." Also, it was required "to provide a series of measures to implement to the maximum extent to remove radioactive contamination (their compulsory transportation and disposal but not through washing out by water) by providing a series of construction works to reduce radioactivity release outside the Chernobyl working site and other contaminated areas."

Thus, in May 1986 a list of required measures to protect the Pripyat, Dnieper, and Desna rivers included the following actions:

- Provide urgent measures for monitoring surface and groundwater.
- Set up an interagency group in the Academy of Sciences to create the database and predict the radiation situation.
- Accelerate waterline construction from the Desna River to provide clean water to Kiev.
- Increase water supply volume by two to three times for the groundwater sources in the Kiev region and take measures for urgent drilling of artesian wells in the towns.

These measures were based on worst-case predictions of water quality in Kiev, Kanev, and Kremenchug reservoirs due to extremely high levels of contamination in the Dnieper River, though the worst case could hardly have occurred. However, the proposed measures needed to support the vital activities of people and industry even in worst-case post-accident scenarios.

Thus, within a month after the accident, the first predictions of possible contamination of the Dnieper water system were made, the approximate estimates of consequences for water resources were obtained, and the first concept of long-term measures was developed to reduce the scale of radioactive contamination of the Dnieper water system.

- May 29, 1986. By decree of the Communist Party of Soviet Union Central Committee and the Soviet Ministers of Ukraine, the State Committee of Hydrometeorology was commissioned to provide:
    - regular predictions of rainfall
    - systematic observations of radioactive contamination of surface flow from the Pripyat River basin in the CEZ and Kiev reservoir
    - estimates of the efficiency of the measures taken to reduce the level of radioactivity in the water, suspended substances, and bottom sediments in the rivers and other water bodies.

By the same decree, several ministries and agencies of the former Soviet Union and Ukraine were assigned to fix the soil surface in the basin and banks of the Pripyat, Uzh, and other rivers in the Chernobyl zone to prevent erosion of contaminated soils and sedimentation of radioactive substances in the water bodies by using sorbents by July 1, 1986. The sorbent substances, vessels, and other means needed for implementing monitoring and water protective measurements were also provided.

- May 30, 1986. The Head Sanitary Doctor of the former Soviet Union approved “Temporary permissible levels of the content of radioactive substances in food products, drinking water” as $1x10^{-8}$ Ci/L for total beta activity.
- May 31, 1986. The Kiev Executive Committee adopted a decision of compulsory purification of wastewater at sanitary treatment locations. Discharge locations of clarified radioactive disposal of transported radioactive silts were set up.
- May 31, 1986. Following unsuccessful attempts to use ceolite as a sorbent due to the inefficiency of ceolites that were discharged into the river, a decision was made to stop using the sorbent for decontaminating the Pripyat River and Kiev Reservoir. The Ministries of Water Management and Geology of Ukraine proposed constructing underwater dams and channel pits for intercepting radioactive silts (river sediment) in the Pripyat River and Kiev Reservoir.
- June 10, 1986. Actions were taken to prevent radioactive runoff from the Chernobyl site. The work was to be completed in five to ten days. Construction began on a filtering wall in the soil around the

Chernobyl site, artesian wells on the cutoff and bank drainages, and the bank drainage around the cooling pond and catchment collector.

- June 10, 1986. According to the Resolution of the Ministers of Ukraine No. 322, a commission was created for radiological monitoring and observation of the environment. A decision was also made concerning the assigned functions among the ministries and agencies of the Ukraine on radiometric and dosimetric monitoring. The Ministry of Water Management of Ukraine was commissioned to control the level of radioactive contamination in catchments of the main channels and irrigated systems. This ministry was also commissioned to develop a series of measures to prevent secondary contamination of surface water bodies and groundwater. There were two official monitoring systems for radioactive contamination of surface water bodies in the country: those of the State Committee of Hydrometeorology of the former Soviet Union and those of the Ministry of Water Management of Ukraine.
- July 1, 1986. The Ministry of Water Management of Ukraine informed the government that in May and June 1986, water protection measures were provided to control radioactivity in the drinking water in all catchments. The following actions were also carried out:
  - Reserve schemes were prepared for water supply for Kiev and other towns in the Dnieper area.
  - Fifty-eight new artesian wells were drilled in Kiev and its suburbs; 200 artesian wells were established in the Kiev region.
  - Two water supply lines were constructed and placed into service for Kiev from the Desna River, as was an emergency water line from Lake Verblyuzh'je.
  - A submerged dam was built in front of the dike of the Kiev hydropower plant.
  - Construction of submerged pits-dams began in the Pripyat River near Chernobyl Town and Otashev village. Shipping began on Kiev Reservoir to traverse Zeleny Mys settlement.
  - Temporary dams were constructed along the right bank of the Pripyat River, and operation of 26-km-long anti-flooding dams was renewed on the left bank.
  - Construction of filtering dikes imbedded with natural sorbents began on June 20, 1986 in small rivers in the CEZ, and all work was planned to be completed by October 15, 1986.
- July 8, 1986. An order was issued by the Soviet of Ministers of Ukraine for radioecological predictions of hydrosphere conditions in September and October 1986.

- August 26, 1986. The Decree of the Central Committee and the Soviet of Ministers of Ukraine No. 302 was issued wherein the ministries and agencies were assigned "timely and with complete efforts" to provide measures to prevent radioactive contamination of the Dnieper reservoirs. It stated that the maximum total radionuclide concentration (that is, total beta activity) should not be permitted to go above $1x10^{-8}$ Ci/L in the Pripyat River.
- September 15, 1986. A prediction of the condition of the Dnieper River was prepared by the monitoring working group, as were proposals to decrease radioactive contamination of the Dnieper River. In particular, it was recommended to:
  - provide optimal water discharge operation at Kiev Reservoir (Ministry of Water Management and Ministry of Energy)
  - develop radioecological prediction of consequences and optimal conditions for irrigating agricultural lands with Dnieper River water (Ministry of Water Management and State Committee of Agricultural Industry)
  - stop commercial fishing in Kiev reservoir based on predictions of water contamination (Chief Fishery of Ukraine)
  - take maximum measures to reduce ship movement in Kiev Reservoir and the Desna River to prevent bottom sediment re-suspension and bank erosion.
- November 1, 1986. State Committee of Hydrometeorology of the former Soviet Union (Institute of Applied Geophysics in Moscow) was entrusted to coordinate all work on "control and prediction of radioactive contamination of natural environment" according to the Decree of the Soviet of Ministries.
- February 25, 1987. First predictions were made on expected Pripyat River flooding and possible radioactive contamination of the Desna River.
- March 1987. Construction of the last four submerged dam traps was completed on the Pripyat River near Ivanovka Village with excavation of more than $4.5x10^6$ m^3 of sand from the river bottom.
- April 22 to May 1, 1987. The conclusion of the experts of the Academy of Sciences of Ukraine, Ministry of Defense, Ministry of Water Management of Ukraine, State Committee of Hydrometeorology, was prepared on the efficiency of hydro-engineering construction of overflow or filtering dams with a core of clinoptillolite in the 70-km zone of the Chernobyl plant. The study showed their inability to intercept radioactive flows and reduce

radioactive contamination of Kiev Reservoir. The decision was made to stop operating most of them after the flood of 1987.

- September 12, 1987. The Chernobyl Department for Operation by Water Protection Construction (ChDOWC) was established. This organization was given broad authority for construction and operation during implementation of water protection measures.
- December 4, 1987. Ministry of Health of Ukraine gave permission to the State Committee of Agricultural Industry of Ukraine to use water from the Dnieper for irrigating crops. The decision was justified by optimistic predictions of the Academy of Sciences of Ukraine concerning "possible contamination levels of the Dnieper reservoirs in full water years" and final conclusions by the Ministry of Health supporting the previously assigned standards for permissible radioactivity in water used to irrigate crops.
- February 8, 1988. "Programme of works for 1988–1990 of the Permanent Commission at the Soviet of Ministers of the Ukrainian SSR on elimination of the consequences of accidents, catastrophes and natural calamities" was adopted for decontamination and sanitary-hygienic measures that should be fulfilled by "radiological control, scientific studies and other works directed to elimination of the consequences of the Chernobyl accident." Particularly, the Academy of Sciences of Ukraine, the State Committee of Hydrometeorology of Ukraine, and the Ministry of Water Management of Ukraine were assigned "to provide the scientific and research for decontamination of reservoirs of the Dnieper cascade (first of all, Kiev Reservoir) and to submit appropriate proposals" in 1988. According to this program, these agencies were assigned "to provide highly effective water protection measures in the Chernobyl zone and adjacent regions during intensive snow melting in the river basins of Kiev, Chernigov, Zhytomir and Cherkassy regions" in the springs of 1988 to 1990. The task was also set up "to study effects of protective and fencing dams on the radionuclide levels of groundwater in 30-km zone of the Chernobyl NPP."
- March 4, 1988. At the meeting of the Permanent Emergency Commission at the Soviet of Ministers of Ukraine on the problems of usage of the Chernobyl 30-km zone, a decision was made "to conduct qualitative observations of the radiogeochemical composition and the radioactive level of subsurface water," and also to perform a specialized survey of the contamination of the Pripyat floodplain and, if necessary, provide decontamination measures. The ChDOWC should provide maintenance and operation of all water protection

construction in the Chernobyl zone, according to the contracts with PA "Combinat."

- During the emergency phase of the acute post-accident period (1986 spring and summer), the measures taken were mainly administrative:
  - reduced intake of ^{131}I through drinking water
  - increased use of groundwater in towns and, if possible, minimal use of contaminated surface water; additional purification of drinking water by treatment plants
  - construction of additional wells for groundwater.

During the early intermediate phase, from the summer of 1986 to 1988, protective ground-filling dams were constructed for several kilometers along the Pripyat River to keep contaminated surface runoff from reaching the towns of Chernobyl and Pripyat. Several special bottom pits (Chernobyl pit at Ivanovka and Otashev villages downstream of the river and along the shipping channel of Kiev reservoir near Strakcholesje village) were built at the bottom of the river to intercept contaminated sediments during the first summer and spring after the accident and the winter of 1986–1987. However, these pits could intercept only 5 to 7 percent of the total radionuclides adsorbed on the surface of suspended river sediments. The pits collected the coarser sediments. The silted and most contaminated fractions of suspended matter were transferred with the water and not trapped in the pits; they were deposited in Kiev reservoir, where flow velocity was below the critical velocity to support suspended sediments in the water. The effectiveness of the sediment traps was analyzed in detail by Voitsekhovich et al. (1989).

Previous studies (Voitsekhovich et al. 1995; Botchkov et al. 1988) showed that hydro-engineering (hydraulic) and expensive measures such as local banking of the river, damming of small rivers, dredging of pits in river bottoms, and other attempts to keep radioactive water from the watersheds or to intercept radionuclides in the mouths of rivers by engineering means were unable to control water quality in Kiev reservoir. They were only able to partially regulate runoff from the contaminated catchments to the rivers and thus reduce the maximum radioactive contamination of water in the areas downstream of the Chernobyl nuclear plant.

In the first year after the accident, the main source of concentrated radionuclides was filtering runoff (seepage) from the Chernobyl cooling pond and surface runoff from the plant and sanitary treatment. An attempt was made to isolate the contaminated drainage waters of the cooling pond from the Pripyat River. A special system of drainage wells combined with a large-diameter pipeline were constructed around the cooling pond to pump contaminated drainage water back into the cooling pond. This drainage system was never used because of uncertainty about the consequences of its operation, despite the fact that its construction was expensive (Voitsekhovich et al. 1995; Kazakov 1995). During 1989 and 1990, another drainage channel with a water receiver was constructed to pump seepage back to the cooling pond.

In 1986 and early 1987, more than 130 special filtering dams containing ceolite were constructed on the small rivers to adsorb radionuclides. During the entire period of operation, their effectiveness at intercepting radionuclides and flow regulation of small rivers was studied (e.g., Botchkov et al. 1988; Ivanuchkina et al. 1989; Kopeikin 1990). By 1987 their ineffectiveness was recognized, and the decision was made to destroy most of them. The low efficiency of filtering dams is further discussed in Section 6.6.2.

In 1987, one of the measures for reducing the consequences was disposing of contaminated materials at temporary radioactive waste sites. The wastes included soil, vegetation, construction materials, small buildings from the Chernobyl site, and trees from the heavily contaminated Red Forest that died from radiation. These measures were deemed necessary to protect people cleaning up after the accident at and near the site and the workers at the nuclear plant (Bariyakhtar 1995). The larger tree trunks and bushes removed from the forest were buried in sandpits and trenches without special barriers to keep radioactivity from seeping into the groundwater. After several years, significant local contamination of the groundwater had occurred and created a long-term problem of restoring the natural environment in this region (Dzhepo et al. 1994; Kopeikin 1994; Zhilinskiy et al. 1997). Many other countermeasures implemented in the first post-accident years were planned and carried out with the sincere conviction that they were needed. Unfortunately, most of them turned out to be ineffective.

Most of the measures were implemented without an assessment of their efficiency because there was no time to study and gain experience. Expediency was the driver, along with the emotional state of the people and the decision makers in the first years after the accident. Most of the measures were too expensive and too limited in scope to produce the desired results. Unfortunately, archival and published data do not provide an accurate determination of the costs. Experts give different estimates of the direct and indirect costs for water protection in the CEZ and adjacent regions from 1986 to 1988; the sum exceeds 50 million U.S. dollars. It is even more difficult to estimate the prevented exposure dose to the people of Ukraine from the measures that were implemented in the emergency and intermediate phases; however, the prevented collective dose equivalent of water protection measures implemented between 1986 and 1988 cannot be more than about 500 man-sv^{-1}. The cost for 1 man-sv^{-1} was estimated to be more than U.S.

$100,000 for all the water protection measures (Voitsekhovich and Panasevich 1998). However, economic factors were not considered when selecting countermeasures; the predominant factors in making decisions at that time were socio-psychological and political justification.

6.4.3.2 Early intermediate phase (summer 1986 to 1988)

In the first months after the accident, the greatest fear was rainstorms—and protecting oneself from radioactive rain and radioactive dirt that would wash off the slopes above Chernobyl. Therefore, earthen embankments were bulldozed along the rivers. These structures were built to prevent most of the dirt from moving through to the river. They were covered with a polyethylene film and soil. In addition, in the summer of 1986, several long dikes were constructed on the southwest bank of the Pripyat River in Chernobyl to catch contaminated runoff from the city. This was not effective because all the land area could not be covered, and the runoff could not be controlled. These were some of the first water protection measures around Chernobyl. Many protective structures were built outside the CEZ; among them the 17- and 3-km-long embankments for the Uzh and Teterev rivers.

In the first years after the accident, isolating the Chernobyl cooling pond from the Pripyat River was considered one of the major tasks. A special drainage and well system was built around the cooling pond to catch infiltrating radioactive groundwater but did not operate because of uncertainty about the consequences of its use. Indeed, pumping water from the wells back into the cooling pond might cause problems with water balance and dissolved salts in the cooling pond. The cost of construction and maintenance was very high, and in 2000 and 2001 the system was removed.

Drainage systems of wells were built in the CEZ, bored at a depth of 20 to 30 m (327 m total) with deep well pumps and pressure lines joined by the main collector. The drainage water was to be released into the cooling pond. This reserve system is kept in complete readiness, although there has been no need to use it. To keep groundwater from the river and cut off drainage, a drainage screen was designed for the cooling pond.

Another action taken during this period was building a slurry wall and a series of drainage wells to prevent underground migration. In September 1986, a vertical antifiltration screening wall of ball clay 2.8 km long and 33 m deep ("a wall in the ground") was built on the border between the main power plant building and the cooling pond (Dzhepo et al. 1995). The Pripyat municipal sewage drainage system was moved to the cooling pond to reinforce the foundation plate under Unit 4. Wells were drilled for Chernobyl industrial site drainage, and radioactive waste disposal sites were built.

Sixteen wells were bored at the industrial site to regulate groundwater levels. Some liquid radioactive wastes were sunk in the cooling pond via a stretch of canal built during nuclear plant construction. Studies showed that migration of radionuclides in the subsurface water was too slow for these wells to be effective. The slurry wall and wells could not prevent contamination of surrounding groundwater, and the project was terminated.

During 1986 and early 1987, over 130 special filtration dams were built using ceolite (clinoptolite) sorbing screens. These filtration dams detained radionuclides while letting the water through at a large number of tributaries of these rivers and canals to a total length of 4.9 km. The ceolite filtration dams captured short-lived radionuclides effectively during the summer of 1986, but soon their capacity decreased dramatically because the pores were blocked by suspended matter and for other unforeseen reasons. Subsequent studies indicated that only 5 to 10 percent of entrained ^{90}Sr and ^{137}Cs was adsorbed by the ceolite barriers in the dams. The river current through the dams would need to be controlled to obtain better performance.

A special technology was required for preparing ceolite and other natural sorbent materials; this greatly increased the costs. Moreover, the streams that were dammed contributed only a few percent to the total flow of the Pripyat and Dnieper drainage basins. After the 1987 spring flood, construction of new dams was terminated and the decision was made to destroy most of the existing dams. By 2000 only about 10 dams were still in use. In 2005 a special action plan on complete demolishing of the dams and small river canals restoration has been developed and its implementation is planned in 2006. However, the dams that cut into the highly contaminated river bays and old river channels away from the main river remained in use and are used to regulate fluxes of the high contaminated water to the river (Figure 6.7).

During the first summer after the accident, several canal-bed traps were dredged in the Pripyat to widen the river and reduce its velocity. This was an attempt to increase sedimentation of suspended radioactive particles. However, subsequent studies showed these traps also were ineffective (Voitsekhovich et al. 1989; Zheleznyak 1995). The suspended radioactive particles were too small to settle in a large natural river such as the Pripyat with large discharges and turbulent flow conditions.

At the time there was a sincere desire to do something to save millions of people and to protect a large portion of land, but knowledge and experience were lacking. That awareness came later. At that time, within the CEZ, heroic and selfless attempts were being made to save the situation.

Figure 6.7. View of one dam with ceolite windows for filtering the water from the contaminated Yanov Bay as it flows into the Pripyat River in the Chernobyl near zone (2000)

To operate the protective facilities in the CEZ, a special water management department was created in July 1987. In 1993, the Administration of the CEZ renamed it the State Specialized Industrial Water-Protective Enterprise. Its purposes are to:

- operate protective installations and systems intended to reduce radionuclide outflow from contaminated areas to the Pripyat River
- design and conduct measures for preparing the water protective installations for flood events
- provide hydrological water protective measures
- maintain and control wells decreasing the water table and observation wells
- bore and maintain experimental and observation wells
- attend nonfunctional wells
- maintain wells at resettlers' ("self-settlers") residences.

6.4.3.3 Later intermediate phase (1988 to 1994)

A new phase of hydrological remediation began after the 1988 summer flood. This was the first flood after the nuclear accident to have high levels of river water covering much of the contaminated floodplain and introducing secondary ^{90}Sr contamination of the river. After this flood, a special study was made of runoff processes from the contaminated floodplain near the Chernobyl nuclear plant (Zheleznyak 1995). Surface water hydrologic modeling showed that a realistic worst-case scenario with the highest possible radionuclide concentration was a spring flood with a maximum discharge of 2,000 m^3/s. Such a flood has a probability of 25 percent occurrence per year (or once in four years). Computer simulations of the flooding indicate that if radionuclides in the floodplain were isolated from the river, the ^{90}Sr concentration in the river would decrease by two to four times (Voitsekhovich et al. 1994).

Six years after the Chernobyl accident, a systematic approach was taken to develop a new concept of water protection. One of the priorities was construction of a water protection dam on the northeast (left) bank of the Pripyat River. This was the first countermeasure that worked. The idea was very simple: limit contact between the contaminated soil of the floodplain and the river. Various strategies were considered. In one scenario the contaminated soil was to be completely removed to a place that does not contact the water. In another, the floodplain was to be covered with a suitable waterproof layer. Unfortunately these options were too expensive. The simplest and most effective version was to isolate the most contaminated sites in the floodplain with an alluvial sand dike and to filter the river water through the dike and return it with pumps, if needed. These were combined with other general environmental protection work in the CEZ, such as planting trees, land remediation, recultivation, and others. The dike construction was completed before spring 1993.

During 1993 summer rainfall, 1994 winter ice jams, and 1999 spring flooding, this water protection system proved its effectiveness and justified the completion of a limited set of the countermeasures. Also, computer predictions to select the right strategy were confirmed (Voitsekhovich et al. 1994; Bilyi et al. 1994), as presented here. To assess the effectiveness of the dike on the northeast (left) bank, the unsteady, two-dimensional, finite element hydrodynamic RMA-II code (Wang and Connor 1975) and the sediment-contaminant transport FETRA code (Onishi 1981) were applied to the Pripyat River and the approximately 3-km-wide, 10-km-long floodplain with and without the dike. The radionuclide contamination levels in the area as well as the earthen dike are shown in Figure 6.8.

Like the TODAM code (Onishi et al. 1983) as discussed in Section 2.4, the FETRA code simulated the migration (transport, deposition, and resuspension) of sediment and dissolved and particulate radionuclides with their interactions (e.g., radionuclide adsorption/desorption and sediment-sorbed radionuclide migration), as well as river bed changes.

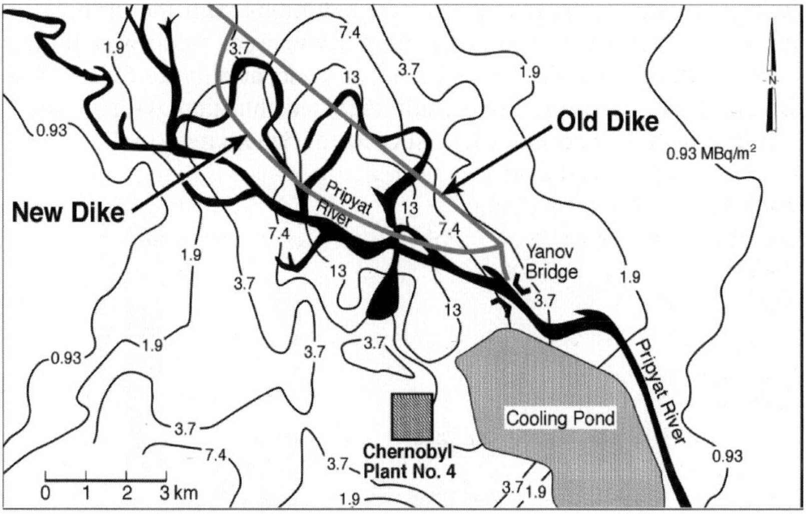

Figure 6.8. ^{90}Sr distributions on the ground and in the Pripyat River floodplain near the Chernobyl nuclear power plant

A four-year flood condition (2,000 m^3/s river discharge) was imposed to cover the entire floodplain with a depth of water similar to the January 1991 flood. Figure 6.9 presents the predicted flow and radionuclide distributions without the dike. FETRA predicted that the maximum ^{90}Sr concentration in the river would be more than 20 times normal. The predicted ^{90}Sr level at Yanov Bridge (at the downstream end of the floodplain) was up to 10 Bq/L, in good agreement with the 9~11 Bq/L observed during the 1991 flooding.

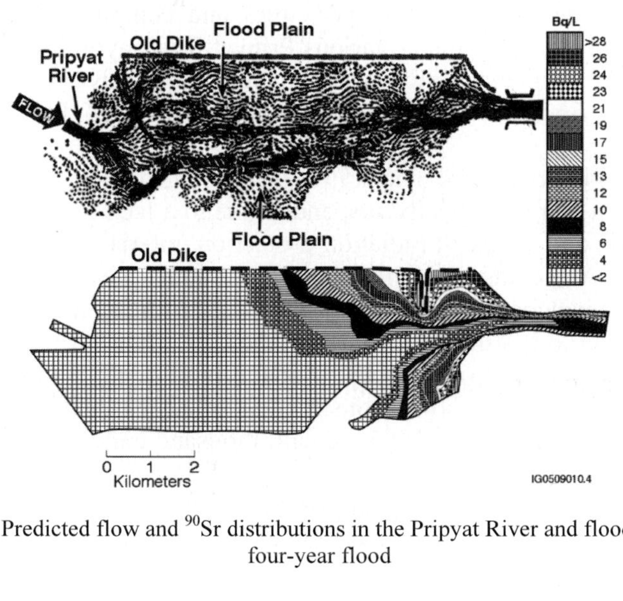

Figure 6.9. Predicted flow and ^{90}Sr distributions in the Pripyat River and floodplain under a four-year flood

Figure 6.10 shows predicted ^{90}Sr concentrations with the northeast (left) dike on the floodplain eliminating flooding of the area. According to the simulation, the dike effectively lowered the ^{90}Sr concentration to 5.5 Bq/L in the river at the Yanov Bridge, about half the concentration without the dike. Thus, the earthen dike proved to be a successful remediation tool.

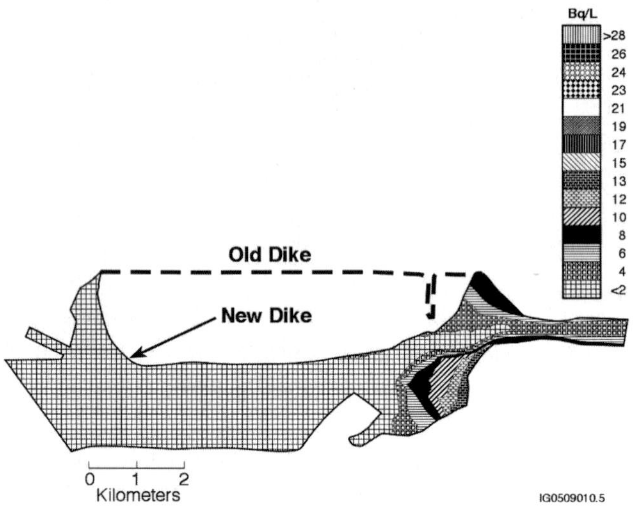

Figure 6.10 Predicted ^{90}Sr concentrations in the Pripyat River with the earthen dike in the floodplain under a four-year flood

6.4.3.4 Later phase (1994-1999)

Based on radiation protection principles and computer simulations for water protection strategies under various Pripyat River hydrologic conditions, a comprehensive remedial action plan for the Dnieper River was established in 1994. Most priorities and actions were achieved by 2002 (Figure 6.11).

However, between 1994 and 1999, water remediation activities at the Chernobyl plant were practically suspended due to a lack of funding for their implementation and a lack of radiation protection criteria on water protection intervention. Achieving cost-effective remediation measures is uncertain without fully developed criteria for cleanup. At present, a large amount of radioactive material is still concentrated in the CEZ. For instance, the Pripyat River floodplain has 40 to 50 percent probability of being inundated by river water each spring. Even after construction of an earthen dike along the northeast bank of the river in 1993, several thousand curies (1 Ci = 37 GBq) of ^{90}Sr and ^{137}Cs still remain in lowland soils. A large amount of radioactive waste is also stored in temporary disposal sites in areas regularly flooded in spring or in contact with groundwater flowing toward the river.

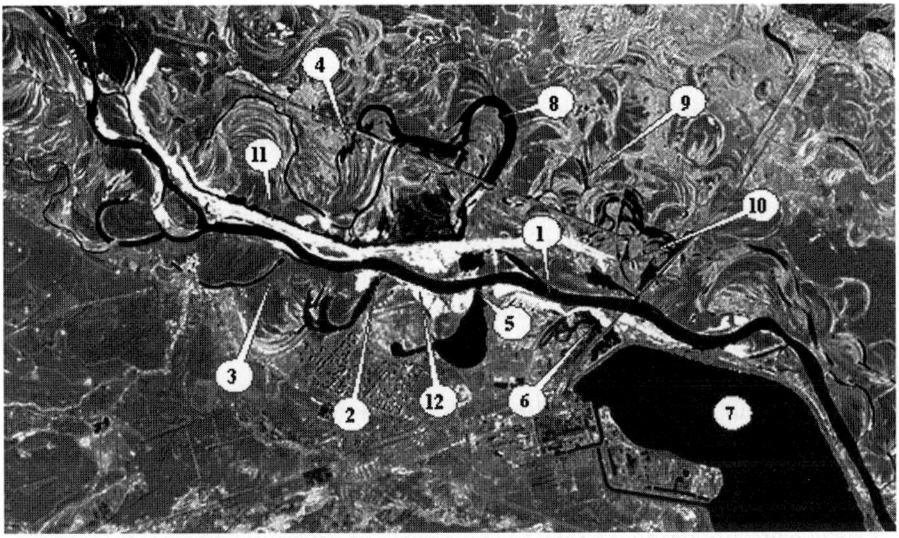

Figure 6.11. Water protective construction in the Chernobyl near zone; 1, Pripyat River (dredged during 1993-1998); 2, Semikhodi Bay (cut off from river by dam in 1986); 3, Shepelitchi Bay (cut off from river in 1986); 4, Glubokoe and Vershina lakes (most contaminated); 5, Pripyat Bay (closed by dam in 1986; 6, Azbutchin Lake (covered by sand taken from river in 1996-1997); 7, Chernobyl cooling pond and drainage system; 8, old canals joined in hydrotechnical system with pumping station to control water elevation; 9, polder system; 10, water elevation control gate No. 7; 11, water protective dike and drainage canal on left bank (1992); 12, water protective dike on right bank (temporary installation built in 1999 and completed in 2003).

The technological feasibility of controlling the existing sources of radioactive contamination in a large catchment area is limited, but an understanding of the problem may be an important result of the first water protection stage at the Chernobyl site. Thus, the main conclusion of the earlier studies is that countermeasures should be optimized, accounting for the averted human doses and economic costs.

Simulation also shows that, similar to the northeast dike, a dike along the southwest bank of the Pripyat River could prevent radionuclide washout from the floodplain. The main reason for a new southwest-bank dike was to isolate extremely contaminated soils and waste disposal sites in this area from inundation during a high flood.

Except for monitoring, there were no planned actions to control secondary radioactive contamination of water resources in Belarus and Russia (except for research programs and initially planned remediation actions by former Soviet Union authorities before 1991). In fact, fish contamination predictions from the ECP-3 study (the Commission of European Community's study conducted in 1991–1996) (Sansone and Voitsekhovich 1996) indicated that the radiocesium concentration in the muscles of predator fish might still be rather high (5 to 10 kBq/kg wet weight) even in 2016. Therefore, the top predators such as pike and perch were banned as fishery products. Potassium was thought to reduce future uptake in fish, but a sensitivity analysis showed potassium was effective only in very oligotrophic lakes with potassium levels below 1 mg/L. Banning these fish was the only countermeasure that could reduce the dose to the population. Other studies (e.g., Grinzhevskiy et al. 1997) reached similar conclusions on remediation of fish-producing lakes in Ukraine (Ivankiv, Ladyzhichi and others), Belarus (Svyatskoe Lake and others), and Russia (Lake Kozhanovskoe, Svyatoe, and others).

6.4.3.5 Present stage of activities (1999–present)

The southwest (right)-bank flood control dike was designed before the spring of 1999 as the first step in preventing floods in the area. But it was not completed before the 1999 spring flood, and 15,000 sand bags were placed on the river bank by hand. Thanks to the efforts of hydraulic engineers, flooding of the southwest bank floodplain was prevented and overall flooding of the southwest (right) bank avoided in areas of waste disposal with high levels of radioactivity. Experts estimated that these measures alone prevented 10 to15 man-sv of collective dose. The southwest-bank protective dike was completed in 2002.

The more contaminated northeast (left)-bank floodplain near the CEZ remained dry during the 1999 floods. A computer simulation showed that this prevented ^{90}Sr washoff of up to 200 Ci into the Dnieper reservoirs. If we consider that the probability of such floods is once every five to seven years, the benefit of this measure is clear. Completion of all walls/dikes in the area will prevent an additional 200 man-sv over 70 years.

One of the key problems under consideration by the Chernobyl Administration authority is to develop cost-effective strategies and elaborate designs for cleaning up (or restoration of) the contaminated bottom sediments in the cooling pond since the last reactor was shut down in December 2000.

For instance, if no water is pumped from the river to the cooling pond, it will be drained in a few years, and about 60 percent of the bottom with its highly contaminated sediments would be exposed, becoming an additional radiation exposure source for people who work in the Chernobyl zone. The problem is what to do with the cooling pond bed. There is no clear answer to this problem. The following strategies are under consideration:

- Do nothing
- Immobilize the sediments in situ (e.g., screening, phytostabilization)

- Remove the sediments for subsequent treatment, conditioning, and disposal (or storage)
- Completely remediate (including water treatment and retrieval and disposal of sediments).

Potential combinations of these strategies would also be considered, for instance, removal of heavily contaminated exposed sediments and in situ immobilization of the remaining waste. Preliminary remediation targets would be identified for the most important radionuclides, such as ^{137}Cs (for resuspension pathway), ^{90}Sr (for groundwater pathway), and actinides.

Remediation strategies would be selected for particular areas or levels of contamination. The volumes and inventories of generated waste streams as a result of retrieval would be assessed and the resulting wastes classified according to an IAEA methodology and waste acceptance criteria for the existing or planned Ukrainian facilities. Strategies will address actions for each stage of Chernobyl nuclear power plant decommissioning and respective water requirements. They would be intended to reduce the following (potential) risks from:

- groundwater and atmospheric dispersion exposure pathways
- direct exposure and doses to the operators
- human intrusion
- surface erosion and other environmental impacts.

The risk needs to be considered and estimated. Estimated risk may be so small that the simplest strategy such as "do nothing" or "just provide some phytostabilization and anti-erosion measures at the exposed bottom" might be chosen. However, the selection must be based on a cost-effective analysis.

Candidate technologies may be identified by first considering a broad spectrum of possible individual methods for capping, waste treatment, encapsulation, storage, and disposal. Some methods may be suitable only for specific waste types, whereas others are more generally applicable. Overall, the analysis should present viable solutions (or components of the solution) for identified strategies. Viability will depend on the specific conditions of the Chernobyl site, the contamination level of the surrounding territory, options for waste storage and disposal available in Ukraine, and the feasibility of a particular technique. Preliminary screening should be done to narrow the list of appropriate technologies that would address the requirements of each strategy. Selected technologies are likely to address the following elements:

- multilayer capping (by sand, clay, or soil)
- biological immobilization (e.g., grass)
- retrieval, offsite treatment/conditioning, and storage/disposal
- maintenance of water level of the cooling pond
- immobilization and/or consolidation of exposed sediments
- monitoring requirements
- dredging (hydraulic or mechanical)
- in situ disposal, including solidification or transfer of sediments to deeper parts.

Each option needs to be screened on the basis of satisfying the following:

- fundamental principles and detailed regulations (including international guidelines and Ukrainian regulations)
- planning considerations
- underlying technical approach (e.g., whether the technology is appropriate for a particular type of waste)
- cost.

The best strategy could possibly be "do minimum" and be the cheapest way to prevent potential wind resuspension (erosion) from the cooling pond bottom, with some phytostabilization approaches using special clones of willows and grasses. Results from phytoremediation studies in the international "PHYTOR" project provide some optimism (Viktoriova et al. 1999).

Further measures include the following:

- continue water regulation of wetlands reclamation system at the CEZ
- complete design and implement optimal actions focused on restoration of the cooling pond
- expand monitoring around radioactive waste disposal sites
- provide reliable monitoring and control of transuranic materials due to surface water and groundwater transport beyond contaminated area
- prevent expansion of radionuclide groundwater transport beyond present locations in the waste disposal site by constructing engineering and geochemical barriers around waste disposal sites.

Research designed to underpin future engineering decisions is being conducted in the Chernobyl zone under the framework of international projects. Chernobyl has become an important scientific location, attracting scientists from many countries to research and test remediation technologies for contaminated land and water, leading to the restoration of natural landscapes.

Water protection problems are difficult enough outside the context of radiation protection. For example, the best way to fight fire may be to flood the peat bed and forests in the floodplain. But once the fires are extinguished, radioactive washoff increases. What is the best way to restore the water balance of this area? The whole ecosystem is deteriorating. The reclamation channels are overgrown, and the land is becoming boggy. When the land was dry, there was a forest. Now the trees are dying and the peat bog is returning.

After 20 years of research there are still more questions than answers. Perhaps most answers are to be found not in the sphere of technology but rather in economics. The problems of the CEZ have not been solved. We must use the experience acquired during our scientific research.

6.5 Application of Cost-Benefit Analysis to Optimal Strategy Selection for Aquatic Countermeasures

Revising the present strategy of water protective measures is justifiable due to the difficult economic situation in Ukraine and the new regulations for "Permissible levels of radionuclide content in the food products and drinking water" (DU-97 1997) and "Radiation Safety Standards of Ukraine" (RSSU-97 1998). According to the present concepts, the optimization and selected technologies to control secondary radioactive contamination sources of the Dnieper water system (most are within the CEZ) should be determined based not only on dose but also on economic and social factors. Most of the economically developed countries follow a similar strategy of optimization of radiation protection. The principle of optimization of water protection activity is to provide the maximum needed benefits at the minimum cost. Optimization is most important for radiation protection of the Ukrainian population with small exposure doses.

The same principles must determine the strategy of water protection measures in the CEZ, considering that contaminated water bodies, floodplains, and watersheds are potential sources of secondary radioactive contamination. According to Ukrainian standards (RSSU-97) for normal operation of nuclear power plants, these sources are out of regulatory control when the impacts on the population meet the following criteria:

- Annual effective dose of a person through all exposure pathways will not exceed 10 μSv per year without active management of contamination source.

- Annual collective effective dose through all internal exposure pathways will not exceed 1 man-sv without active management of a radiation source or optimization of antiradiation protection proves that no active management is the best decision.

However, these criteria are irrelevant to post-accident situations. A decision to terminate or continue the water protection activity should be made based on criteria from RSSU-97, providing that the dose impact of a radiation exposure source on the population would not exceed 1 mSv per year for chronic internal and external exposures for more than 10 years or 5 mSv per year for the first two years and 15 mSv for the first ten years.

But when the source with small individual doses of exposure contributes to a large collective dose (e.g., water from the Dnieper River) and the contamination source can be controlled, it is necessary to use the criteria of collective dose and radiation risk. In previous work by the Ukrainian Hydrometeorological Institute and the Ukrainian Scientific Centre of Radiation Medicine, it was shown that if no additional measures were taken to reduce radioactive runoff outside the CEZ, the expected effective collective dose of Ukrainian population in the regions that use Dnieper River water (i.e., drinking water, fish, and irrigated crops) may reach 3,000 man-sv by 2056 under the statistically mean hydrological regimes of the Pripyat River. It is not so significant to the individual annual dose for a population of 20 million. However, countermeasures to reduce the dose exposure should be selected based on dose reduction efficiency as well as social and economic benefit.

The reality is that source control of secondary radioactive contamination of the water environment cannot be achieved at any cost. The economic and social advantages of the countermeasures are greater than the cost of their implementation or, if not implemented, the harm to public health. The fact is that water protection measures have been provided continuously in the CEZ during the post-accident period.

Following the optimization principles, planning of water protection measures should be directed to gain the most benefit with the least cost. This is true even for reasonable water protection activities such as reducing runoff of contaminated water from the Chernobyl cooling pond, constructing dikes on the river banks, and building geochemical and engineering barriers around the temporary radioactive waste disposal sites.

The above criteria were not used for selecting countermeasures in the past. With very limited information and predictions available at that time, most selections were justified only by social and economic criteria. However, during the final phase of water protection activity presently being considered for the CEZ, an economic evaluation of exposure dose and social efficiency of countermeasures should be considered.

The most sophisticated approach is to define the justifiable economic and social cost of water protective countermeasures. The cost of countermeasures can be estimated for different variants of technical solutions to the problem, and the lowest-cost variant can be chosen to reduce exposure dose. This problem is usually solved at the stage of a feasibility study. But after the feasibility study was approved in 1997 for the CEZ, most experts were not sure a sound solution was chosen due to a lack of clear criteria of rational cost to be paid for reducing the dose to the population (e.g., 1 man-sv) as a result of water protection. Their concern was natural because there was no experience in solving water problems under conditions similar to Chernobyl. In many countries the optimal value of dose reduction by remediation varies from U.S. $1,000 to $10,000 per unit of averted dose (1 man-sv). However, the range of optimal cost of countermeasures reflects the economic conditions of the country, the social factors of its society, and whether the state pays for radiation protection measures. Therefore, the economically and socially justified countermeasure cost must be determined by the government agencies in charge of radiation protection of the population.

The World Health Service has a concept called "Population Health" that indicates that eliminating disease implies physical and psychological well-being. However, if there is a real physiological connection between the effect of reducing water contamination and the health of the population, the psychological response of society to implementation or rejection of a water protective activity in the CEZ cannot be estimated. Nevertheless, its manifestation may produce a "stress-dependent" component of radiation risk that may require major cost for its control.

For instance, hundreds of thousands of the Kiev citizens were panicked in January 1991 by temporary secondary radioactive contamination of the Pripyat River and Kiev reservoir (i.e., radionuclide washoff from the floodplain after flooding). This was sufficient reason to suggest that the cost of measures to reduce radioactive runoff from the CEZ may be higher than the optimal cost. Some consider that an increase in countermeasure cost due to social needs is reasonable for nonaccident radiation protection, but it should be within twice the optimal cost. It is evident that the role of a social factor to justify countermeasures has been increasing in post-accident conditions.

According to the previous assessment by experts headed by Professor Los'y at the Ukrainian Institute of Radiation Protection, the economically reasonable dose component of countermeasure cost does not exceed U.S. $2,000 for 1 man-sv in the present Ukrainian economic conditions. According to RSSU-97, this cost takes into account the GDP per capita as an economical component plus the "willingness to pay" adjustment. However, the same document states that "a monetary value of risk should also consider the cost of compensation for psychological perceptions of risk (psychological or social component). As a rule, in optimization of protection, the economical component is 5 to 10 percent of the willingness to pay adjustment (or psychological cost). If the economical and social justification criteria of water protection measures are estimated in a similar manner, the cost to reduce the anticipated effective collective dose of population exposure would be in the range U.S. $20,000 to $40,000 per man-Sv.

A seminar of European countries held June 29–30, 1998 in Brussels discussed implementing countermeasures for rehabilitation of territories contaminated by the Chernobyl accident. They estimated that the cost of the most effectively implemented countermeasures in agriculture was U.S. $30,000 to $40,000 for each man-sv reduced. This level could be considered the upper limit of effective expenses in optimizing water protection activities in the CEZ. Figure 6.12 shows the cost per man-sv recommended in various developed countries. For instance, in the United Kingdom the value per man-sv are U.S. $10,000 to $30,000 for the public, $75,000 for workers, and U.S. $150 for children. In the United States, this value is estimated up to U.S. $100,000 per man-sv. However, if a GDP-based monetary value is used for the dose reduction, the value adopted for Ukraine should be much lower than those recommended as an optimal value for European developed countries.

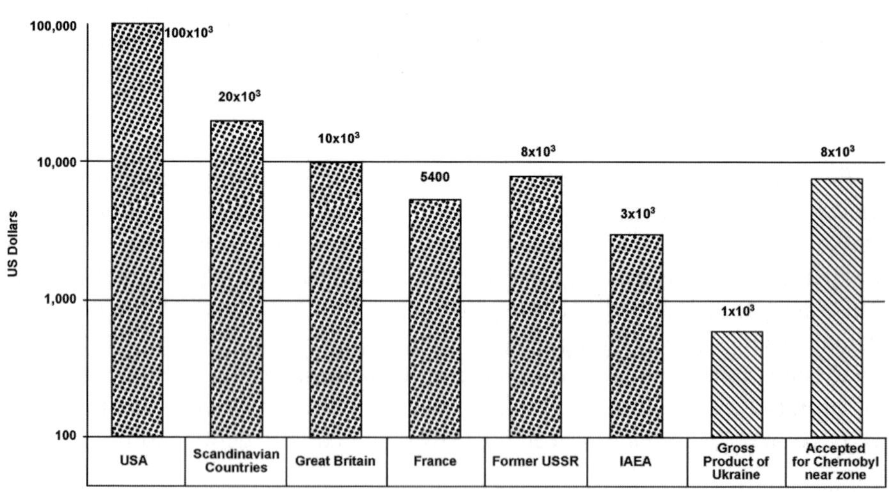

Figure 6.12. Comparative monetary value of dose reduction optimal cost in various countries

It was not a trivial task to apply such an approach to justify water protective flood control dikes at the CEZ because of uncertainty about the probability of floods and dose assessment scenarios. Computer simulations indicate the effectiveness of the flood control dike construction on the northeast and southwest banks (see Figures 6.9 through 6.11) would be much greater under high flood conditions, but this remedial action can be useless during low water levels with no flooding (Voitsekhovich et al. 1997b). Furthermore, their effectiveness will also be low when contamination levels in the reservoirs are already low due to natural factors. The use of available funding to implement measures in the CEZ is suitable in spite of their relatively low contribution to reducing the global health risk from the Chernobyl accident.

The long-term Dnieper water contamination was estimated for occurrence of the highest flooding events in three sequences, as shown in Table 6.1. The dose assessments were then performed for the annual radioactive discharges associated with various flooding sequences simulated for a 70-year reference period (Lepicar and Droz 1999). Because radiological impacts integrated over this long time period are strongly affected by the time distribution of the flooding events, given their magnitude and time of occurrence, a set of nine flooding sequences was extracted from the complete 1,000 data sets. The nine flood sequences shown in Table 6.1 were assumed to be representative of the most probable "best" and "worst" cases.

Table 6.1. Criteria of selection of the nine tested flooding sequences

Occurrence of highest magnitude flooding events in the sequence	**Magnitude**		
	Mean	**Low**	**High**
Uniformly distributed in the time period	Sequence 0	Sequence 3	Sequence 6
Distributed at the end of the time period	Sequence 1	Sequence 4	Sequence 7
Distributed at the beginning of the time period	Sequence 2	Sequence 5	Sequence 8

The dose reduction by the construction of the southwest (right) bank dike is estimated in the range of 200 to 300 man-sv until 2056 for collective doses of 20 million Ukrainians potentially exposed under the most probable hydrological conditions, compared with the no-action case (Lepicar and Droz 1999). This reduction is in addition to the 600 to 700 man-sv dose reduction from the northeast-bank dike that was constructed in 1993.

The loss of life expectancy from radiation-induced cancer is accepted as 16 years, which agrees with the ICRP (1990) estimate (see also Section 6.4.1). A value of U.S. $1,200 is obtained per man-sv based on the 1999 Ukrainian GDP. Adjusting for social factors, the value of a man-sv was recommended as $8,000 to $10,000, as stated above. Justification for the water protection plan may include reducing the cost of the technology, optimizing the remedial and cleanup activity, or obtaining additional outside funding. Even though the remediation cost-benefit criterion may not be met for countermeasure implementation, the cost can still be justified for the measures controlling radioactive outflow from the CEZ and considering the social factors and/or stress for people living in the Dnieper River water use region.

For the protective dike on the southwest bank of the Pripyat River, the optimal monetary value of dose reduction was accepted as about U.S. $8,400/man-sv. This value was obtained by considering the Ukrainian GDP-based value plus the "willingness to pay" adjustment. The total cost of the dike (construction and maintenance during 70 years of operation) at 5 to 8 percent discount rates and the expected collective dose reduction of 150 to 300 man-sv over 70 years would lead to a benefit of dose reduction from U.S. $1.26 to $2.52 million (Lepicar and Droz 1999). The cost-benefit ratio in this case is within 0.5 to 1.2. Potential investment on implementation of this dike construction can be considered cost-effective because the cost-benefit ratio for most scenarios has been less than 1.0. The international study concluded:

- Although a high degree of variability exists (dose estimates and economic valuation), the expected benefits tend to justify the cost of southwest-bank dike construction (cost-benefit ratio around unit).
- Expected secondary positive effects (economic, social benefits) would favor the southwest-bank dike construction.

Other examples of countermeasure expenses were also considered reasonable and socially justifiable, though the cost was higher than optimal. This is because society feels good about and is willing to pay for radiation safety of the environment. Society also faces supplemental expenses of retraining people affected by the Chernobyl accident (such expenses may be too large).

The radiological effects of water protection and further countermeasures should be estimated considering the fundamental tenets of modern science on radiation protection. To provide such an accounting it is essential to estimate the benefits in terms of averted radionuclide fluxes into the Dnieper reservoirs. The absolute effect was estimated by monitoring surface water. Modeling anticipated collective dose of population exposure for all water protection measures under various hydrological regimes of the Dnieper water system has not been performed. Exposure doses were modeled only for the most significant countermeasures (northeast and southwest banks). For all other measures the experts determined the benefits by assuming a direct correlation between the averted radioactive runoff into the Pripyat and Dnieper reservoirs and anticipated collective dose of population exposure. An approximate factor of proportionality between these quantities was determined by modeling some water protection activities as a feasibility study. Prevention of about 7.4×10^{10} Bq (2 Ci) of the main dose-forming radionuclides (^{90}Sr and ^{137}Cs) in releases from the water system was estimated to decrease the anticipated effective collective dose by about 1 man-sv until 2056 under existing relations of these radionuclides in water (Table 6.2).

It is evident that averted transfer of ^{137}Cs and ^{90}Sr into the rivers and their contributions to the anticipated dose exposure would be different for various time periods. Nevertheless, even such rough estimates permit comparisons of the efficiencies of various countermeasures in terms of cost of expected averted risk. The results were obtained from archives of the enterprises involving construction of water protective facilities and estimates of dose effects of implementing water protection measures.

Table 6.2 reveals that despite relatively low construction costs along the southwest bank of the Pripyat River, if the dike had been constructed in May–June 1986 to prevent surface runoff into the river, its effect on exposure dose was not apparent. On the other hand, the exposure doses to workers during its construction in the first weeks and months following the accident would be too high. Thus, such countermeasures have done more harm than good.

The construction of underwater sand pits in the Pripyat River channel would not result in heavy exposure to workers because the work was under water in areas of low surface contamination. Nevertheless, the cost of building the sand pits (dredging about six million m^3 of soil) was too high; that is, no less than 3 million U.S. dollars, although the labor costs increased by two to three times in the Chernobyl zone during 1986-1987. However, the Ukrainian Hydrometeorological Institute and the State Committee of Water Management of Ukraine estimated that the sand sediment traps accumulated only about 8 to 10×10^{11} Bq of ^{137}Cs and ^{90}Sr between 1986 and 1990. They mainly intercepted contaminated particles of river silt that could hardly contribute to population exposure. Thus, the real cost for expected prevention of 1 man-sv (operational expenses were unforeseen) is U.S. $200,000, which appears to be too high.

In 1986 and 1987, 131 water protection dikes were built on small rivers and creeks of the CEZ, using up to 1×10^8 m^3 of soil. Furthermore, the dikes were filled with up to 34,000 $m^3$of ceolite raw materials to inhibit radionuclide runoff. During the whole operational period, these ceolite dikes absorbed only 1.0×10^{10} Bq to 3.7×10^{10} Bq of ^{137}Cs and less than 3.7×10^{10} Bq of ^{90}Sr. The construction costs of these dikes, most of which were dismantled before 1988, exceeded U.S. $10 million. The maximal averted collective dose was estimated as no more than 10 man-sv. Therefore, the cost of this countermeasure in units of "risk cost" was more than $1,000,000 per man-sv.

Table 6.2. Estimated dose and cost-efficiency of measures in CEZ to reduce risk of water use from Dnieper reservoirs

(from Voitsekhovich and Panasevich 1998)

Year	Water protection measure	Averted dose since counter-measure implemented	Anticipated dose to be averted in 1998-2056	Cost of countermeasures in US $ without (*) and with (**) operational costs	Cost of dose reduction in US $1,000 per man-sv	Notes
1986	Construction of levee on Pripyat River bank	Benefit not obtained	None	up to 1×10^5		Exposure dose higher than effect of countermeasure
1986-1987	Construction of channel sediment traps on Pripyat River	10-15 man-sv.	None	up to 3×10^6	more than 200	Up to 1991 open cuts filled with sand; localized about 30 Ci of radioactivity
1986-1987	Construction of dikes and ceolite filters in small rivers and creeks	4-5 man-sv for 1986-1987 (up to 1988 most dikes were destructed)	None (excludes dams on Pripyat Bay)		$1\text{-}1.3 \times 10^3$	Specific cost of dike effect on rivers/creeks is U.S. $25-30,000/man-sv(*)
1987-1988	Construction of drainage pipe system along cooling-pond	Not functioning	Not operational	up to 2×10^6	Effect not determined	
1993-1998	Alternative drainage system in streams to filter	up to 5 man-sv	Further function not determined	up to 5×10^5 (**) for 5 years of operation	about 100	In 1994 about $1.48\ 10^{11}$ Bq returned to pond; in 1997, less than $3.7\ 10^{10}$ of Sr
1993	Construction of left bank protective dike	120-150 man-sv	500-700 man-sv	up to 1.5×10^7 (**) for 5 years and up to 2.5×10^7 of work (**) to 2056	35-40	Anticipated effect estimated with dike operation expenses for 60 years
1998	Construction of right bank protective dike	-	200-250 man-sv	up to 4×10^3 (*) and $6\text{-}7 \times 10^6$ of work(**) to 2056	30-35 (**)	Anticipated effect estimated with operation expenses for 60 years

The dam built on the southwest bank was relatively effective for the Pripyat, Semikhody, and Shepelychi to separate the creeks and springs from the main river flow. The creeks were heavily contaminated, and radionuclides were washed out during flooding. Separating them from the main flow prevented more than $18.5x10^{12}$ Bq of ^{90}Sr and ^{137}Cs from entering the river. No less than 10 percent of the radioactivity localized on the bottom of these water bodies and could contribute to radioactive contamination of the water ecosystems and additional exposure to the population using the water.

A drainage canal with a large-diameter tube was built along the cooling pond but never operated and may have been ineffective. Direct and indirect costs for construction and anticipated operation were estimated to be at least U.S. $2 million. The alternative drainage complex was constructed in 1994 to intercept and filter drainage water and return it to the cooling pond; this was more effective but not effective enough. For instance, about $2x10^6$ m^3 of contaminated water with activity up to 10 Bq/L (200–270 pCi/L), corresponding to $1.85x10^{10}$ Bq, was recycled in 1997. It was decided that further operation after 1998 would not be reasonable, and it was terminated.

From 1993 to 1998, construction of an anti-flooding dike with pumping stations to regulate the flow of the most contaminated part of rivers and creeks on the northeast bank was the most effective measure. This dike worked well during 1993 summer flooding and 1994 winter ice-jamming on the Pripyat, preventing probable washoff of radionuclides into Dnieper reservoirs of more than $7.4x10^{12}$ Bq of ^{90}Sr and $1.11x10^{12}$ Bq of ^{137}Cs plus transuranics. The estimated effect of the averted dose due to the northeast (left)-bank hydraulic engineering complex constituted 120 to 150 man-sv. The effect of this countermeasure during its operation (up to 60 years) was to reduce the anticipated collective dose to the population near the Dnieper by 500–600 man-sv. The money for construction and operation for the next 60 years is U.S. $25 to $30 million. Nevertheless, considering the social significance, the cost does not exceed the optimal level, and further reduction of annual operational expenses in the future may even increase its economic efficiency.

6.6 Case Studies of Countermeasure Applications at the Chernobyl Affected Zone

6.6.1 Rehabilitation of Water Bodies in Remote Areas

After the Chernobyl accident, considerable attention was paid to reducing the dispersal of radionuclides from contaminated catchments and water bodies from the Chernobyl plant into the reservoirs downstream in the Pripyat and

Dnieper rivers. In Ukraine, more than 4,000 lakes and ponds were exposed to radioactive contamination. Many have complex water management assignments. Only 1 to 3 percent of the water bodies are used for fishing. About 1 to 6 percent are used for drinking water and up to 10 percent for reclamation and irrigation. The other 60 to 80 percent have various uses.

A serious systematic study of these water bodies was started only after 1991, when the acute problem had subsided. The majority of water body rehabilitation in the contaminated regions was to restore fishing. Measures to restore fishing have also improved recreation and other types of water use in the contaminated water bodies.

According to a database of radioactive contamination of water bodies in Ukraine (Samoilenko et al. 1997), abnormally high levels of contamination were observed in the Kiev, Zhytomir, and Rovno regions, especially in the first years, despite the relatively favorable overall situation. The levels of contamination of most water bodies were low, and no countermeasures were required to reduce them. The radioactive contamination of the bottom of some water bodies has a limited impact on normal fishing, e.g., at Kiev, Ivankov, Chernigov, and Zhytomir fish-production plants. Fish hatcheries were suspended in many ponds after the accident; the water was drained and the pond beds became overgrown, preventing restoration. Therefore, developing good technologies for fishing is still needed in the contaminated regions of the Ukrainian Polessye even 20 years after the accident.

From 1995 to 1997 the Institute of Fishery of the Ukrainian Academy of Agrarian Sciences of Ukraine issued recommendations for reclamation of the ponds where fishing was suspended (Grinzhevskiy et al. 1997). This document summarized the experience gained in Ukraine and proposed further organizational and technological methods to restore the normal process of fish breeding in the contaminated territories. It proposed to use the ponds mainly for breeding annual plant-eating fish to sell to other fish farms and biennial fish with a special fattening technology for Kiev reservoir to increase its fish productivity. The key recommendation was to breed annual fish to minimize radioactive contamination of fish. Using the contaminated water bodies without draining the water from ponds under triennial and more frequent cycles, commercial fish were proposed to be bred under polyculture to reduce the ^{137}Cs in the fish to 20–30 percent of those grown under monoculture, and ^{90}Sr to only 3–8 percent due to more rapid renewal of forage. The optimal technology of fish breeding would be denser planting of individual fish. For example, if the population density of carp hybrid fry increased from 600 to 1,400 per hectare, the contamination levels of ^{137}Cs would be reduced by two times in commercial fish. Artificial feeding of young fish provided a significant benefit as well, reducing radioactive contamination of commercial fish by 10 to15 percent. Under a full regime of feeding for a month before the commercial catch, the contamination was reduced up to 40 percent.

Some studies present the chronology of radioactive contamination of fish in the water bodies in the affected zone and the effect of total mineralization on accumulation of ^{137}Cs in fish (Kryshev et al. 1988; Grinzhevskiy et al. 1997; Kudelskiy et al. 2001). They show that reduction of water mineralization by 1.5 to 2 times increases by twice the radioactive contamination in the muscles of fish as well as the rate of accumulation. Using mineral fertilizers and lime in the water bodies as a countermeasure to reduce the rate of radionuclide accumulation in fish is well known (IAEA 1994). Such measures were also used in the water bodies of the Ivankiv district of the Kiev region. Experimental studies of the efficiency of potassium and lime countermeasures for fish breeding ponds found that the effective amount of mineral fertilizers was 2 mg/L nitrogen and up to 0.5 mg/L phosphorus in water. Potassium fertilizers are applied to the water within fishery standards. The positive effect has been observed at 10 kg/ha of potassium oxide (K_2O). The best results of ^{90}Sr and ^{137}Cs reduction in fishery lakes were obtained by a mixture of mineral fertilizers (potassium, phosphorus and lime) of 40 to 50 kg/ha. Liming of fish ponds increases calcium, which is close to ^{90}Sr in chemical properties. Therefore, its competition with ^{90}Sr through water trophic chain transfer has a positive effect—reducing ^{90}Sr accumulation in benthos and contributing to purification of the pond from sedimentation by forming a carbonate insoluble phase of ^{90}Sr.

The limits on fishing and fish consumption increase radiation protection (Kryshev et al. 1988; Ryabov et al. 1996). For example, from 1986 through 1992, fishing was banned in the upper Kiev reservoir. Some administrative countermeasures for the lakes in Russian regions adjacent to Ukraine and Bryansk were also introduced as the areas most contaminated by the Chernobyl accident.

Russian and Ukrainian scientists paid the most attention to Kozhanovskoe and Svyatoye lakes (a remote zone of radioactive fallout), which are about 300 km from the Chernobyl plant. Fallout of ^{137}Cs on the catchments of these lakes constituted 0.62 and 1.1 MBq/m^2. The characteristics of these lakes are relatively weak adsorption of ^{137}Cs by their catchment soils and bottom sediments of organic origin. As a result, during post-accident years the ^{137}Cs concentration in a soluble form constituted the same order as that in the most contaminated lakes in the 10-km zone around Chernobyl. Six to seven years after the 1986 accident, the content of ^{137}Cs in muscles of fish remained high at 104 Bq/kg (carp, in natural moisture) to 105 Bq/kg (perch). This amount of ^{137}Cs in fish exceeded the maximum permissible levels of contamination adopted by the Standards of Russia and Ukraine for radioactivity in fish. In these lakes commercial fishing and peat working continued until 1991, and only after wide publicity and recommendations based on monitoring results did the government authorities ban fishing and limit peat working in the lakes. Thus a considerable decrease in the food contribution to exposure dose was realized for the population from lake fish consumption. This was especially important for fishermen and their families, for whom fish consumption was a main source of internal exposure.

However, limiting or banning activities is not effective if the people have no alternative but to violate the ban and the State is unable to fully replace this vital element of the diet. In some cases, like the Ivankiv fish farm, the quality of the water ecosystem might be restored by changing the fish production technology and operation of fish-breeding lakes. No other water bodies in populated areas of Ukraine have such extremely high contamination as Lake Kozhanovskoe. Nevertheless, the temporary limits on fishing in the fish-breeding ponds in Kiev and Zhytomir areas were established until authorization can be given for fishing based on observations.

No special countermeasures were provided in Belarus, but under the International Project "AQUASCUR" on Lake Svyatskoye, an attempt was made to reduce the levels of ^{137}Cs contamination in fish by using potassium fertilizers. About 20 tons of potassium fertilizer was unloaded on the ice in the winter and melted into the lake. As a result, the potassium content in water increased by over an order of magnitude. However, a significant long-term decrease in ^{137}Cs content has not been achieved, though the effect was somewhat different in predatory and non-predatory fish. The studies are still in progress, and it is too early for a final conclusion. Similar measures would probably give only local and short-term effects for the natural fattening of fish. The methods of artificial fish breeding with uncontaminated feed showed the best results in extracting ^{137}Cs from fish.

Despite the relatively limited effects of similar measures to reduce radioactive contamination of fish, these studies are still needed to decrease water pollution in the contaminated lakes. The decisions concerning "interference" are on principles of protection (see Chapter 5) or social and economical acceptability of the countermeasures for ecosystem restoration. The remediation methods in regions of low contamination may be analyzed by work performed by the European countries for areas affected by the Chernobyl accident. In Sweden, Norway, and Finland, considerable success was achieved in developing technologies to restore lakes exposed to post-accident contamination. These Scandinavian countries were the first to notice radioactive fallout in Western Europe on April 27 to 29, 1986.

The contamination of the soil in the catchments reached 50 to 70 kBq/m^2 (average contamination was 10 to 30 kBq/m^2) in Sweden. The maximum concentrations exceeded 1 Bq/L in Swedish and Finnish lakes in 1986 and 1987. In Sweden more than 90 lakes were monitored, and data from 41 kinds of fish showed ^{137}Cs levels at or higher than the European standard for fish of 600 Bq/kg (wet weight) (Hakanson 1992). Naturally, attempts were made to decrease the contamination level in fish in these lakes despite the relatively low exposure level. In Sweden, consumption of fish from the contaminated lakes was a main contributor to the exposure of a critical population. Thus, the countermeasures implemented to decrease fish contamination in Swedish and Norwegian lakes are worth describing here.

In 1986 and 1987, the "Lime-Mercury-Caesium" project implemented various reconstruction measures in 41 Swedish lakes (Hakanson et al. 1996). These measures were expected to reduce cesium contamination and perhaps toxic metal contamination as well. The following measures were used to address efficiency and ecological availability:

- liming of acid soils around the lakes and catchments
- anti-erosion and other countermeasures to reduce washoff of ^{137}Cs and other toxicants from the catchments into the lakes
- partial liming of the lakes and other ameliorants like phosphate and potassium fertilizers to decrease biological accessibility of ^{137}Cs and ^{90}Sr in the lake waters
- addition of clay and other additives into the water bodies to increase sedimentation effects
- intensive harvesting of fish to extract a portion of ^{137}Cs in the full balance of ^{137}Cs in lake ecosystem.

In addition, the various kinds of fish demanded an individual approach to choose relevant additives (Hadderingh et al. 1996a). The Swedish experience led to the following conclusions:

- Applying potassium fertilizers increased potassium content from 10 to 20 meqv/L. However, it was difficult to increase the potassium content significantly in lakes with rapid water exchange and thus impossible to significantly block the bio-availability of ^{137}Cs. However, where the long-term concentration of potassium was more than 5 meqv/L, ^{137}Cs was reduced by up to 10 percent per year in such fish as redeye and similarly throughout the food chain. In some lakes ameliorants reduced the stable ^{137}Cs accumulation in predatory fish like pike by 5 percent a year when conductivity was above 1.5 mg/L.

The hydrologic regime and sedimentation processes played a significant role in lake restoration. Decreasing contamination levels was insignificant and slow in shallow lakes and catchments consisting mainly of organic soils. In deep lakes with a significant river inflow having suspended substances, the self-purification occurred much faster than the expected rates of water and fish purification by additives. The effectiveness of hydroameliorants was affected by water acidity. Thus, the U.S $4-million Swedish study showed that, even with the proper choice of countermeasures, the maximum effect might be to decrease lake water contamination by only 5 to 10 percent of natural self-purification of the ecosystem for some fish.

There are many other studies on this topic (Hadderingh et al. 1996b; Helling et al. 1996). For example, aquarium experimental studies were conducted at the Institute of Hydrobiology of the Academy of Sciences of Ukraine to test the effectiveness of potassium ameliorants to remove ^{137}Cs from some food-fish. The Ukrainian Hydrometeorological Institute also conducted laboratory experiments of kinetics of adsorption and desorption of ^{137}Cs on the clayey minerals of the bottom sediments and availability of ^{90}Sr in water. These studies as a whole may conclude that potassium countermeasures to decrease contamination levels are not always simple. In some cases, adding potassium enhances ^{90}Sr desorption from the bottom sediments, intensifies the secondary contamination of water, and causes other indirect impacts to ecological quality. This complex process, in turn, was observed to increase rates of ^{137}Cs removal from fish.

The limited experiences of the countermeasures for decreasing the radioactivity levels in lake water and biota showed that fertilization additives had no significant benefits. Applications of countermeasures would have significantly greater effects if the measures are provided not once but on a regular basis, for example, every two years. However, multiple injections of additives to increase the reclamation efficiency lead to higher costs, decreasing their economic efficiency. This results in a decreased level of justification and reasonableness of countermeasure implementation. The most effective measures are those that (1) reduce washoff of radioactive substances into water bodies, (2) decontaminate or remove contaminated bottom sediments, and (3) reduce radionuclide migration from the source through the water ecosystem. However, the appropriate countermeasures also require economic justification.

6.6.2 Natural Sorbents for Decontaminating Surface Water

Using mineral sorbents for wastewater purification to remove various pollutants including radionuclides is well known in municipal and water management practices. These water purification technologies and other coagulation and biological purification methods are used for settling and filtering. Sorbents based on mineral raw materials that can extract radionuclides from the water may be recommended for decontamination. But mineral sorbents for water protection practices require special conditions for preparation and development of special technologies. Such requirements limit their application to natural waters compared with industrial and laboratory conditions.

The following are some factors limiting their use for decontamination of radioactively contaminated runoffs:

- variable chemical composition of natural water (mineralization, composition of main ions, acidity, redox potential, availability of organic complex-former, etc.)
- large flow rate of natural water moving through the filters with sorbent for decontamination
- peculiarities of raw sorption materials, etc.

This section discusses examples of mineral sorbents used for decontamination of river and drainage runoff from the territories near the Chernobyl nuclear plant They were used to purify waste water in the settling basins of the sites of sanitary treatment, other contaminated areas, and the Chernobyl zone.

6.6.2.1 Use of natural sorbents at the Chernobyl site

Because of their physical and chemical characteristics, mineral sorbents based on clayey minerals such as ceolites, bentonites, and others were widely used in the Chernobyl zone to reduce the amount of radionuclides flowing into natural water bodies. Long before the Chernobyl accident, geochemical studies showed that minerals like illite, glauconite, vermiculite, and ceolite extracted soluble ^{137}Cs and other radionuclides from river water (Kopeikin et al. 1988; Gradev et al. 1978). Sorbents of mineral raw materials containing ceolite and vermiculite can also extract ^{90}Sr under specific conditions (Spitsyn et al. 1958). Ferric oxide and activated charcoal were used to decontaminate construction materials in the nuclear industry. Good results were obtained in reducing radioactive contamination by filtering runoff water through sorbents of natural organic components (e.g., sawdust, peat).

Bentonite (a clayey material with a high cation exchange capacity) is widely used for purification of liquid radioactive waste at the Chernobyl nuclear plant. The mineral sorbents were attractive because of their low cost. For example, during the 1970s and 1980s the U.S. nuclear industry treated 4,000 m^3 of waste annually with clinoptillolite zeolites. Despite the relatively high cost of preparation, loading, and replacement of used materials, the cost to purify water is estimated at about U.S. $0.3 per ton. In the former before 1986, clinoptillolites of Sokirnyansk (Ukraine) and Dzegvinsk (Georgia) deposits were widely used for the same purpose.

After the Chernobyl accident, the bentonite and palygorskite clay of the Cherkassk deposit, in addition to the above deposits, was recommended. Geochemical studies and the characteristics of mineral sorbents of various deposits in Ukraine and other regions of the former Soviet Union (Kopeikin et al. 1988; Minikh et al. 1989) predetermined their use for decontamination of contaminated water in the CEZ.

The raw mineral material of the Sokirnyansk deposit in sufficient amounts for water purification was supplied to the Chernobyl plant during the summer of 1986. It was mainly used to decontaminate radioactive runoff at sanitary treatment sites, the drain system and water in the filtering fields of the plant, industrial site, and the town of Pripyat. Depending on the accumulated contamination level, fresh material was provided regularly and later disposed at a disposal site. Presently, it is difficult to estimate accurately the amount of radioactivity prevented from entering the water bodies of the CEZ. According to the State Specialized Enterprise on Decontamination and Radioactive Waste Management Complex, the approximate radioactivity of sorption material disposed in the main sites and repository sites constituted 600 to 800 Ci. Thus, the total potential radioactivity prevented from flowing into surface and subsurface waters by the mineral sorbents used in the Chernobyl zone is on the order of a thousand curies for the post-Chernobyl decade. Two-thirds of that amount was trapped in spent sorption raw materials used in 1986 and 1987. This amount could have entered the Pripyat River without the decontamination effort and was a significant part of the annual contribution of all radioactive flows from the Chernobyl zone into the Pripyat River during that period.

At that time, ceolite was used to purify the natural river and stream flows from radioactive contamination by dropping crushed zeolite from barges into the flows in June 1986. Special filtering appliances were constructed to enhance the effect but did not work well. Reasons and conditions for natural mineral sorbents used in the water protection of the CEZ are discussed below.

6.6.2.2 Efficiency of sorbents used for purification of river flow

After the Chernobyl accident, methods of solids settling and radionuclide sorption were proposed to reduce radioactive contamination of surface water. In 1986, 136 temporary water protection "blinds" and overflowing dams were built on several brooks and small rivers in the CEZ and adjacent areas. The reduced flow rate led to partial sedimentation of radionuclides attached to suspended solids and radioactive dust in the upper reaches of the dams. The sorbent imbedded in the dams should have extracted radionuclides dissolved in water. Zeolite from the Sokirnyansk deposit (in the Transcarpathian Mountains) was used as the filtering material. These dams were built in the summer of 1986 through the winter of 1987.

After building the first set of filtering dams in 1986, Kurchatov Institute, the Department of State Committee of Hydrometeorology of the USSR, and the Academy of Sciences of Ukraine began estimating their efficiency to reduce radioactive contamination of small rivers in the CEZ. For example, observations (Table 6.2) on the Braginka River showed the visible effect of dams catching radioactivity transferred by the suspended solids during the first post-accident months of 1986. This effect was obtained by reducing turbidity of water in the lower reaches of the dams. Cesium isotopes in the water were reduced by 10 to 20 times. For short-lived radionuclides, e.g., ^{95}Zn, ^{95}Nb, ^{141}Ce, and ^{144}Ce, concentrations were reduced by hundreds and thousands of times compared with the lower river reaches. Other rivers and channels of the CEZ showed similar efficiencies at radionuclide interception (e.g., Ivanushkina et al. 1989; Sobotovich and Olkhovik 1989).

However, this positive effect of filtering dams was observed only in their initial operation during the summer of 1986. By 1987 spring flood time, the efficiency of water purification by the filtering dams decreased to a minimum and even to zero. Moreover, the total beta activity of water taken in lower reaches of the dams in the Sakhan River was even higher than it was before filtering through the dams (Botchkov et al. 1988; Ivanushkina et al. 1989). This occurred because filtering prisms of dams were quickly sealed by deposits and transformed into "blind" dams from filtering dams.

In the upper reaches of the dams, reservoirs were formed and considerable areas of radioactive contaminated territories were flooded. At that time, an extra amount of radioactivity started to enter the rivers due to additional washoff from the flooded territories. As a result, the water protection effect of filtering dams abruptly decreased. The washed-off soils began to redeposit on the river bottoms. Up to 5-cm-deep silt deposits accumulated in large flooded areas, forming additional highly contaminated zones. Surface water and groundwater mixed in many parts of the flooded territories, and secondary radioactive contamination began. With the spring floods, a portion of radioactive substances previously intercepted by zeolites in the dams began to wash off, reducing the water protection features. The data showed that these dams held only 3 to 5 Ci of ^{137}Cs during the 1987 flood, while the radioactive fluxes in all small rivers were about 100 Ci (Botchkov et al. 1988).

Thus, the efficiency of filtering dams with clinoptillolite zeolites in the rivers was too low. Considering the negative effects of overflowing and secondary contamination of surface and subsurface waters in the summer of 1987, a decision was made to destroy most of the temporary water protection dams. By 1988 only 13 of 136 dams built in 1986 were still operating. The conclusions on the efficiency of temporary filtering dams with zeolite base are the following:

- The proposal to use zeolites (clinoptillolites) from the Sokirnyansk deposit during summer/autumn of 1986 was timely and justified. The water protection effect of the dams and sorption properties of the loaded material had been significantly reduced since the autumn of 1986, and during the spring flood of 1987 their efficiency was extremely low, so the decision to destroy them was sound.
- The low sorption efficiency of clinoptillolites in the river occurred because the raw material was not prepared properly before loading it into the dams. The inherent urgency of zeolite delivery to the CEZ caused sorbent to be loaded without adequate drilling and purification. As a result, a large number of rocks, including a clayey fraction, were found in the sorbent. During construction of the dams, ceolite was drilled without controls and the clayey component was transformed into dust. Zeolite was not properly scattered to meet recommended size fractions. All of these problems resulted in fast blocking of the pores of the filtering dams, quickly reducing their water protection ability after just several months of operation.
- The experts concluded that filter construction should have a cassette-type construction to remove the spent sorbent. The flow through dams with sorbents should be at the lowest possible rate (optimally less than 0.3 m/s). Sorbent contamination could reach up to $1x10^{-6}$ to $1x\ 10^{-5}$ Ci/kg in some water flows. The large volumes of sorbent used, which should be in a category of solid radioactive waste, led to a disposal problem. The cost of disposal was estimated at up to U.S. $\$1,000/m^3$. Such high prices for large volumes of radioactive waste have cast a doubt on the efficiency of the sorbent technology for purifying radioactively contaminated natural flows.

In spite of these problems, it has been recommended to use Sokirnyansk ceolites for purification of some highly radioactive discharge water and for decontamination of small but highly contaminated flows.

6.6.2.3 Decontamination of flows in the regulated hydraulic structures

Despite these problems, at the September 20, 1988 meeting of the Permanent Working Group of the State Committee of Hydrometeorology of the USSR, it was decided to test sorbents from other regions of the country in the CEZ and to recommend the most effective technologies for water protection. In autumn of 1988, Professor V.A. Kopeikin delivered sorbents from various deposits to the town of Chernobyl (Kopeikin and Plenkin 1988). They were tested by the Ukrainian branch of the Central Scientific and Research Institute of Comprehensive Use of Water Resources. The coordination of work on the program "Sorbents" was assigned to the Permanent Working Group of the State Committee of Hydrometeorology in Chernobyl Town.

Under this program in 1988 and 1989, a list of raw material inventory in the country was produced. A series of sorbents was loaded into a discharge channel of polder system specially built (Unit No. 7) on the northeast (left) bank of the river and were evaluated for their ability to purify the radioactive flow under conditions as close as possible to natural ones (Lisichenko 1990). Laboratory studies of sorption properties were also performed using highly contaminated water from Goluboy Ruchey (a small tributary of the Pripyat in the Chernobyl near zone) with water contamination having beta activity of $1x10^{-9}$ to $1x10^{-8}$Ci/L. Sorbents of Sokirnyansk deposit (clinoptillolites), pink bentonite from Gumbriysk deposit of Zkhaltubo in Georgia, brown bentonite from Biklyansk formation in Tataria, green bentonite from Sarigyuksk deposit in Armenia, palygorskite clay in the Cherkassy region, montmorilonite clay in the Moscow and Kaluga regions, zeolites from Idatsk deposit in Azerbaijan, and clinoptillolites in the Chita region and others were chosen for testing.

Fractions of sorbents were split into granules 3 to 5 mm and 5 to 7 mm in size and tested under laboratory conditions. The rate of water flowing from Goluboy Ruchey into a sorbent column was regulated within 0.4 to 0.8 m/s. Based on the tests, clinoptillolites of Sokirnyansk (Ukraine) and Kholinsk deposits in Chita, Russia, ceolites of Idagskoe deposit in Georgia, and Cherkassy polygorskite clay were accepted. The rest of the rocks were sodden in water, changed their structures and lost their filtering properties. Therefore, they were judged not suitable. The selected ceolites were tested in a natural flow at Unit No. 7 of the discharge channel of the polder system (Lisichenko 1990). The initial cassette-type filter had broken stones. At the second stage the rates were regulated in the range 0.05 to 0.9 m/s.

These studies concluded that the effectiveness of sorption increased with a reduced flow rate through the broken stones and ceolites. The maximum sorption effect of ceolites was observed at a flow rate of about 0.05 m/s and reduced total beta activity by 80 percent. At an increasing flow rate of up to 0.1 m/s, the efficiency was decreased to 30 percent, and at 0.2 m/s and faster it extracted no more than 20 percent of activity. The sorbent efficiency stabilized at 40 to 50 percent when the contaminated water passed through at the optimal flow rate for 30 to 40 hours. Flows with specific activity of 10^{-8} to 10^{-7}Ci/L were purified very effectively (the extraction efficiency reached 90 percent and greater). However, for very high levels of water contamination, even several percent of residual contamination presented a serious problem. Because the sorbent could be used for decontamination in only limited water bodies, it was considered not to be effective.

The efficiency of a similar water purification technology did not exceed 20 percent for ^{90}Sr for no more than 10,000 m^3 of water passing through 1 m^3 of sorbent at the flow rate of about 0.05 m/s. Further increase in water volume through the sorbent filtering box reduced the efficiency.

Thus, the natural sorbents used in water protection at the CEZ were effective only under controlled conditions of changeability of cassettes with the sorbent and for specific levels of radioactive flows. The main consideration to use relatively inexpensive raw materials for water purification is preparing suitable application conditions for achieving high efficiency. An efficiency analysis for some other measures, e.g., building of channel dams and water regulation of contaminated river flow in reservoirs, was performed with mathematical models of the Pripyat River and Kiev Reservoir.

6.7 References

Baryakhtar VG (Ed.). 1995. *Chernobyl Catastrophe.* Naukova Dumka, Kiev, (in Russian).

Beasly IT. 1984. "Transuranic Radionuclides in the Columbia River Sources, Inventories and Geochemical Behavior." *Proc. of Symposium on Environmental Research for Active Elements.*

Bilyi IY, OV Voitsekhovich, Y Onishi, and R Greivs. 1994. "Modeling of ^{90}Sr Washoff from Flood Plain Sites of Pripyat River During Flooding of 25 Percent Probability of Exceeding." *Isotopes in Water Management. Proc of Int. IAEA Symp.*, IAEA, Vienna, pp. 9-10.

Botchkov LP, AI Nikitin, OV Voitsekhovich, AI Shereshevskiy, SM Vakulovskiy, VA Borzilov, EG Tertyshkin, VB Chumichev, and ES Hersonskiy. 1988. "Efficiency of Water Protective Construction in Chernobyl NPP Effect Zone." *Radiation Aspects of Chernobyl Accident,* 1(6):373-377, Obninsk (in Russian).

Bulgakov A, OV Voitsekhovich, and V Kanivets. September 2001. "Modeling of long-term dynamic of radiecaesium in rivers of Chernobyl accident influenced areas." *Proc of International Congress on the radioecology-ecotoxicology of continental and estuarine environments,* pp. 3-7. Aix-Provence, France.

Deville-Cavelin G, H Biesold, C Brun-Yaba, and V Chabanyuk. September 2001. "The consequences of the Chernobyl accident. First results in the radioecological project of the French-German initiative. *Proc of International Congress on Radioecology-Ecotoxicology of Continental and Estuarine Environments.* Aix-Provence, France.

DU-97. 1997. "Permissible Levels of Cs and ^{90}Sr Contents in Food Products and Drinking Water." *State Sanitary Norms, Rules, Hygienic Standards*. Kiev, Ukraine (in Ukrainian).

Dzhepo SP, AS Skalskiy, DA Bugai, VV Gudzenko, SA Mogilny, and NI Proskura. 1994. "Ground Tests of Radionuclide Migration at the Radioactive Waste Temporary Localisation Site Red Forest." *Problems of Chernobyl Exclusion Zone*. Kiev, Issue 2, pp. 77-84 (in Russian).

Dzhepo SP, AS Skalskiy, DA Bugai, et al. 1995. "Hydrogeological effects of the principal radioactive waste burial sites adjacent to the Chernobyl NPP (review). The state of the Chernobyl nuclear power plant unit 4." *Proceedings of international symposium*. OECD, Zeleny Mys, Chernobyl, Ukraine, pp. 370-382.

Gradev GD, MV Milanov, YD Prodan, et al. 1978. "Study of sorption features klinoptilolite, vermiculite, and bentonite aiming at decontamination of radioactive drain waters." *Nuclear Energy*, No. 7, pp. 30-41 (in Russian).

Grinzhevskiy VP et al. (Eds.). 1997. "Recommendations on Restoration of Ponds Excluded from the Fishery Process in the Result of Chernobyl Accident." Institute of Fishery, Kiev Ukraine (in Russian).

Hadderingh RH, O Nasvit, MCV Carreiro, IN Ryabov, and VD Romanenko. 1996a. "Processes and parameters governing accumulation and elimination of radiocaesium by fish." *The Radiological Consequences of the Chernobyl Accident*, pp. 137-147.

Hadderingh RH, O Nasvit, IN Ryabov, G van Aerssen, V Prostantinov, and NV Belova. 1996b. "Field Studies of Fish-Size Effect in ^{137}Cs Accumulation." *Modeling and study of the mechanisms of the transfer of radioactive material from terrestrial ecosystems to and in water bodies around Chernobyl*, pp. 84-94.

Hakanson L. 1992. "Remedial measures against radioactive cesium in Swedish lake fish after Chernobyl." Dept. of Physical Geography, Birkhauser Verlag, Basel, Switzerland.

Hakanson L, JE Brittain, L Monte, U Bergstrom, and R Helling. 1996. *Modeling of Radiocesium in Lakes. Sensitivity and Remedial Strategies*. Elsevier Science Limited, p. 25.

Helling R, RH Hadderingh, and O Nasvit. 1996. Application of 'Lakeco' Lake Model to Explain the Size-Effect Phenomenon Observed in Fish. *Modeling and study of the mechanisms of transfer of radioactive material from terrestrial ecosystems to and in water bodies around Chernobyl*, pp. 158-167.

International Atomic Energy Agency (IAEA). 1994. *Manual on Application of Countermeasures in Agriculture in Case of Accidental Release of Radionuclides into the Environment*. IAEA, Vienna.

International Commission on Radiological Protection (ICRP). 1990. "Recommendation of the International Commission on Radiological Protection." ICPR Publication 60, Pergamon Press, New York.

Ivanushkina NI, GP Ryabtseva, and VN Zuzanskaya. 1989. "Efficiency of Dams at Small Rivers." *Measures on Reduction of Migration and Disposal of Radioactive Waste. Proc. of Chernobyl-88, Scientific and Technical Meeting*, Vol. 6, pp. 17-26 (in Russian).

Izrael YA (Ed.). 1990. "Chernobyl: Radioactive Contamination of Natural Environment." Hydrometeoizdat, Leningrad (in Russian).

Kanivets VV and OV Voitsekhovich. 2000. "The bottom sediment radioactive contamination of the Chernobyl NPP cooling pond." *Proc. UHMI*, Vol. 248, pp. 154-171.

Kashparov VA, SM Lundin, and SI Zvarich. 2003. "Territory Contamination with the Radionuclides Representing the Fuel Component of Chernobyl Fallout." *The Science of The Total Environment*, Vol. 317, pp. 105-119.

Kazakov SV. 1995. *Management of Radiation State of NPP Cooling Ponds.* Technika, Kiev, Ukraine (in Russia).

Kopeikin VA. 1990. *Use of Natural Sorbents of the USSR in the Process of Radionuclide Removal from the Waters of the Zone of Liquidation of the Consequences of Chernobyl NPP Accident.* Kazan, No. 37, Vol. 1, Book 2 (in Russian).

Kopeikin VA. 1994. "Geochemical Consequences of Chernobyl Catastrophe. Problems of Chernobyl Exclusion Zone, Issue 2. Kiev, Ukraine, pp. 128-138 (in Russian).

Kopeikin VA, AP Plenkin, AI Azizov, and VM Fedotov. 1988. *Assessment of source of raw natural sorbents of the USSR and assessment of possibilities of its application in the process of radionuclide removal from waters in the zone of liquidation of Chernobyl accident consequences.* Kazan, VNIIgeolnerud (in Russian).

Kopeikin VA and AP Plenkin. 1998. *Source of raw natural sorbents of the USSR and assessment of possibilities of its application in the process of environmental protection under conditions of radioactive contamination.* Kazan, VNIIgeolnerud (in Russian).

Kryshev II, TG Sazykina, IN Ryabov, OL Zarubin, and VK Chumak. 1988. "Dynamics of Radioecological Processes in Water Bodies of Accidental Track of Chernobyl NPP." *Radiation Aspects of Chernobyl Accident,* Vol. 2, pp. 3-8. Obninsk (in Russian).

Kudelskiy AV, JT Smith, and AA Petrovich. 2001. "An experiment to test the addition of potassium to a non-draining lake as a countermeasure to ^{137}Cs accumulation in fish." *Proc. Intern. Congress on the radioecology-ecotoxicology of continental and estuarine environments.* Aix-en-Provence, France.

Laptev GV and OV Voitsekhovich. 1993. "Experimental Studies of the Radionuclide Flood Washout from the Pripyat River Floodplain Soils." *Proceedings of the Ukrainian Hydrometeorological Institute*, Vol. 245, pp.127-143, Kiev, Ukraine (in Russian).

Lepicar S and P Droz P (Eds.). 1999. "Chernobyl exclusive zone protective dike (UNDP expertise of justification)." *Collective dose reduction and cost-benefit analysis.* Geneva, Switzerland.

Lisichenko GV. 1990. *Use of Natural Sorbents for Localisation of Radioactive Contaminations in Chernobyl NPP Zone.* Report on Research Contract with SPA Pripyat. Kiev, Ukraine (in Russian).

Maringer FJ, M Tschurlovits, and D Rank. 1997. "Recent radioecological investigations in the Austrian part of the Danube River." *Freshwater and estuarine radioecology.* Elsevier, The Netherlands, pp. 249-255.

Minikh VV, PV Shvydko, SV Gorbunov, and SG Alekseev. 1989. "Natural clay deposits from Cherkassy to be used as an effective sorbent in Chernobyl Exclusion Zone." *Measures on reduction of migration and disposal of radioactive waste.* Proc. of Chernobyl-88, Scientific and Technical Meeting, pp. 252-264 (in Russian).

Nikiforov AS, VV Kulichenko, and MI Zhiharev. 1985. *Neutralisation of Liquid Radioactive Waste.* Energoizdat, Moscow (in Russian).

Nikipelov BV, EI Mikerin, GN Romanov, DA Spirin, YB Kholina, and LA Buldakov. 1990. "The Radiation Accident in the Southern Ural in 1957 and Elimination of its Consequences." Proc. Int. Symp., November 1989. IAEA-SM-316/55, Vienna, pp 373-403.

Onishi Y. 1981. "Sediment and Contaminant Transport Model." *Journal of Hydraulics Division.* American Society of Civil Engineers, 107(HY9):1089-1107.

Onishi Y, ST Yokuda, OV Voitsekhovich, and MJ Zheleznyak. 2003. "Evaluation of and Air Condensation and Ventilation Needs of the Chernobyl New Safe Confinement." *Proc. 48th Annual Meeting of Health Physics Society*, San Diego.

Pileev AS and VA Tischenko. 1988. "Accidental Situations at NPPs in USA 1985-1986." *Nuclear Technique Abroad*, No. 4, pp. 22-26 (in Russian).

RSSU-97. 1998. "Radiation Safety Standards of Ukraine; State Hygienic Norms." Ukrainian Centre of State Sanitary and Epidemiological Supervision of Ministry of Health of Ukraine, p. 121. Kiev, Ukraine (in Ukrainian).

Rust J and L Wiver. 1980. "Safety of Nuclear Energy." Atomizdat, Moscow (in Russian).

Ryabov I, N Belova, L Pelgunova, and RH Hadderingh. 1996. "Radioecological phenomena of the Kojanovskoe Lake." *Radiological consequences of the Chernobyl accident*, pp. 213-217.

Samoilenko VN, YS Tavrov, and OI Nasvit. 1997. "Creation, Running and Development of Cadaster of Radioactive Contamination of Water Bodies of Local Water Use." *Radiogeoecology of Water Bodies of Chernobyl Accident Effect Zone. Chernobylinterinform, Vol. 1: Monitoring of Radioactive Contamination of Natural Waters of Ukraine*, pp. 96-126 (in Russian).

Sansone U. and OV Voitsekhovich (Eds.). 1996. *Modeling and Study of the Mechanisms of the Transfer of Radionuclides from the Terrestrial Ecosystem to and in Water Bodies around Chernobyl.* Final report, ECP-3 Project, EUR 16529 EN. European Commission, Luxemburg, Belgium.

Smith J, OV Voitsekhovich, A Konoplev, and A Kudelsky. 2005. *Radioactivity in Aquatic Systems, Chernobyl Catastrophe and Consequences*, J Smith and N Beresdorf (Eds.). Springer, pp. 139-181.

Sobotovich EV and YA Olkhovik. 1989. "To the issue on protective features of zeolite dams." Chernobyl-88, *Measures on reduction of migration and disposal of radioactive waste,* pp. 26-40 (in Russian).

Spitsyn VI and VV Gromov. 1958. "Study of regularities of radioactive strontium sorption on montmorillonite and fixing it by decrepitation method." *Nuclear Energy,* 5(4):446-452 (in Russian).

Ternovskiy IA, GN Romanov, EA Fedorov, EN Teverovskij, and YB Kolina. 1989. *Radioactive cloud trace formation dynamics after the radiation accident in the Southern Urals in 1957: migration processes.* IAEA SM 316/553, International Atomic Energy Agency, Vienna.

Tolochko PI (ed.). 1996. *Chornobyl Tragedy. Document and Materials.* Naukova Dumka, Kiev (in Ukrainian).

TRANSAQUA. 2003. "Transfer in Aquatic Ecosystems." Final report of the French-German Initiative for Chernobyl, CD database, IRSN, Cadarach, 2003.

United States Geological Survey (USGS). 1991. *Radioactivity in bottom sediment of the Clinch and Tennessee rivers.* U.S. Geological Survey, Tennessee Valley Authority, pp. 57-78.

United States Navy. 1996. *Arctic Nuclear Waste Assessment Program Summary, FY-1995.* ONR 32296-16, Office of Naval Research, United States Navy, Arlington, Virginia.

Viktoriova NV, OV Voitsekhovich, BV Sorochinskiy, H Vandenhove, AV Konoplev, and IG Konoplev. 1999. "Phytoremediation of Chernobyl contaminated lands." *Radiation Protection Dosimetry,* Vol. 92, pp. 59-64.

Voitsekhovich OV. 1997. "Development and Prospects of Water Protective Activity in the Chernobyl NPP Exclusion Zone." *Bulletin of Ecological State of Exclusion Zone,* 4(9):36-45 (in Ukrainian).

Voitsekhovich OV. 2001. *Water quality management in the areas influenced by the Chernobyl accident.* Kiev, Ukraine (in Russian).

Voitsekhovich OV, VV Kanivets, and AI Shereshevskiy. 1989. Evaluation of the river sediment trip effectiveness caught. *Proc. of UHMI,* Vol. 228, pp.60-68.

Voitsekhovich OV, VV Kanivets, and GV Laptev. 1993. "Analysis of formation of radioactive contamination of the Dnieper hydrosystem five years after Chernobyl accident." *Proc. of Ukrainian Hydrometeorological Institute,* Issue 245, pp.106-125 (in Russian).

Voitsekhovich OV, MJ Zheleznyak, and Y Onishi. 1994. *Chernobyl Nuclear Accident Hydrologic Analysis and Emergency Evaluation of Radionuclide Distributions in the Dnieper River, Ukraine, During the 1993 Summer Flood.* PNNL-9980, Pacific Northwest National Laboratory, Richland, Washington.

Voitsekhovich OV, OI Nasvit, IP Los'y, and VV Berkovskiy. 1995. "Assessment of water protection operations to minimize the risk from using water from the Dnieper water system at various stages after the accident at the Chernobyl nuclear power plant." *Environmental impact of radioactive releases.* International Atomic Energy Agency, Vienna, pp.549-557.

Voitsekhovich OV, VV Kanivets, and GV Laptev. 1997a. "Contemporary State of Radioactive Contamination of the Water Bodies of the Exclusion Zone of Accident Influence." *Radiogeoecology of Water Bodies of Zone of Chernobyl Accident Effect. Kiev: Chernobylinterinform, Volume 1. Monitoring of Radioactive Contamination of Natural Waters of Ukraine,* pp. 60-96 (in Russian).

Voitsekhovich OV, OI Nasvit, IP Los'y, and VV Berkovsky. 1997b. "Present thoughts on the aquatic countermeasures applied to regions of the Dnieper river catchment contaminated by the 1986 Chernobyl accident." *Freshwater and estuarine radioecology.* Elsevier, pp. 75-87.

Voitsekhovich OV and EL Panasevich. 1998. "On Dose and Social and Economical Expediency of Modern Water Protective Activity in Chernobyl NPP Exclusion Zone." *Bulletin of Ecological State of Exclusion Zone*, №12, pp. 3-8 (in Ukrainian).

Wang JD and JJ Connor. 1975. *Mathematical Modeling of Near Coastal Circulation*. Report No. 200, RM Parsons Laboratory, Massachusetts Institute of Technology, Cambridge, Massachusetts.

Whicker FW, DJ Niquette, and TG Hinton. 1993. "To remediate or not: A case history." *Environ. Health Physics*. Research Enterprises Pub. Segment, Richland, Washington, pp. 473-485.

Zheleznyak MI (Ed.). 1995. *Primary Water Protective Measures for Protection from Radiation Contamination of Surface and Underground Waters in the Chernobyl NPP Exclusion Zone*. Institute of Mathematical Machines and System Problems of Ukrainian National Academy of Sciences, Kiev (in Russian).

Zhilinskiy VV, VM Antropov, and AI Habrika. 1997. "Inventory of Long lived Radionuclides." *Bulletin of Ecological State of Exclusion Zone,* 4(9):29-35 (in Ukrainian).

Chapter 7
Water Protection Measures for Radioactive Groundwater Contamination in the CEZ

D. A. Bugai, A. S. Skalskiy, and S. P. Dzhepo
Institute of Geological Sciences, Kiev, Ukraine

This chapter presents results of simulations of radionuclide migration and filtration processes in subsurface water as well as radioactive groundwater contamination risk assessments for the CEZ. For decisions on groundwater protection criteria and countermeasures, a cost-benefit analysis methodology was adapted to address hydrogeologic problems and used to account for uncertainty in groundwater flow parameters of contaminant transport modeling. Cost-benefit analyses are presented for groundwater remediation measures related to ^{90}Sr migration into Pripyat Town's water supply wells.

7.1 Health Risks for Potential Residents in the CEZ Due to Radionuclide Migration in Groundwater

This assessment assumed that hypothetical residents use the unconfined aquifer for potable water. Contaminated groundwater consumption risks were compared with risks from external irradiation due to environmental contamination and consumption of agricultural products cultivated on lands contaminated with radionuclides.

The time frame of radionuclide migration in groundwater is much longer than that for radionuclide migration in contaminated surface soil. Both affect direct radiation exposure and contamination of foods. It is not evident which component of radiation risk from these pathways dominates at a given time; risks from external radiation and agricultural product contamination decrease exponentially with time due to radionuclide decay and downward migration in soil. On the other hand, Chernobyl-released radionuclide concentrations in the groundwater at the time of radioactive fallout on the soil surface are assumed to be zero. Groundwater contamination and associated risk increase with time as the radionuclides migrate through the unsaturated zone and accumulate in the aquifer below. Therefore, the maximum radionuclide concentrations and risks will occur some time later than the moment of deposition on the soil surface. This delay in reaching maximum levels is

Y. Onishi, O.V. Voitsekhovich and M.J. Zheleznyak (eds.), Chernobyl – What Have We Learned?, 203–245.

controlled by the protective properties of the geologic environment (e.g., thickness of the unsaturated zone, infiltration recharge rate) and kinetics of radionuclide leaching from fallout and radioactive fuel particles. After reaching the maximum, the risk caused by drinking contaminated groundwater begins to decrease due to radioactive decay and reduced influx from lessened or exhausted radionuclide sources on the soil surface.

This section discusses model assumptions, subsurface water models, dosimetry models, dose assessment, conversion equations from dose to risk, results of groundwater contamination, and dose estimates.

7.1.1 Methodology

The modeling assumptions were selected such that groundwater contamination risks represent upper-bound estimates, while estimated risks from external radiation and contaminated agricultural product consumption represent lower-bound estimates. These assumptions are as follows:

- Only ^{90}Sr migration is considered. Other long-lived radionuclides released from Chernobyl (^{137}Cs and plutonium isotopes and their daughter products) are not assessed because previous investigations (Voitsekhovich 1997) showed that they migrate slowly due to relatively low geochemical mobility, causing relatively minor health problems compared with ^{90}Sr.
- Model parameters for reasonable conservatism were used to calculate the upper bound of ^{90}Sr groundwater concentrations using an analytical model developed by the Institute of Geological Sciences (Bugai et al. 1996b).
- For assessing internal effective doses from consumption of contaminated water, we used a model based on "activity-dose" conversion coefficient developed by the U.S. Environmental Protection Agency (EPA) (Eckerman et al. 1988).
- ^{137}Cs external dose levels from consuming contaminated agricultural products are similar to ^{90}Sr concentrations. Plutonium isotopes and ^{90}Sr, mainly internal sources of radiation, are represented at lower-bound estimates. Dose assessments of ^{137}Cs use the model developed by the Moscow Biophysics Institute.
- We only considered ^{137}Cs for assessing external doses and effective doses resulting from consumption of contaminated agricultural products. The ^{137}Cs concentrations in the CEZ soil are roughly equal to the ^{90}Sr concentrations. Because ^{90}Sr and plutonium isotopes were not accounted for, our dose assessment from external irradiation and

effective doses from consumption of agricultural products would produce the lower-bound estimates.

- For assessing the effective dose from ^{137}Cs, we used a model developed by the Moscow Biophysics Institute.
- We used the conversion coefficient recommended by ICRP (1992) to estimate carcinogenic radiation doses.

7.1.1.1 Model for radionuclide migration in groundwater

For the screening assessment of ^{90}Sr migration in groundwater, we used a simple analytical model (Bugai et al. 1996b) that estimates aquifer depth-averaged concentrations of radionuclides in groundwater. It was developed by combining the IGS model for radionuclide transport in an unsaturated zone and the Hoeks model (1981) for contaminant transport in a saturated zone. The model accounts for advective transport but neglects solute diffusion (dispersion). In relatively permeable unconsolidated sediments composing the unconfined aquifer in the CEZ, advective transport almost always dominates over diffusive (dispersion) transport (Hoeks 1981; Massman and Freeze 1987); therefore, the last assumption is commonly used in screening groundwater assessments (e.g., Massman and Freeze 1987; Jury et al. 1987). The model schematization is shown in Figure 7.1.

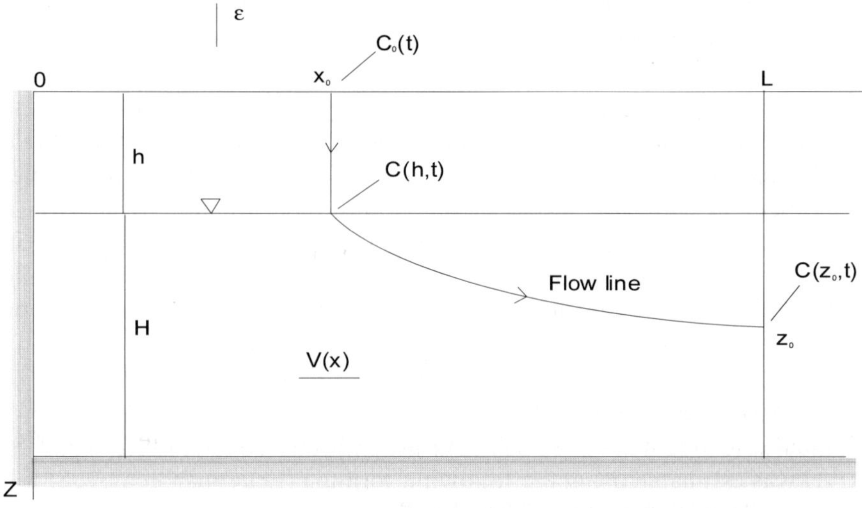

Figure 7.1. Radionuclide transport model flow domain (from Bugai et al. 1996a)

Here, h is the thickness of the unsaturated zone (m), H is the thickness of the saturated zone (aquifer) (m), and L is distance to water divide (m). We assumed that ^{90}Sr leaching from radioactive fallout ("hot" particles) was described by the first-order reaction equation. Radionuclide concentration in the infiltrating water was calculated based on a conservative assumption that all radionuclides leached from fallout migrate downward with the infiltrating water. The resulting equation for radionuclide concentration in groundwater averaged over the aquifer depth is as follows:

$$C_{GW}(t)=\frac{A_0}{(n+\rho K_d)H}exp(-\lambda t)\{1-exp(-K_L(t-t_v)\},$$
$$t_v=\frac{h(\theta+\rho K_d)}{\varepsilon} \qquad (7.1)$$

where

- A_o = initial surface contamination with ^{90}Sr at time t=0 (Bq/m^2)
- n = porosity of aquifer (dimensionless)
- ρ = density of aquifer sediment (kg/m^3)
- K_d = a distribution coefficient of radionuclide for the system (m^3/kg)
- λ = a radioactive decay constant (yr^{-1})
- K_L = first-order rate constant describing radionuclide leaching from fallout particles (yr^{-1})
- t_v = time of radionuclide migration in the unsaturated zone accounting for the retardation due to sorption (yr)
- θ = moisture content in unsaturated zone (dimensionless)
- ε = an infiltration recharge rate (m/yr).

Equation 7.1 is valid for $t > t_v$, otherwise $C_{GW}(t)$=0. The complete derivation is presented in Bugai et al. (1996a). The model has a simple physical interpretation; the term $A_o/(n+\rho\times K_d)H$ represents the groundwater concentration in case of instantaneous dissolution of the total surface inventory of radionuclides in the aquifer. The first exponent-multiplier in equation 7.1 accounts for reduction of activity due to radioactive decay, while the last multiplier accounts for kinetically limited leaching of radionuclide from the fallout particles.

7.1.1.2 Dose assessment model

Annual effective dose from drinking groundwater contaminated with ^{90}Sr (mSv/yr) was estimated using the following formula (Eckerman et al. 1988):

$$CED_{GW}^{(a)}(t)=C_{GW}(t)IRIDF \qquad (7.2)$$

where

- $C_{GW}(t)$ = ^{90}Sr concentration in groundwater calculated by equation 7.1
- IDF = coefficient for consumed activity-dose conversion for ^{90}Sr (mSv/Bq).

IR = consumption rate of drinking water (m^3/yr)

Parameter *IR* was assumed to be 2 L/day (EPA 1991). For the *IDF* coefficient we used the value 3.9×10^{-5} mSv/Bq (Eckerman et al. 1998).

Using the depth-averaged radionuclide concentration calculated from equation 7.1 in equation 7.2 implies that a screen is installed in the water supply well over the entire aquifer depth. External, internal, and total annual effective doses from ^{137}Cs were estimated using the dosimetry model:

$$CED^{(a)}_{EXT}(t) = 0.0044\, A_{CS}\, (0.7 \exp(-0.3\, t) + 0.3 \exp(-0.024\, t))$$

$$CED^{(a)}_{INT}(t) = (0.0044\, A_{CS} + 0.66)\, (\exp(-0.05\, t) + 0.43 \exp(-0.35\, t)) \qquad (7.3)$$

$$CED^{(a)}_{TOTAL}(t) = CED^{(a)}_{EXT}(t) + CED^{(a)}_{INT}(t)$$

where A_{CS} is surface inventory of ^{137}Cs (kBq/m^2). Risks of acquiring fatal cancer due to exposure to low doses of ionizing radiation were estimated with a conversion factor of 5×10^{-5} $(\text{mSv})^{-1}$ (ICRP 1992).

7.1.2 Radionuclide Transport and Risk Assessment

Calculations were performed for the following two scenarios:

- Scenario 1, CEZ average hydrogeologic and radiological conditions
- Scenario 2, extreme radioactive contamination condition corresponding to a temporary radioactive waste site (RAWTLS) in the Red Forest in the Chernobyl near zone. This worst-case scenario assumes that disposed radioactive waste materials have direct contact with groundwater and does not include contaminant attenuation due to radioactive decay during ^{90}Sr transport in the unsaturated zone.

Parameter values for ^{90}Sr migration in groundwater are summarized in Table 7.1. Moderately conservative values to predict the upper bound of ^{90}Sr concentration in the aquifer were selected for unsaturated zone thickness, infiltration recharge rate, and distribution coefficient.

The modeling results are presented in Figures 7.2 and 7.3. According to Scenario 1, the depth-averaged ^{90}Sr concentration reaches a maximum value of 5.2 Bq/L in groundwater 40 years after the accident. This level is 2.6 times greater than the Ukrainian drinking water standard of 2 Bq/L set by Permissible Levels–97 (DU-97). For Scenario 2, the ^{90}Sr maximum concentration would be 670 Bq/L, 15 years after the accident. This concentration is 335 times greater than the drinking water standard (see Figure 7.2).

Table 7.1. Parameter values used for assessing risk to hypothetical residents of CEZ from groundwater contamination

***Scenario 1*: Average conditions of the CEZ**		
Surface inventory of ^{90}Sr, A_o	kBq/m^2	1100
Surface inventory of ^{137}Cs, A_{Cs}	kBq/m^2	1200
Thickness of the unsaturated zone, h	M	1.5
***Scenario 2*: Conditions in Red Forest RAWTLS (worst-case scenario)**		
Surface inventory of ^{90}Sr, A_o	kBq/m^2	0.8×10^5
Surface inventory of ^{137}Cs, A_{Cs}	kBq/m^2	1.2×10^5
Thickness of the unsaturated zone, h	M	0
Parameters identical in Scenarios 1 and 2		
^{90}Sr leach rate constant, K_L	yr^{-1}	0.14
Radioactive decay coefficient, λ	yr^{-1}	0.023
Infiltration recharge rate, ε	m/yr	0.2
Moisture content in unsaturated zone, θ	-	0.1
Soil density, ρ	kg/m^3	1.65
Distribution coefficient of ^{90}Sr, K_d	m^3/kg	0.002
Saturated thickness of aquifer, H	m	20
Porosity, n	-	0.3

Conservatively estimated risks for residents of the CEZ from drinking this contaminated groundwater would be $1\text{x}10^{-5}$/yr under Scenario 1 (average radiological and hydrogeologic conditions). For Scenario 2 (the worst case), this risk would be $1\text{x}10^{-3}$/yr, 20 times higher than the risk under Scenario 1. It will take 150 and 250 years of Scenarios 1 and 2, respectively, for the groundwater risks to be reduced to the $1\text{x}10^{-6}$/yr level.

For both scenarios, risks from contamination of soil with ^{137}Cs would be approximately 10 times greater than that caused by ^{90}Sr migrating to the groundwater system, as shown in Figure 7.3. Thus, these model results indicate that radioactive contamination of groundwater may be a health risk to residents of the CEZ for 150 to 300 years. However, soil contamination with ^{137}Cs in the CEZ poses a much greater health risk to the residents.

7.2 Economic Risk of Radioactive Contamination of Pripyat Town Water Supply Wells

This section describes simulation of ^{90}Sr migration to Pripyat Town water supply wells and a cost-benefit analysis of possible groundwater protection measures. Pripyat Town's water well field is in the Chernobyl near-zone. These water wells exploit the confined aquifer in Eocene deposits and are the source of industrial and drinking water for the Chernobyl nuclear plant.

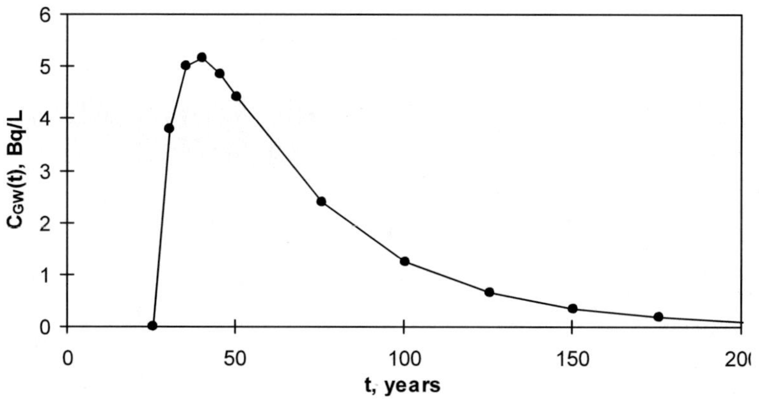

(a) Forecast for scenario no.1

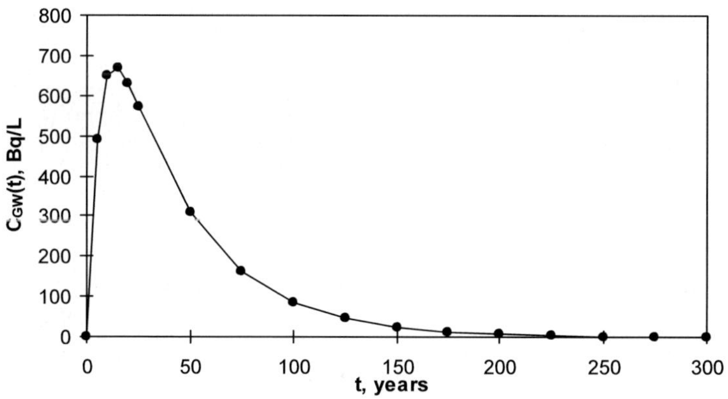

(b) Forecast for scenario no.2

Figure 7.2. Predicted ^{90}Sr concentrations in groundwater of CEZ

Hydrogeologic conditions and radiological monitoring data from Pripyat Town groundwater supply wells are discussed in detail in Voitsekhovich (1997). The confined aquifer in the Eocene sands of Buchak and Kanev suites is separated from the overlying unconfined aquifer in Quaternary sandy deposits by a low-permeable (hydraulic conductivity $1x10^{-3}$–$1x10^{-2}$ m/d) layer of Eocene clays and marls of Kiev suite 10 to 15-m thick. The depth of wells is approximately 80 m. Surface contamination with ^{90}Sr fallout (key radioactive contaminant in groundwater migration) in the well field constitutes $3x10^3$ to $7x10^3$ kBq/m^2 (100 to 200 Ci/km^2). According to 1991–1995 monitoring data, trace quantities of ^{90}Sr and ^{137}Cs were observed episodically in wells at about 0.01 Bq/L and are at the sensitivity limit of radiometric measurement. This causes concern that in time wells might become contaminated beyond the permissible levels of extracted groundwater.

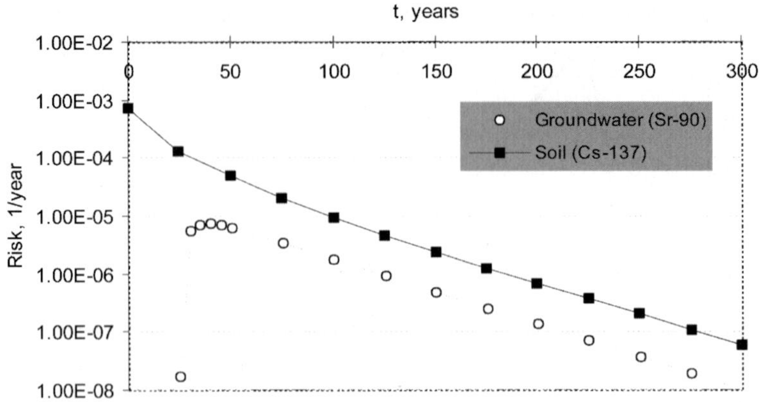

(a) Forecast for scenario no.1

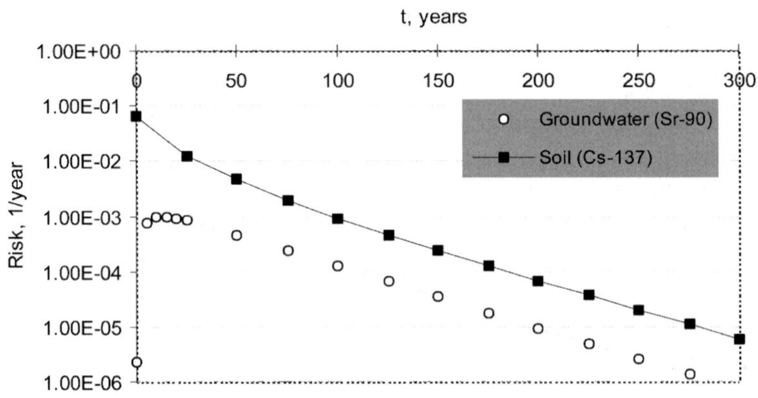

(b) Forecast for scenario no.2

Figure 7.3. Risks from radioactive contamination of soil and groundwater in the CEZ

In 1986 special engineering measures were undertaken to protect the Pripyat Town water supply wells. The southern drainage curtain and Pripyat Town bank drainage were constructed, totaling approximately 100 drainage wells that were intended to (1) improve protection of the Quaternary aquifer as the last source of water recharge to the Eocene aquifer, and (2) decrease the hydraulic head in the Quaternary aquifer and thus reduce the intensity of water exchange between the two aquifers. These drainage systems were never operated, but the possibility of their activation was repeatedly discussed in the early post-accident period by CEZ authorities.

Because the Pripyat Town well field is the sole source of the Chernobyl water supply, a modeling analysis of radionuclide migration to water wells and development of a sound protection strategy for the water supply are of particular interest. To choose a rational protection strategy for the groundwater supply, we used a cost-benefit analysis methodology that was adapted for groundwater problems, accounting for the uncertainty in the parameters of groundwater flow and contaminant transport (Bugai et al. 1996a).

Using the Monte-Carlo method, we estimated a probability that ^{90}Sr concentration in water wells exceeds the permissible level and estimated economic risk (probabilistic damage) associated with contamination of the groundwater supply. With these probabilistic estimates, we discuss the necessity of water protection measures for the groundwater supply and outline priority tasks for further hydrogeologic investigations.

7.2.1 Methodology of Cost-Benefit Analysis Applied to Groundwater Problems

The main provisions of the cost-benefit analysis methodology applied to groundwater problems were originally developed by Freeze and co-workers (Massman and Freeze 1987; Freeze et al. 1990). The methodology is based on combined use of the following three mathematical models:

- Model for decision-making based on a economic objective function considering risk-cost-benefit relationships
- Groundwater simulation model
- Model describing the uncertainty of data and parameters of the groundwater simulation model.

The decision-making model is intended for comparing alternatives (strategies) for managing the groundwater (e.g., water wells) and selecting the optimal alternative. In general, the objective function for risk-cost-benefit relationships has the following form:

$$\Phi = \sum_{t=0}^{T} \left[B(t) - C(t) - R(t) \right] / (1+i)^t \tag{7.4}$$

where

B(t)	= benefits
C(t)	= costs
R(t)	= risks (in monetary terms) in year *t* under the condition of implementation of one or another alternative
T	= a duration of prediction (approximately 50 years, as a rule)
i	= a bank discount rate.

The risk is determined as anticipated costs in a case if the selected strategy fails to satisfy the relevant (e.g., technological) requirements:

$$R(t) = P_f(t)\, C_f(t)$$

where

$P_f(t)$ = probability of failure

$C_f(t)$ = costs (economic damage) associated with failure in year *t*.

The optimal alternative (strategy) maximizes the objective function. The main problem is to estimate the probability of failure $P_f(t)$ for the compared alternatives (strategies). For this the groundwater simulation model is used. The simulation was performed in a stochastic (probabilistic) framework to account for uncertainties that are inherent to a mathematical model of a groundwater object. Uncertainties in parameters and input data are described by the probabilistic models. The methodology and examples of its practical application are described in Massmann and Freeze (1987), Freeze et al. (1990), Massmann et al. (1991), and James et al. (1996).

The cost-benefit analysis methodology applied to the Pripyat Town water wells can be simplified. Similar to examples described in Freeze et al. (1990), direct benefits from water protection measures are absent. We also assumed that costs related to failure of groundwater supply, C_f, do not depend on time and that capital costs to implement water protection measures would exceed operational costs. Under these conditions, equation 7.4 simplifies to

$$\Phi = -C - \sum_{t=0}^{T} \left[P_f(t) C_f\right] / (1+i)^t \qquad (7.5)$$

where C is the capital costs of particular alternative water protection measures. For convenience, one may remove the minus sign and consider this problem as cost-risk minimization.

The simplest approach is a “do-nothing” alternative, that is, no measure is implemented. For this case, C=0 and the objective function may become

$$\Phi^{(*)} = \sum_{t=0}^{T} \left[P_f(t) C_f\right] / (1+i)^t \qquad (7.6)$$

On the other hand, in the case of ideal failure-free functioning of water protection measures, $P_f(t)$ in equation 7.5 is equal to zero. Therefore, for any strategy assuming intervention and accomplishment of countermeasures the value of objective function would be greater than or equal to

$$\Phi(^{**}) = C \quad (7.7)$$

where C is capital costs for the strategy considered.

Therefore, as a first approximation, the need for water protection measures may be estimated by matching equations 7.6 and 7.7 and selecting the alternative for which these two objective function values are minimized. In other words, the costs of implementing countermeasures shall not exceed the anticipated (probabilistic) damage from contamination of the water wells (Bugai et al. 1996a). As pointed out by Voitsekhovich (1997), the time horizon (interval) of prediction T for more than 50 years is hardly justified in groundwater problems from economic considerations and relevant uncertainties. In this study we selected T=70 years with some reservation.

7.2.2 Mathematical Model for Subsurface Water System of Pripyat Town Water Supply Wells

We balanced to account for the most important features of the actual hydrogeologic system of the Pripyat Town groundwater supply and to simplify the mathematical representation of the problem to a reasonable degree. Therefore, the functioning of water wells was simulated using a rather simple numerical model. We assumed a steady-state groundwater flow regime and radial symmetry of groundwater inflow to the well (that is, we used a cross section-averaged, two-dimensional formulation). The model domain scheme is shown in Figure 7.4.

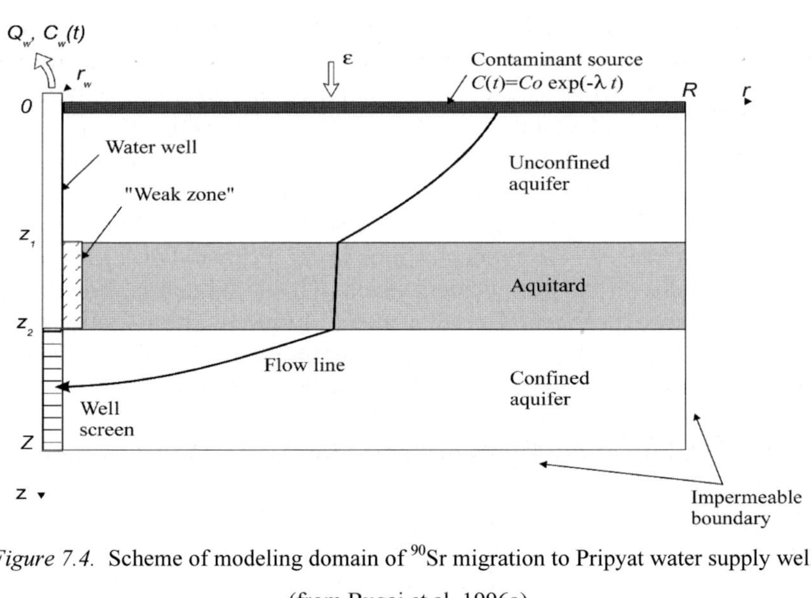

Figure 7.4. Scheme of modeling domain of ^{90}Sr migration to Pripyat water supply wells (from Bugai et al. 1996a)

When modeling the radionuclide migration in groundwater, we accounted for one-dimensional advective transport of ^{90}Sr along groundwater flow path lines, taking into account the equilibrium reversible radionuclide sorption by aquifer sediments and radioactive decay. Radionuclide concentrations at the migration source (at the level of the groundwater table) was set by

$$C(t) = C_o \exp(-\lambda t)$$

where

C_o = a known initial ^{90}Sr concentration at time t=0

λ = radioactive decay constant.

The ^{90}Sr migration in an unsaturated zone from the soil surface contaminated by fallout to the groundwater table was not simulated. By changing C_o values within the reasonable limits based on monitoring data and expert judgment, we can use this approach to obtain estimates of the contaminant concentration in water wells very simply for different scenarios. For instance, setting C_o based on conservative (worst case) assumptions, one may obtain and upper estimate of possible ^{90}Sr concentrations in a well.

Monitoring data indicated that trace quantities of ^{90}Sr and ^{137}Cs existed in groundwater supply wells in the post-accident period. To explain this, a hypothesis was proposed for the existence of a near-well weak zone with higher permeability in the marl aquitard layer separating hydrogeologic horizons (see Figure 7.4). This weak zone could be formed while installing water wells if a hole adjacent to the well casing was not properly sealed and served as a pathway for rapid (facilitated) transport of radionuclides to a generally well-protected confined aquifer. The possibility of such weak zones for deep groundwater supply wells is indicated by the results of some experimental tracer studies (personal communication, V. M. Shestopalov, IGS). In their experiments, the tracer (NaCl) injected near the well in the shallow unconfined aquifer was quickly found in the well extracting water from a deep confined aquifer. This hypothesis is also discussed in Voitsekhovich (1997) to explain the increased ^{90}Sr concentrations in the Pripyat Town groundwater supply compared with the pre-accident level. This argument is, however, not decisive, and the experimental results allow different interpretations. The monitoring data are also scarce and sometimes unreliable. Thus, we simulated the subsurface migration with and without weak zones.

7.2.3 Uncertainty of the Hydrogeologic Simulation Model Parameters

The model parameters of the Pripyat Town water supply were split into two groups, known and uncertain. The known (deterministic) parameters include those characterized reliably and those less important to model results.

These parameters and their best estimated values are listed in Table 7.2. The second group includes parameters for which reliable estimates are not readily available and that, in our opinion, represent the main uncertainty factors for predicting ^{90}Sr migration into wells. These include ^{90}Sr concentrations in the upper part of the unconfined aquifer (parameter C_o) and distribution coefficients describing ^{90}Sr sorption by sediments of the aquifer and the separating layer. These parameters were assigned the uniform probability distributions presented in Table 7.3. Values were based on site-specific data and literature sources (Bugai et al. 1996b).

Table 7.2. Deterministic parameters of model for Pripyat Town groundwater supply wells

Parameter	**Value**
Infiltration recharge rate, mm/yr	200
Well diameter, mm	320
Pump rate of well, m^3/d	500
Radius of filtration region, m	540
Radius of the "weak zone", m	0.5
Unconfined aquifer	
Saturated thickness, m	20
Hydraulic conductivity , m/d	10
Porosity, %	30
Soil density, kg/m^3	1.65
Separating layer	
Thickness, m	10
Hydraulic conductivity, m/d	0.01
Porosity, %	5
Rock density, kg/m^3	2.0
Confined aquifer	
Thickness, m	20
Hydraulic conductivity, m/d	5
Porosity, %	30
Rock density, kg/m^3	1.65

Table 7.3. Upper and lower limits assigned to the uncertain model parameters with uniform probability distributions

Parameter	**Lower**	**Upper**
^{90}Sr concentration in the source of migration, Bq/L	10	1000
Distribution coefficient of ^{90}Sr for unconfined aquifer, mL/g	0.5	5
Distribution coefficient of ^{90}Sr for separating layer, mL/g	0.5	20

In addition to parameter uncertainty, our analysis assumed a geologic uncertainty (Freeze et al. 1990), the possible existence of a near-well weak zone with higher permeability in the separating marl layer, as discussed in

Section 7.2.2. The radius of this zone was assumed to be 0.5 m and hydraulic conductivity 10 m/d, the same as a conservative estimate of the unconfined aquifer. The possibility of a weak zone was assumed at 50 percent.

In view of the long-term nature of prediction, the drinking water standard of ^{90}Sr in groundwater was also considered to be an uncertain parameter. In 1995, VDU-91 was the controlling regulation on temporary permissible levels (established by Ukraine's Radiation Protection Commission after the Chernobyl accident). According to VDU-91, the permissible ^{90}Sr concentration in drinking water was 3.7 Bq/L (100 pCi/L). The VDU-91 requirement is more stringent than the previous Soviet *Standards of Radiation Safety–76/87* (NRS-76/87) regulation allowing ^{90}Sr concentration of 14.8 Bq/L (400 pCi/L) in drinking water. We assumed that the trend to lower the permissible concentration by Ukrainian regulatory authorities may continue in the future. For example, in the United States the drinking water standard is 0.3 Bq/L (8 pCi/L). Thus, we assumed that the drinking water standard of 8 pCi/L will replace the current 100 pCi/L value in the future with 50 percent probability.

7.2.4 Simulation of Groundwater Filtration to Wells and Estimated Adequacy of the Hydrogeologic Model

The hydraulic head distribution and flow path lines in the model domain are presented in Figure 7.5 for the scenario without the weak zone (Bugai et al. 1996a). The hydrodynamic network for the scenario with the weak zone is similar. An exception is a small near-well zone where the vertical interflow (i.e., component *Vz* of seepage flow rate) is 30 times higher than the adjacent "solid" aquitard area due to the high gradient between the unconfined and confined aquifers and the high permeability of the weak zone (Figure 7.6). However, due to the small area of the weak zone (amounting to only ≈ 10^{-5} percent of the well recharge area), the total interflow through it is as low as ≈ 0.6 percent of the total water well debit rate, according to modeling results.

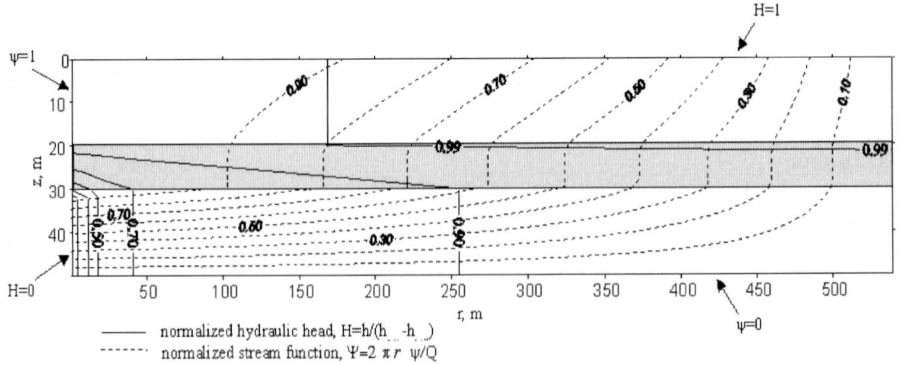

Figure 7.5. Hydrodynamic groundwater flow network to the water well

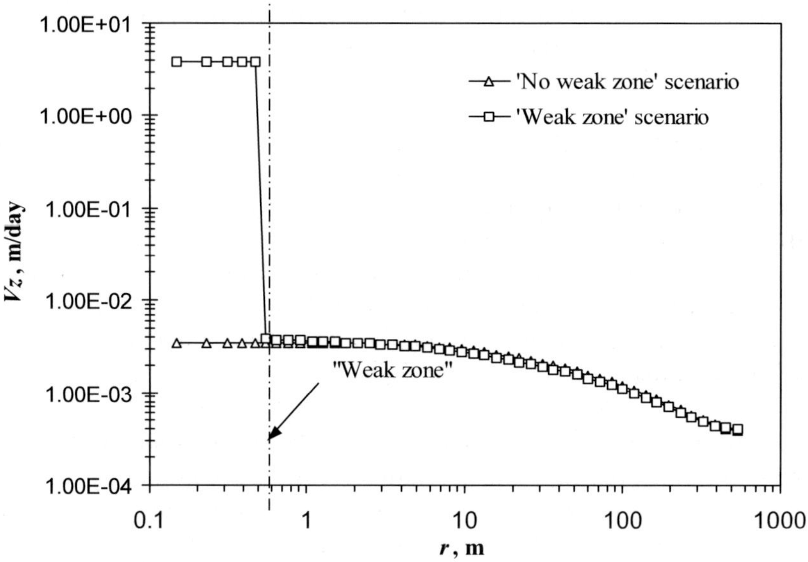

Figure 7.6. Vertical interflow through separating layer (from Bugai et al. 1996a)

To evaluate how well the model reproduces the actual groundwater system of Pripyat Town, we estimated the age of water extracted from the well. The water migration time for different flow lines ranges from 12 years to about 100 years. The average time (for flow line $\psi=0.5xQ_w$) is 30 years in the unconfined aquifer, three years in the separating aquitard layer, and 17 years in the confined aquifer, totaling approximately 50 years. The model result agrees well with the tritium dating of the Pripyat Town water supply (Voitsekhovich 1997).

Another model validation was provided by data on the piezometric level difference between the confined and unconfined aquifers at water supply wells in the pre-accident period. According to the Institute of Industrial Technologies (Moscow), this difference was 18 to 20 m. For filtration with a well extraction rate of Q_w=1000 m^3/d (pre-accident situation), the predicted difference is about 12 m. Thus, the model results matched well considering the simple formulations. This supports selected parameter values, especially the hydraulic conductivity value of 0.01 m/d for the separating aquitard layer. Compared with a similar sediment type, a relatively high permeability value used for the marl layer may be indicative of fracturing (Davis 1969).

7.2.4.1 Conservative prediction

Prior to performing the Monte-Carlo simulation, we simulated ^{90}Sr migration to the Pripyat Town water supply wells for the extreme (most conservative) radionuclide sorption value: the ^{90}Sr distribution coefficient, K_d was set to 0.5 mL/g for all hydrogeologic layers. This simulation provided interesting interpretations.

The maximum concentration of ^{90}Sr in the wells for the next 70 years was predicted to be 0.0025 C_o. In the worst case for C_o=1000 Bq/L (Table 7.3), the ^{90}Sr concentration in the well may reach up to 2.5 Bq/L, less than the VDU-91 standard but higher than the alternative specification (U.S. standard) of 0.3 Bq/L. Therefore, if permissible concentration remains equal to VDU-91 in the future, the failure of water wells will not occur for 70 years. The maximum ^{90}Sr concentration in the water wells would not exceed 0.3 Bq/L in the next 70 years if C_o<90 Bq/L.

7.2.4.2 Probabilistic prediction and economic risk assessment

The probability of water supply failure (^{90}Sr concentration in the water well exceeds permissible levels) was estimated using the Monte Carlo method. In this method a computational experiment generates random sets of input model parameters that conform to preset probability distributions, and for each parameter set a contaminant transport simulation was performed (model implementation). Thus, model implementation data are generated that further serve to determine probabilistic characteristics of the calculated output parameters of the model.

In the Monte Carlo assessment, we performed 50,000 implementations of the model. Figure 7.7 shows the resulting conditional probability of the Pripyat Town water supply failure, assuming that the permissible concentration of ^{90}Sr is 0.3 Bq/L. As discussed above, there is no chance that the calculated concentrations exceed the permissible concentration of 3.7 Bq/L. The cumulative conditional probability of water well failure during the next 70 years under the scenario of the weak zone and the probability of exceeding the permissible concentration of 0.3 Bq/L is approximately 30 percent. Without the weak zone, it is as low as 0.05 percent.

The probabilistic estimates of water well failure $P_f(t)$ (Figure 7.7) allow a calculation of an objective function value for the alternative that no water protection measures are undertaken (equation 7.6). Another relevant parameter required for the estimation is the cost of water well failure, C_f. Obtaining an accurate estimate of the possible cost for a particular adverse event in a subsurface water application is a complex problem (Massmann and Freeze 1987). We used the approximate expert estimate, which is accurate within an order of magnitude.

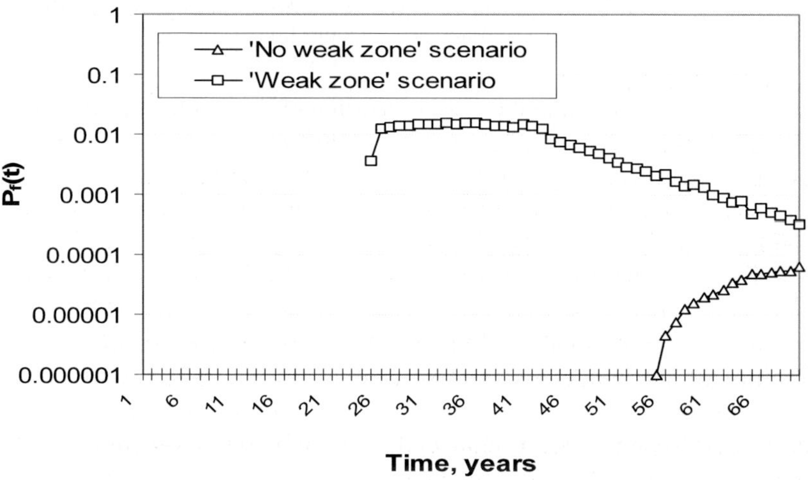

Figure 7.7. Conditional probability of water well failure assuming that permissible ^{90}Sr concentration is 0.3 Bq/L

We assumed that in case Pripyat Town water supply wells fail, it would be necessary to construct a similar water supply of 20 wells, exploiting the same aquifer in an alternative location. In this case, the damage (cost of equipment/installation of the new water supply) would be

$$C_f = 20\times \ \$\ 40{,}000 \times 2 = \$1{,}600{,}000$$

where the first multiplier accounts for the number of wells, the second term represents the expert estimate of equipment costs for a water supply well with a depth of 80 m, and the last roughly accounts for the additional costs related to exploration, construction of water management utilities, and other costs.

Using these values with the earlier estimate of $P_f(t)$ and assuming a 10 percent bank discount rate as suggested in Freeze et al. (1990), we obtained

$$\Phi^{(*)} \approx \ \$5{,}000$$

Because any large-scale water protection measures at the Pripyat wells (excavation of surface contamination, activation of protective drainage systems, etc.) would cost considerably more than the estimated economic risk of U.S. $5,000, such measures are not economically justified.

According to model results, water well failure may take place sooner than 25 years. An interesting question would be how the economic risk would behave in the short term. The first column in Table 7.4 shows the value of the

Table 7.4. Present (t=0) and future (in t=10 years and t=25 years) economic risks of contamination of Pripyat Town water wells as a function of possible maximum ^{90}Sr concentration in the source of contamination

Time	**Upper bound for concentration of ^{90}Sr in Quaternary aquifer (Bq/L)**		
	1,000	500	200
t=0	\$5,000	\$2,000	\$100
t=10 years	\$12,000	\$5,000	\$400
t=25 years	\$51,000	\$23,000	\$1,500

objective function (equation 7.6) in 10 and 25 years, assuming that the uncertainty of the model parameters is unchanged. The economic risk of Pripyat Town water well contamination in 25 years would increase by about a factor of 10 (\$51,000 U.S.), primarily due to the banking discount rate.

Because the risk grows considerably, it is appropriate to conduct a short-term study to reduce the uncertainty of the contaminant transport model parameters. The parameter that may be relatively easily constrained based on routine monitoring is the ^{90}Sr concentration in the unconfined aquifer, C_0. The second and third columns of Table 7.4 show values of economic risk, assuming the upper bound of C_0 is limited by 500 and 200 Bq/L, respectively. Decreasing risk with reduced uncertainty in the C_0 parameter justifies radiological monitoring of the unconfined aquifer near the water supply wells.

This decision model for the Pripyat Town groundwater supply wells was reevaluated using a more sophisticated, three-dimensional, hydrogeologic simulation model and more comprehensive parameter values (Smith and Gaganis 1998). This model predicted the cumulative probability of contamination of Pripyat Town water wells to be 20 times less than this study did. However, the revised estimates of the probability of contamination do not significantly affect selection of the preferred management option. Therefore, the simplified hydrogeologic model described above provides an adequate basis for evaluating the necessity of protective countermeasures for the Pripyat Town water wells (Smith and Gaganis 1998).

7.3 Radionuclide Transport to the River Network by Subsurface Runoff

Hydrologic migration of radionuclides from the CEZ to the Pripyat River and on to the Dnieper River is the most significant pathway of offsite radionuclide transport. Hydrologic migration is a source of radiation risk to tens of millions people living in the Dnieper basin. The most important long-term problem is posed by water migration of ^{90}Sr, which has higher mobility in soil and surface water than ^{137}Cs. Strontium-90 will be the main dose-forming radionuclide for the Dnieper River exposure pathway over the next 70 years and will contribute up to 80 percent of the collective dose from the aquatic pathway (Berkovsky et al. 1996).

Runoff of ^{90}Sr from the CEZ contributes 60 to 70 percent of Pripyat River contamination (Voitsekhovich 1994; Derevets et al. 1996). It was determined that the main source of Pripyat River contamination is the floodplains releasing radionuclides during flooding events (e.g., spring high water and winter ice jams). Under such conditions, direct washout (desorption) of radionuclides occurs from the contaminated soil to the surface water. The role of floodplains as a source of contamination is expected to decline with time due to natural processes (vertical migration of radionuclides in soil) and water protection measures (construction of protective dikes around the most contaminated floodplains). However, radionuclide transport via groundwater is expected to increase with time because of vertical migration from the contaminated soil surface through the unsaturated zone to the groundwater system and lateral migration from radioactive waste disposal sites.

Radionuclide transport via groundwater to the river network in the CEZ is discussed in the next section. Long-term radionuclide transport by groundwater to the Pripyat River from the RAWTLS in the Chernobyl near zone is also estimated. In addition, the screening assessment of radionuclide subsurface runoff (seepage) from distributed radionuclide sources (i.e., contaminated river catchments of the CEZ) was conducted.

7.3.1 Assessment of Current Radionuclide Transport in Groundwater to the River Network

Strontium-90 transport to the Pripyat River by the groundwater base flow within the CEZ was estimated to be less than 0.06 TBq/yr (1.6 Ci/yr) in 1992. The model used 1.7×10^{8} m^3/yr as base flow seepage to the river, obtained from a quasi-three-dimensional regional groundwater flow model of the CEZ. Radionuclide transport in groundwater from the water catchments of the Chernobyl near zone was also estimated in 1996 in the same way.

The estimated radionuclide transport rate in groundwater is 10^{-3} Ci/yr, as shown in Table 7.5. The slow groundwater transport is due to the fact that vertical migration of radionuclides to the groundwater system has been constrained by the barrier capacity of the unsaturated zone. Thus, the contamination in the unconfined aquifer is in the early stages.

Table 7.5. Transport of ^{90}Sr and ^{137}Cs to surface water bodies in the Chernobyl near zone by groundwater in 1996

Surface water reservoir	Rate of groundwater (m^3/d)	Average concentration in groundwater (Bq/L)		Transport Bq/day Ci/yr	
		^{90}Sr	^{137}Cs	^{90}Sr	^{137}Cs
Semikhody Inlet	500	0.14	0.13	71000 0.0007	64000 0.0006
Pripyat Inlet	1850	0.145	0.047	290000 0.003	87000 0.0009
Azbychin Lake	1200	0.185	0.14	222000 0.002	171000 0.0017

7.3.2 Long-Term ^{90}Sr Transport from Radionuclide Sources in the Chernobyl Near Zone

Long-term ^{90}Sr migration in subsurface water to surface water bodies was predicted for the right (southwest) bank of the Pripyat River in the Chernobyl near zone. This area is heavily contaminated by radioactive fallout, and water protection measures are planned (Skalskiy and Bugai 1996). Two ^{90}Sr migration sources were considered: surface contamination of soil by radioactive fallout (distributed sources) and point sources of RAWTLS and the Chernobyl industrial site (Figure 7.8).

This site includes the catchments in Pripyat Inlet, Semikhody Inlet, and an area where groundwater discharges to Azbuchin Lake. The Pripyat Inlet catchments encompass the point sources of migration—RAWTLS Red Forest, Stroibaza, Yanov, and Neftebaza. The groundwater from the Chernobyl industrial site and Shelter also discharges to Azbuchin Lake. The Pripyat Inlet connects directly to the Pripyat River, while Azbuchin Lake is an isolated reservoir with no direct connection to the Pripyat (see Figure 7.8). The most important distributed sources of radionuclides are the catchments of Pripyat Inlet and Azbuchin Lake, and the most important point sources are RAWTLS Stroibaza and Chernobyl industrial site.

For ^{90}Sr migration modeling for groundwater, we used the Performance Assessment Groundwater Analysis of low-level Nuclear waste (PAGAN) model developed for safety assessment of near-surface low-level radioactive waste burials (Chu et al. 1991). The PAGAN model and associated reports were provided to the Department of Monitoring of Geological Environment of Institute of GeoScience by Sandia National Laboratories in Albuquerque, New Mexico, USA, under an exchange program, Collaborative Research in Sectorial Policy, of the U.S. National Academy of Sciences in 1996.

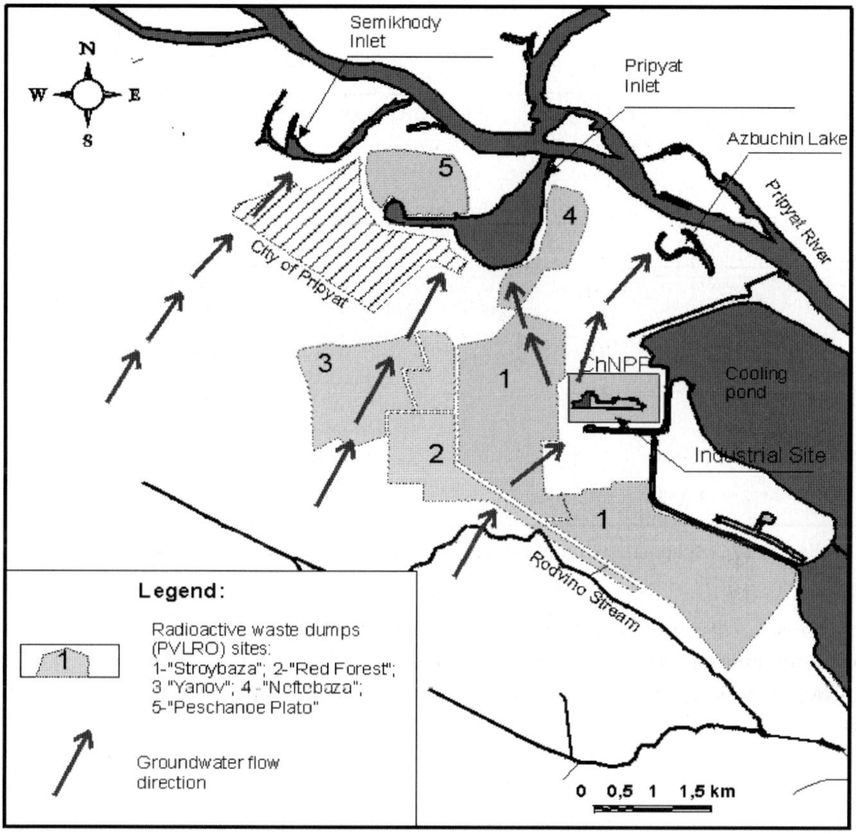

Figure 7.8. Chernobyl near zone

PAGAN incorporates analytical models of radionuclide migration in groundwater to predict (1) vertical migration from a rectangular area (the contamination source) near the soil surface into the aquifer and (2) lateral transport within the aquifer. Radionuclide migration in the unsaturated zone from source to groundwater table is treated as a one-dimensional process. The time of migration in the unsaturated zone accounting for sorption is assigned as the delay time parameter. Radionuclide transport in the aquifer is simulated by a three-dimensional advection-diffusion equation (one-dimensional advection transport in the direction of a filtration flow, and three-dimensional dispersion in longitudinal and transverse directions). It is assumed that the aquifer is uniform for hydraulic and sorption properties and has a constant thickness. The groundwater flow is assumed to be one-dimensional and is characterized by a seepage rate that is constant in time and space. The schematization of migration is presented in Figure 7.9.

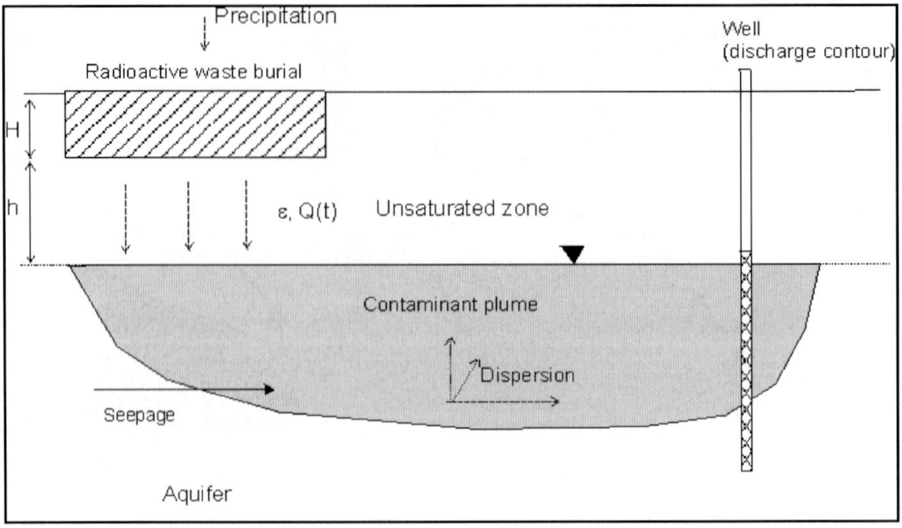

Figure 7.9. Schematization of radionuclide migration assumed in the PAGAN code

To describe the time-varying dynamics of radionuclide input to a hydrogeologic system (i.e., unsaturated zone and aquifer), PAGAN uses the following exponential model of the source:

$$Q(t)=Q_o \times \exp(- \alpha \times t) \qquad (7.8)$$

where

$Q(t)$= an integral rate for the radionuclide influx from the entire rectangular source to a hydrogeologic system (Bq/yr)

Q_0 = initial radionuclide flux (Bq/yr)

α = decay rate (yr^{-1})

t = time (yr).

Hydrogeologic schematization for modeling the Chernobyl near zone is listed in Tables 7.5 and 7.6.

Relatively large catchments (Pripyat and Semikhody Inlets and Azbuchin Lake) were divided into areas with constant unsaturated zone thickness (Table 7.6). Catchments and RAWTLS sites were simulated as contamination sources with a uniform distribution of radioactive inventory within each one. Delay time for ^{90}Sr penetration to the aquifer due to migration through the unsaturated zone was estimated using the formula:

$$t_v = h\,(\theta + \rho\, K_d)\,/\varepsilon$$

Table 7.6. Schematization of conditions of radionuclide migration sources

Source	Discharge contour	Dimension of catchments (m x m)	Distance from the centre of catchments to discharge contour (m)	Average gradient of groundwater flow	Thickness of unsaturated zone (m)	Inventory of ^{90}Sr (TBq [Ci])
Pripyat Inlet catchments, zone 1	Pripyat Inlet	2000 x 1000	500	0.003	5	37 (1000)
Pripyat Inlet catchments, zone 2	-	2000 x 500	1250	-	3	22(600)
Pripyat Inlet catchments, zone 3	-	2000 x 2000	2500	-	1	52(1400)
Azbuchin Lake catchments, zone 1	Azbuchin Lake	1000 x 1000	500	0.0043	1	11(300)
Azbuchin Lake catchments, zone 2	-	1000 x 200	1100	-	3	7.4(200)
Azbuchin Lake catchments, zone 3	-	1000 x 600	1500	-	5	8.5(500)
Semikhody Inlet catchments, zone 1	Semikhody Inlet	2000 x 1500	750	0.003	5	5.5(150)
Semikhody Inlet catchments, zone 2	-	2000 x 2000	1750	-	3	1.8(50)
Semikhody Inlet catchments, zone 3	-	1000 x 1000	3000	-	1	11(300)
RAWTLS Red Forest	Pripyat Inlet	2000 x 2000	2500	0.003	0	74(2000)
RAWTLS Yanov	-		2000	0.003	2	15(400)
RAWTLS Stroibaza	-	*1500 x 1200*	*1500*	*0.003*	*2*	*222(6000)*
Industrial Site of Chernobyl NPP	Azbuchin	500 x 500	1500	0.0043	5	59(1600)

Note: 1 TBq=10^{12} Bq.

where

h = the thickness of unsaturated zone (m)
ρ = the soil density (kg/m^3)
K_d = distribution coefficient of radionuclide (m^3/kg)
θ = moisture content in the unsaturated zone (dimensionless)
ε = infiltration recharge rate (m/yr).

Hydrogeologic and migration parameters for ^{90}Sr transport simulation were conservatively selected assuming the worst-case conditions (Table 7.7). Thus, these values should be considered upper bounds.

The radionuclide influx from the rectangular source area S (m^2) to the aquifer is given by:

$$Q(t) = C(t) \times \varepsilon \times S \tag{7.9}$$

where C(t) is radionuclide concentration in the infiltration recharge (Bq/m^3).

Assuming that ^{90}Sr leaching from fuel particles is described by the first-order equation with coefficient K_L, the concentration $C(t)$ may be estimated by (Bugai et al. 1996b):

$$C(t)=\frac{K_L A_0}{\varepsilon}\exp(-(\lambda+K_L)t) \tag{7.10}$$

where

K_L = $\ln(2/T_{se})$ (yr^{-1})
$T_{se.}$ = the half-life of ^{90}Sr leaching from fuel particles (yr)
A_o = concentration density of surface contamination (Bq/m^2)
λ = radioactive decay (yr^{-1}).

Table 7.7. Parameters for assessing ^{90}Sr transport via groundwater to surface water

Parameter	**Value**
Unsaturated zone	
Infiltration recharge rate, mm/yr	200
Volumetric moisture content in unsaturated zone, dimensionless	0.1
^{90}Sr distribution coefficient for the unsaturated zone, mL/g	1
Soil density, kg/dm^3	1.65
Aquifer	
Saturated thickness, m	20
Hydraulic conductivity , m/d	10
Porosity, dimensionless	0.2
Coefficient of longitudinal dispersion, m	5
Coefficient of transverse dispersion, m	0.5
^{90}Sr distribution coefficient for saturated zone, mL/g	0.5
Soil density, kg/dm^3	1.65

Combining equations 7.9 and 7.10, we learn that radionuclide influx to the aquifer may be described by exponential model (equation 7.8), where

$$Q_o = K_L \times A_o \times S, \text{ and } \alpha = \lambda + K_L$$

Note that product $A_o \times S$ is the initial inventory of ^{90}Sr in the migration source. In our radionuclide transport modeling analyses, we assumed that $T_{se} = 10$ yr, that is, K_L=0.07/yr (Bobovnikova et al. 1990; Dolin et al. 1990).

Strontium-90 inventory within catchment areas of the Chernobyl near zone was based on surface contamination maps from the Ukrainian Institute of Hydrology and Meteorology (1992). Inventory of radioactivity in RAWTLS Red Forest was based on data from the Institute of Industrial Technologies (Il'ichev et al. 1992). Inventory of ^{90}Sr in RAWTLS Yanov, and Stroibaza was estimated by experts and measured data. Inventory at the Chernobyl industrial site was estimated using data from Belyaev et al. (1990) that indicates 0.3 percent of the complete load of Chernobyl Unit 4 fuel was deposited at the industrial site and the total ^{90}Sr inventory was 540,000 Ci. Model parameters and results of radionuclide transport through the subsurface water pathway are presented in Tables 7.7 and 7.8 and Figure 7.10.

The maximum predicted ^{90}Sr concentration in the groundwater at seepage discharge points is 215 to 340 Bq/L, which is two orders of magnitude higher than the permissible ^{90}Sr concentration in drinking water, according to NRS-76/87. Thus, the contaminated groundwater incoming to the surface waters is regarded as liquid radioactive waste (based on classification of SPORO-85), considering its radionuclide characteristics. However, the discharge rate of the contaminated groundwater flow is relatively low. Thus, despite the high ^{90}Sr concentration in groundwater, its transport from these radionuclide sources to Pripyat Inlet and Azbuchin Lake in the future would be relatively low, less than several 10s of GBq/yr (several Ci/yr).

The maximum travel time for subsurface ^{90}Sr transport to surface water bodies ranges from 33 to 145 years after the accident. The maximum cumulative transport from all these sources is estimated at 130 GBq/yr (3.5 Ci/yr) in approximately 100 years, or 0.02 percent per year of the total inventory within the contaminated catchments. The total amount of radionuclides transported over 300 years is estimated as 15 TBq (420 Ci), or 3 percent of the total initial inventory of radioactivity in the catchments (Figure 7.10).

Table 7.8 and Figure 7.10 do not account for ^{90}Sr transport from RAWTLS Neftebaza. Other assessments showed that the maximum ^{90}Sr transport from this RAWTLS may be 0.75 GBq/yr (0.02 Ci/yr) in the next 20 years (Mishutina et al. 1995).

Table 7.8. Prediction of ^{90}Sr transport to groundwater in the surface reservoir of the Chernobyl near zone

Source	Discharge contour	Max. ^{90}Sr in groundwater at discharge contour (Bq/L)	Max. transport GBq/yr (Ci/yr)	Term of maximum. transport (yr)	Integral transport for 300 years TBq (Ci)
Catchments of Pripyat Inlet	Pripyat Inlet	280	63 (1.7)	65	4.2 (115.7)
Catchments of Azbuchin Lake	Azbuchin Lake	340	63 (1.69)	33	3.9 (107.1)
Semikhody Inlet Catchments	Semikhody Inlet	28	6.3 (0.17)	65	0.4 (11.1)
RAWTLS Red Forest	Pripyat Inlet	16	3.3 (0.09)	200	0.23 (6.3)
RAWTLS Yanov	Pripyat Inlet	5	0.22 (0.006)	220	0.015 (0.4)
RAWTLS Stroibaza	Pripyat Inlet	215	81.4 (2.2)	110	5.3 (144)
Industrial Site of Chernobyl NPP	Azbuchin Lake	230	37 (1)	145	1.4 (37.2)
Total			129.5 (3.5)	110	15.4 (421.8)

Note: 1 TBq=10^{12} Bq, 1 GBq = 10^{9} Bq.

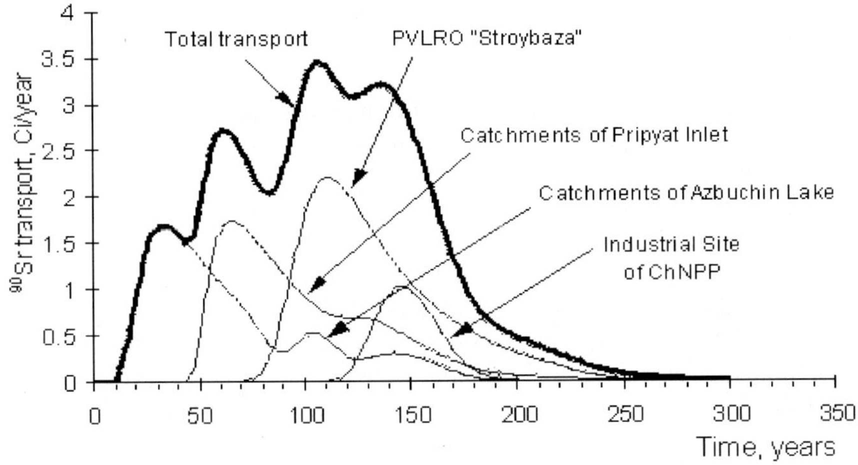

Figure 7.10. Prediction of ^{90}Sr transport via groundwater pathway to surface water in Chernobyl near zone

The low flow rate of the Pripyat River at 97 percent probability is 60 m^3/s in an open channel. The ^{90}Sr concentration increment in the Pripyat River due to subsurface migration from the radionuclide sources will not exceed 0.1 Bq/L (2 pCi/L) at this flow rate. Therefore, contrary to certain statements (Sobotovich and Bondarenko 1995), hydrogeologic migration of ^{90}Sr from RAWTLS sites will not result in catastrophic amounts to the Pripyat River. According to these results, ^{90}Sr subsurface migration modeling results for the Pripyat River southwest (right)-bank catchments of the Chernobyl near zone should not be extrapolated to the entire CEZ.

7.3.3 Screening Assessment of ^{90}Sr Subsurface Water Migration from the Contaminated Catchments of the CEZ to Surface Waters

The following are some favorable factors for migration of radioactive contaminants to the surface waters (small rivers, melioration drainage channel) in the CEZ. Predominant infiltration of precipitants to deeper soil layers and in minimum surface runoff can reduce the potential surface runoff.

- Mineral and lithologic compositions of soils characterized by low capacity to retain (adsorb) radionuclides
- Developed network of watercourses and melioration channels, providing good drainage conditions for the territory.

The subsurface migration pathway would apparently determine contamination of surface waters in a relatively long-term perspective (30 to 50 years and longer), when the washout of radioactivity by surface runoff will be minimized due to vertical penetration of fallout radionuclides. Under such conditions contamination of surface waters is expected to reach maximum value during the low-flow period, when groundwater base flow accounts for almost 100 percent of runoff from the catchments.

In the next 50 to 100 years, subsurface water migration of radioactive fallout to the surface water would occur mainly from 200- to 300-m-wide watershed zones immediately adjacent to the surface waters. The effect of more distant catchment areas will be minimal due to long groundwater travel time and retardation by sorption. Thus, hydrophysical and migration regimes of floodplain soils play a key role in both radionuclide washout processes by surface runoff and subsurface migration processes. Unfortunately, these processes have not been studied adequately, and uncertainty in migration parameters and predictions are significant.

We discuss the ^{90}Sr subsurface transport to the surface waters with the analytical model (equation 7.1) described in Section 7.1.1. We assume that the ^{90}Sr concentration in the surface waters in the low-flow period is the same as that in groundwater. Subsurface transport of ^{90}Sr to the river network, $R(t)$ (Bq/yr), was calculated using the following formula (Bugai et al. 1996b):

$$R(t) = C_{GW}(t)\, Q_{GW}$$

where

$C_{GW}(t)$ = ^{90}Sr concentration in groundwater (Bq/m^3) calculated by equation 7.1

Q_{GW} = groundwater runoff (seepage) to the river network (m^3/yr).

Groundwater runoff (seepage) to the river network was estimated using the following formula:

$$Q_{GW} = P\, S\, K_r\, K_{gw}$$

where

P = average annual rainfall (m/yr)

S = catchment area (m^2)

K_r = runoff coefficient for the river basin (dimensionless)

K_{gw} =share of the groundwater flow in the total runoff (dimensionless).

Average ^{90}Sr surface contamination in the CEZ is 1100 kBq/m^2. This value was used for the migration assessment. Model results were extrapolated to the entire CEZ (2600 km^2 catchment area) to estimate total groundwater transport. The assessment was performed for conservative, baseline, and optimistic scenarios. Results and parameter values are shown in Table 7.9.

Table 7.9. Parameters and results of screening assessment of ^{90}Sr transport via groundwater pathway to the CEZ river network

Fixed Input Parameters			
Surface contamination of catchments by ^{90}Sr, kBq/m^2	1100		
^{90}Sr leaching rate from hot particles, yr^{-1}	0.14		
Radioactive decay rate, yr^{-1}	0.023		
Thickness of the unsaturated zone, m	1.5		
Thickness of aquifer, m	20		
Moisture content in unsaturated zone, dimensionless	0.1		
Soil density, kg/m^3	1.65		
Porosity, dimensionless	0.3		
Annual total rainfall, mm	650		
Runoff coefficient from catchments, %	18		
Area of catchments, km^2	2,600		
Variable Input Parameters			
	Conservative scenario	Baseline scenario	Optimistic scenario
Infiltration recharge rate to groundwater, mm/yr	250	200	150
^{90}Sr distribution coefficient , L/kg	1	2	4
Share of the groundwater base flow in the total runoff, %	50	30	25
Results of Prediction			
Maximum concentration of ^{90}Sr in surface water in low-flow period, Bq/L	13	5.1	1
Travel Time of maximum transport, year	25	40	80
Maximum cumulative transport from catchments in CEZ,[a] GBq/yr (Ci/yr)	2,000 (54)	460 (12)	75 (2)
(a) Rough illustrative estimate.			

The values of hydrological runoff coefficients were adopted based on the data of Bilyavsky (1971) and Bogomolov et al. (1983). For natural conditions in Kiev and Zhitomir Polesje (woodlands) regions, the runoff coefficient for rivers is 17 to 18 percent, while the total rainfall is 650 to 680 mm (Bilyavsky 1971). Groundwater flow accounts for approximately 25 to 30 percent of total hydrological runoff. In the melioration drainage systems, the groundwater share of runoff may be 50 to 70 percent (Bogomolov et al. 1983). The unsaturated zone thickness and infiltration recharge rate selected in the transport calculations correspond to those of the floodplain sites.

The maximum predicted ^{90}Sr concentration in surface water and integral transport of the radionuclide for the three scenarios vary significantly (see Table 7.9). These results show that groundwater migration under certain conditions may be an essential mechanism for contamination of surface waters in a low-flow period. Let us note again that our assessments are based upon rather crude schematization and the simplified radionuclide migration model. Thus, these results are more of a qualitative than a quantitative nature.

Based on the screening assessment, we believe that the problem deserves further investigation. First, comprehensive field subsurface water investigations should be performed to obtain the required migration parameters and data for model calibration. Second, subsurface migration must be modeled using more complex and adequate subsurface water simulation models.

7.4 Principles of Water Protection Measures in Connection with Groundwater Contamination in the CEZ

This section discusses approaches to water protection measures and relevant regulatory decisions for groundwater contamination in the CEZ. Groundwater contamination is considered to have unfavorable consequences and is analyzed from a pragmatic point of view. The issues considered are:

- General criteria for analysis of groundwater contamination problems
- Practical problems of groundwater contamination in the CEZ, their brief description and analysis
- Scheme for calculation of reference (controlled) concentrations of radioactive and chemical toxicants in the groundwater.

The analysis used a general approach developed for protection and remediation of groundwater contaminated due to industrial activities, uranium mill tailings, and others (EPA 1988; DOE 1996a; IAEA 1997; Rosen and LeGrand 1997; LeGrand 1982; OTA 1991; Govarets and Luykx 1994).

7.4.1 General Criteria for Analysis of Groundwater Contamination

The man-made contamination of groundwater and geologic environment is an undesirable phenomenon, whether it is groundwater contamination by a newly created, environmentally hazardous object or past and existing *de facto* groundwater contamination by inadvertent management of hazardous wastes, leakage, and emergency spills of chemicals. Newly created, potentially environmentally hazardous facilities (e.g., toxic waste burials) should be developed and constructed to preclude contamination of the hydrogeologic environment. In many cases, once groundwater contamination has occurred, remediation of an aquifer is expensive and technically difficult (or even infeasible). Practical experiences in engineering projects aimed at remediation of contaminated aquifers have shown the low efficiency of existing technologies (LeGrand 1982; OTA 1991).

These circumstances require "recovery of quality of the contaminated groundwater at any price" to be substituted in practice with a more pragmatic and rational approach, where water protection measures and countermeasures are balanced with human health and economic risks caused by groundwater contamination (Massmann et al. 1991; Rosen and LeGrand 1997). When estimating the importance of a groundwater contamination problem and developing water protection measures, it is helpful to be guided by the following criteria (in order of priority) (EPA 1988; DOE 1996a; IAEA 1997):

- Risks to health
- Threat to the environment (ecological risks)
- Economic risks (groundwater is a valuable natural resource)
- Socio-political aspects.

The most serious situation requiring intervention occurs when groundwater contamination is a source of risk to human health. Pathways by which contaminated groundwater may reach particular recipients are presented in Table 7.10. The most typical and relevant exposure pathways are using groundwater for drinking and discharging groundwater to surface reservoirs and using that water for potable water or agricultural purposes.

While the contaminated water supply well in principle may be closed and a water well field may be transferred to a new location, the surface reservoir, fixed on the land, cannot avoid contaminated groundwater migrating its direction unless the migration is attenuated by the protective properties of the geologic environment (e.g., sorption, decay, dispersion, etc.) or engineering countermeasures (intercepting drainages, barriers) (LeGrand 1982).

Active water protection measures (countermeasures) are deemed justifiable beyond some threshold level of risk to health. This threshold level of individual lifetime risk ranges from $1x10^{-6}$ to $1x10^{-4}$. [Lifetime risk corresponds to probability of death (or disease) of an individual exposed for 70 years.] The scale level of individual risk-level of intervention, according to Govarets and Luykx (1994), is presented in Table 7.11.

Table 7.10. Groundwater contamination exposure pathways (EPA 1988)

Recipient	Exposure Pathway	Exposed group
Well for potable water supply	- Absorption - Inhalation - Skin adsorption	Residents
Well for lawn irrigation	- Absorption - Inhalation - Skin adsorption	Residents
Agricultural irrigation well	- Inhalation - Skin adsorption	Agricultural workers
Industrial water supply well	- Inhalation - Skin adsorption	Industrial workers
Spring	- Absorption - Inhalation - Skin adsorption	- Site users - Residents
Surface water reservoir	- Casual swallowing - Inhalation - Skin adsorption - Accumulation by aqueous organisms	- Aqueous organisms - Site users -Terrestrial animals consuming aqueous biosphere - Fishing, hunting, drinking surface water
Construction drainage well	- Inhalation - Skin adsorption	Construction workers

Table 7.11. Intervention criteria based on values of lifetime carcinogenic risk (from Govarets and Luykx 1994)

Risk, yr^{-1}	Level of intervention
$>10^{-2}$	Countermeasures are taken; economic factors are of minor importance
$>10^{-6} – 10^{-5}$	Countermeasures are taken accounting for economic and social factors
$<10^{-8} – 10^{-7}$	No intervention is necessary

In general, international recommendations advise compliance with the ALARA (as low as reasonably achievable) principle so risk is minimized as much as possible; however, the expense should not be excessive. The next priority is the situation in which groundwater has a vitally important ecological value. For instance, groundwater being discharged to a surface reservoir provides a living environment for an especially sensitive ecological system, and contamination of this water may result in the disappearance of unique biological species (EPA 1988). Under these circumstances, active water protection measures and countermeasures are urgent from an environmental standpoint, although no immediate threat to human health may be present.

The economic risk of groundwater contamination is connected with the importance of groundwater as a natural resource. In this case, the groundwater is typically used for potable, household, agricultural, and industrial purposes. Economic risk (*R*) is determined as follows:

$$R = P_f \times C_f$$

where P_f is the probability of occurrence of adverse events and C_f is the economic cost (damage) caused by the adverse event (Rosen and LeGrand 1997).

For instance, groundwater contamination may cause failure of water supply wells. This, in turn, may entail economic losses and costs related to facilities and equipment of an alternative water supply source, or costs related to additional purification of the locally extracted groundwater. Assessing the economic risks of groundwater contamination and comparing this risk with the cost of water protection measures allow the cost-efficiency of such measures to be estimated. Water protection measures are economically justified when their implementation costs do not exceed the possible monetary losses from the contamination (Massmann 1991; Rosen and LeGrand 1997; Raucher 1983). The methodology of economic cost-benefit analysis of subsurface water problems is discussed in Section 7.2.1.

When developing water protection policies, socio-political factors such as negative attitudes of the local population, local authority, political parties, environmental organizations, etc., toward the groundwater contamination problem may play an important role.

To assist in selecting the applicable water protection strategy, the EPA developed classification for groundwater resources. Aquifers are subdivided into three classes, as shown in Table 7.12. Water protection measures are the most urgent in the first-class groundwater contamination. Third-class groundwater contamination (particularly Subclass IIIB) is not a priority environmental and/or economic problem.

Table 7.12. U.S. Environmental Protection Agency classification of aquifers (EPA 1988)

Category	**Description**
Category I	Specially valuable groundwater that is poorly protected from contamination due to local hydrogeologic conditions and is described by the following: ▪ groundwater is irreplaceable; no alternative source of water exists for significant population ▪ groundwater is discharged to a surface reservoir providing a living environment for especially sensitive ecological systems, and contamination may result in disappearance of unique biological species.
Category II	Current and potential sources of potable and household water supply (except those belonging to Category I)
Subclass IIA	Current sources of potable and industrial water supply
Subclass IIB	Potential sources of potable and industrial water supply
Category III	Groundwater is not a potential source of potable water and has limited industrial value; saltwater (>10 g/L); groundwater contaminated from a distributed source; aquifer with low water production, etc.
Subclass IIIA	Aquifer hydraulically connected with surface reservoir or with water bearing aquifer, belonging to Categories I or II.
Subclass IIIB	Aquifer not connected hydraulically with surface reservoir or with water bearing aquifer, belonging to Categories I or II.

7.4.2 Practical Problems Related to Groundwater Contamination in the CEZ

Considering groundwater contamination in the CEZ from the perspective of the criteria described above, the following three problems are discussed:

- Contamination of unconfined aquifer in Quaternary deposits
- Contamination of confined aquifer in Eocene deposits
- Discharging the contaminated groundwater to the Pripyat River and its tributaries.

Because the unconfined aquifer is hydraulically connected to the confined aquifer, the contaminated unconfined aquifer is a potential source of contamination of the confined aquifer below. The migration of the contaminated groundwater in the unconfined aquifer also carries radionuclides to the receiving rivers. Thus, the three problems listed above are interrelated.

Risks to health and possible economic damage are the main criteria. We have insufficient data to reliably isolate and estimate ecological risk. Thus, assuming that protection of human health also provides ecological safety, we adopted a simplified approach. Socio-political factors were not considered because their analysis would require a special study. Moreover, socio-political factors for the CEZ may have less importance because of the special administrative status of this territory and the absence of the regular population and local authorities. On the other hand, safety problems at the Chernobyl nuclear plant, Shelter, or radioactive waste burials, and the radiation safety of the CEZ as a whole, receive attention by the international community. Any negative information from the CEZ on the environment and/or increased radiation risks reverberate worldwide. Therefore, they must be addressed in investigations of environmental protection strategies for the CEZ.

For groundwater contamination problems, Rosen and Legrand (1997) recommend using the schematization of "source of contamination – recipient of groundwater." A similar schematization was first used for groundwater contamination in the CEZ by A. B. Sitnikov (IGS). The radionuclide sources of migration are both distributed (i.e., contaminated catchments of the Pripyat River) and point sources (e.g., the Shelter, the cooling pond, and radioactive waste dumps) of radioactivity within the CEZ.

7.4.2.1 Problem 1: Contamination of the unconfined aquifer

According to radiohydrogeologic monitoring data, contamination of the unconfined aquifer is a serious problem; however, it does not present an immediate health or economic risk because the aquifer at Chernobyl is not used for potable or household water. Drinking water in the foreseeable future is excluded from consideration due to the specific administrative status of the territory; i.e., the residential population does not use this water. However, contamination of unconfined aquifer creates preconditions for radionuclide migration to the confined aquifer in Eocene deposits and to the Pripyat River. Based on classifications in Table 7.11, the unconfined aquifer at Chernobyl may be considered in the second category. The following are potential negative effects from contamination of the unconfined aquifer:

- Possible construction workers' exposure to contaminated water from drainage operations during decommissioning of the Chernobyl plant and transformation of the Shelter to an environmentally safe system cannot be excluded (see Table 7.10 and Chapter 8).
- Subsurface water migration spreads radioactive contamination to initially clean soils. Thus the volume of contaminated soil increases. This may be important in selecting and implementing the Green Lawn project, assuming all radioactive waste at the site will be excavated and reburied.

- Disposal of contaminated drainage water during geotechnical operations at the Chernobyl industrial site may be a serious technical and economic problem in the future.

7.4.2.2 Problem 2: Contamination of confined aquifer in Eocene deposits

Subsurface water migration of radionuclides may result in contamination of the confined aquifer, which is separated from the unconfined aquifer by the regionally discontinuous layer of low-permeability marls (carbonate clays). The possibility of contamination increases with existence of hydraulic "windows" in the marl layer, including those under the Chernobyl plant.

A probable recipient of the contaminated water is the Pripyat Town water supply wells. The Eocene confined aquifer is a source of potable and technical (industrial) water supply for the Chernobyl plant and special facilities that continue to operate at the site. The possibility of contamination of Pripyat Town water supply wells would cause certain economic risks. No potential health risks are present, because the water supply wells are monitored, and if they are out of compliance with drinking water standards, well operations would be suspended. Therefore, an analysis on protection of Pripyat Town water supply wells from radioactive contamination should be based on the economic expediency of the countermeasures.

According to the preliminary assessment (Section 7.2), the economic risk of contamination of Pripyat Town water supply wells in the foreseeable future is estimated as low (due to a low probability of the water well contamination), and expensive countermeasures are not justified. However, the assessments did not account for the possible existence of "windows" in the marl layer. Also, a number of model parameters were assumed based on expert judgment rather than reliable site-specific information. Therefore, the problem deserves additional analysis, taking into account updated radiological monitoring data.

7.4.2.3 Problem 3: Discharge of contaminated groundwater to rivers

The radionuclide migration pathway to the rivers is important because the potential recipients of the contaminated surface water are the downstream populations near Kiev Reservoir and along the Dnieper River. In this case a potential health risk exists for a significant number of people because the Dnieper is a main source of water (including irrigation) for southern Ukraine.

Past assessments demonstrate that subsurface water migration of radionuclides from certain point sources in the Chernobyl near zone, particularly RAWTLS, would not result in catastrophic transport of radioactivity to the Dnieper River system (see Section 7.3.2). The residence time of ^{90}Sr before the groundwater discharges to the river is estimated at hundreds of years, which allows for significant radionuclide decay.

According to screening assessments, ^{90}Sr migration to the river network from contaminated catchments under some scenarios may result in high concentrations in small rivers and reclamation systems of the CEZ in low flow periods (see Section 7.3.3). However, because the assumptions are conservative and the contaminated runoff from the catchments of the CEZ is significantly diluted by the Pripyat River and Dnieper cascade of reservoirs, there is no reason to expect catastrophic offsite transport of radioactivity from the CEZ via subsurface migration. However, this assessment has significant uncertainty in contaminant transport model parameters, and continued monitoring of groundwater migration processes in the CEZ is needed. The monitoring data would reduce parameter uncertainties and refine the modeling predictions and risk assessments.

Long-term radiological problems at the Chernobyl site may be caused by subsurface migration of radionuclides from the Shelter and some radioactive waste repositories containing high-level and long-lived wastes (e.g., Compleksny and Podlesny). For these sources, radionuclide delay in migration by several hundred years provided by local geological barriers may not be sufficient to decay migrating radionuclides to safe levels. If contamination of an unconfined aquifer is allowed to reach a hazardous scale (exceeding the protection properties of local geologic barriers), the future isolation (capture) of the contaminated groundwater and protection of the Pripyat River would become a technically complex and expensive problem. Therefore, the Pripyat River contamination due to subsurface radionuclide migration from these sources may be referred to as a "pending" problem.

In view of the paramount importance of ensuring the long-term safety of the water of the Pripyat and Dnieper system as a whole, the problem of possible migration of radionuclides to the Pripyat River through subsurface migration from point and distributed sources should be considered the most important among Problems 1 through 3. Problem 3 should dictate the priority of tasks in developing a hydrogeologic monitoring system for the CEZ. The risk of Pripyat River contamination may serve as a basis for determining reference radionuclide concentrations in groundwater in the CEZ, as is discussed in Section 7.4.3. These reference concentrations would be used as criteria for making relevant administrative (water protection) decisions.

7.4.3 Estimation of Reference Concentrations of Radioactive and Chemical Toxicants in Groundwater

The reference concentration is the numeric value of radionuclide (or other contaminant) content in groundwater that is the basis for making specific administrative decisions (including those of water protection). The recommended scheme for calculating reference concentrations of contaminants in the groundwater of the CEZ is based on the risk of contamination of surface waters due to contaminated groundwater discharge to surface water.

The starting point is the threshold concentration of a radionuclide in the Pripyat River water at the point of groundwater seepage from the CEZ. For instance, in 1991 the Chief Sanitary Officer of Ukraine set the ^{90}Sr threshold concentration in the Pripyat River at Chernobyl Town at 50 pCi/L. In general, threshold concentration in surface water should be determined based on risk assessment of the water from the Pripyat River and downstream water bodies under relevant exposure scenarios. The threshold concentration of a contaminant in the groundwater can be obtained by back-calculation, taking into account (1) mixing of groundwater discharge with surface water and (2) attenuation of contaminant concentration in the groundwater during migration from its source to the seepage points due to the protective properties of the hydrogeologic subsurface environment. This method has been used in similar assessments by EPA and the U.S. Department of Energy (DOE 1996b).

The contaminant threshold concentration in groundwater in the source zone of contamination is obtained by

$$C_{TC,\ ground} = C_{TC,\ surf} \times K_1 \times K_2$$

where

$C_{TC,\ surf}$ = TC in surface water

K_1 = coefficient for groundwater mixing with surface water

K_2 = coefficient for attenuation of concentration during migration from contamination source to seepage point due to protective properties of the hydrogeologic environment (Figure 7.11).

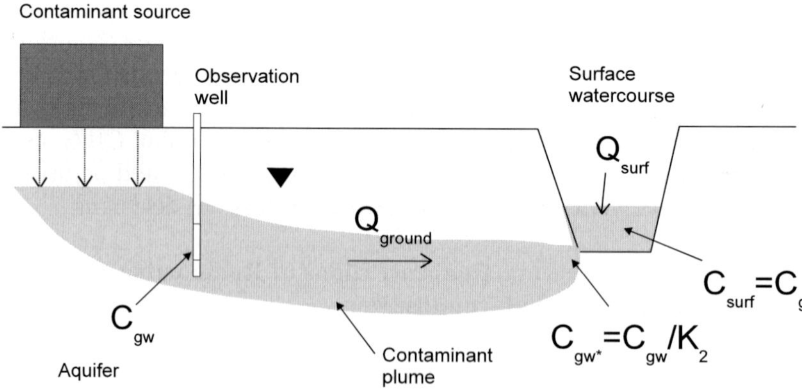

Figure 7.11. Scheme for estimating reference concentration of contaminants in groundwater

The coefficient K_1 may be given by

$$K_1 = Q_{\text{surf}}/Q_{\text{ground}} \times (1 - C_{0,\text{sur}}/\ C_{TC,surf})$$

where

Q_{surf} = flow rate of surface water
Q_{ground} = contaminated groundwater discharge to surface water
$C_{0,\text{sur}}$ = background concentration of contaminant in surface water (resulting from upstream sources).

The last equation assumes that Q_{surf} is much greater than Q_{ground}.

Coefficient K_2 should be determined by site-specific mathematical modeling of subsurface migration. For screening assessments, simplified conservative assessment schemes may be used; e.g., considering attenuation of radionuclide concentration in groundwater due only to radioactive decay and disregarding the hydrodynamic dispersion, the coefficient K_2 can be calculated by

$$K_2 = \exp(-\lambda\ t_{\text{migr}})$$

where λ is a radioactive decay constant and t_{migr} is the travel time of radionuclide in subsurface environment from the contamination source to the seepage location at the river.

The radionuclide concentration exceeding the reference concentration in groundwater monitoring wells at a contaminated site may serve a basis for implementing active countermeasures to mitigate subsurface migration. Along with the threshold concentration levels and determining the necessity of intervention, it is appropriate to define a number of intermediate levels serving for initiation of additional investigations and observations when concentrations approach critical values.

It may be difficult to estimate the background concentration in surface water ($C_{0,sur}$) resulting from upstream sources, especially with multiple sources of surface water contamination. As an alternative, modeling and/or experts may be used to estimate contamination of the Pripyat River.

7.5 Conclusions

Conservatively estimated risks for hypothetical residents of the CEZ from the groundwater as a drinking water supply would be less than $1\text{x}10^{-5}$/yr under average radiological and hydrogeologic conditions in the CEZ. For the Red

Forest radioactive waste dump site, risk from radioactive contamination of groundwater is up to 10^{-3}/yr and 20 times the drinking water standard. It will take 150 and 250 years, respectively, for the risk caused by using groundwater for drinking water to be reduced to 1×10^{-6}/yr. Soil contamination in the CEZ by ^{137}Cs is a priority dose-forming factor in comparison with groundwater contamination.

The results of stochastic simulations of ^{90}Sr migration to the Pripyat Town water wells show the probability of significant radioactive contamination (exceeding permissible levels) is relatively low. The residence time of ^{90}Sr in a hydrogeologic system of water wells is probably long enough for radionuclide decay to lower the concentration to below the threshold concentration. A low economic risk of water supply contamination indicates that expensive countermeasures are not justified. Modeling of ^{90}Sr migration from RAWTLS sites and some other radionuclide sources in the Chernobyl near zone shows that ^{90}Sr transport from these sources to surface waters would be relatively low, not exceeding $n \times 10$ GBq/yr ($n \times 1$ Ci/yr). The travel time of maximum subsurface transport ranges from 33 to 145 years after the accident.

The screening assessment of ^{90}Sr migration from the contaminated catchments of small rivers and water reclamation systems of the CEZ suggests that subsurface water migration under certain conditions may be an essential mechanism for radioactive contamination of surface waters in low flow periods.

In general, due to the conservative modeling assumptions as well as the significant dilution of radioactively contaminated groundwater seepage from catchments of the CEZ by the Pripyat River and the reservoirs of the Dnieper cascade, one would hardly expect that subsurface migration would cause catastrophic off-site transport of radioactivity from the CEZ.

These assessments are subject to significant uncertainty in radionuclide transport model parameters. Large uncertainties in radionuclide transport model parameters result in widely varying estimates. Thus it is necessary to continue monitoring the groundwater migration processes in the CEZ for operational control of the situation. The monitoring data will also reduce uncertainties and refine the modeling predictions and risk assessments. Thus, more purposeful field and modeling studies need to be performed.

To ensure the long-term safety of the Pripyat and Dnieper River water, the radionuclide subsurface migration from point and distributed sources to the Pripyat River should be evaluated to determine the possibility of radionuclide concentrations reaching hazardous quantities. The problem should dictate development of a necessary hydrogeologic monitoring system for the CEZ. The risk of Pripyat River contamination may serve as a basis for developing reference concentrations in groundwater of the CEZ, which could be used for making relevant administrative (water protection) decisions.

7.6 References

Belyaev ST, AA Borovoy, YP Busulukov, et al. 1990. "Some Aspects of Post-Accident Works Within the Controlled Zone of Chernobyl NPP." Proceedings. of International Symposium on Recovery Operations in a Nuclear Accident or Radiological Emergency, Vienna, IAEA-SM-316/42, International Atomic Energy Agency, Vienna, pp. 185-202.

Berkovsky V, G Ratia, and O Nasvit. 1996. "Internal Doses to Ukrainian Populations Using Dnieper River Water." *Health Physics*, 71(3).

Bilyavsky GO. 1971. *Natural Resources of Fresh Ground Waters of the Ukrainian Polessye.* Naukova Dumka Publishers, Kiev, p. 128 (in Ukrainian).

Bobovnikova CI, EP Virchenko, AV Konoplev, et al. 1990. "Chemical Forms of Long-Lived Radionuclides and Their Transformation in Soils of the Chernobyl NPP Zone." *Pochvovedenie*, Vol. 10, pp. 20-25 (in Russian).

Bogomolov YG, AV Romanov, and GA Ulyanov. 1983. "Study of the Influence of Water Withdrawal on Changes of Hydrogeological Conditions under Influence of Land Melioration in West Region of USSR (Byelorussia SSR, Near-Baltic region)." *Regime, Balance and Geochemistry of Groundwater of Pripyat Polessye.* Nauka i Tekhnika Publishers, Minsk, pp. 93-107 (in Russian).

Bugai DA, RD Waters, SP Dzhepo, and AS Skalskiy. 1996a. "Risks from Radionuclide Migration to Groundwater in the Chernobyl 30-km Zone." *Health Physics,* Vol. 71, pp.9-18.

Bugai DA, L Smith, and R Beckie. 1996b. "Risk-Cost Analysis of Strontium-90 Migration to Water Wells at Chernobyl Nuclear Power Plant." *Environmental & Engineering Geoscience*, 2(2):151-164.

Chu MSY et al. 1991. *A Self-Teaching Curriculum for the NRC/SNL Low-Level Waste Performance Assessment Methodology.* NUREG/CR-5539 (SAND90-0585), Sandia National Laboratories, Albuquerque, NM.

Davis SN. 1969. "Porosity and Permeability of Natural Materials." *Flow Through Porous Media*, R DeWiest (Ed.). Academic Press, New York, pp. 53-89.

Derevets VV, YP Ivanov, and SV Kazakov. 1996. "Radiation condition of environment in the Chernobyl Exclusion Zone." Proceedings of Chernobyl-94, Fourth International Conference, *Results of eight years of works on mitigation of consequences of the accident at Chernobyl NPP,* Vol. 1. Zeleny Mys, Chernobyl, pp. 4-20 (in Russian).

DOE. 1996a. *Final Programmatic Environmental Impact Statement for the Uranium Mill Tailings Remedial Action Groundwater Project.* Vol. 1. DOE/EIS-0198, U.S. Department of Energy, Grand Junction Project Office, Colorado.

DOE. 1996b. *Performance Evaluation of the Technical Capabilities of DOE Sites for Disposal of Mixed Low-Level Waste.* DOE/ID 10521/3, U.S. Department of Energy and Sandia National Laboratories, Albuquerque, NM.

Dolin VV, GN Bondarenko, and EV Sobotovich. 1990. "Diffusion Mechanism of Migration of Cesium-137 and Strontium-90 from Fuel Fallout of Chernobyl NPP." Reports of the Ukrainian Academy of Sciences, Series 6, *Geology, Chemical and Biological Sciences.* №12, pp. 6-10 (in Russian).

Eckerman KF, AB Wolbarst, and ACB Richardson. 1988. *Limiting Values of Radionuclide Intake and Air Concentration and Dose Conversion Factors for Inhalation, Submersion and Ingestion.* EPA-520/1-88-020, U.S. Environmental Protection Agency, Washington, DC.

EPA. 1988. *Guidance on Remedial Actions for Contaminated Groundwater at Superfund Sites.* EPA/540/G-88/003, U.S. Environmental Protection Agency, Washington, DC.

EPA. 1991. *Risk assessment Guidance for Superfund, Volume 1: Human Health Evaluation Manual/Supplemental Guidance-Standard Default Exposure Factors.* OSWER Directive 9285.6-03, U.S. Environmental Protection Agency, Washington, DC.

Freeze RA, J Massmann, L Smith, T Sperling, and B James. 1990. "Hydrogeological Decision Analysis: 1. A Framework." *Ground Water*, 28(5):738-766.

Govarets P and F Luykx. 1994. "Restoration of Former Nuclear Sites in Europe: A Historical Review." *Proc. of International Symposium on Remediation and Restoration of Radioactive Contaminated Sites in Europe,* pp. 3-22. European Commission, Antwerp.

Hoeks J. 1981. "Analytical Solutions for Transport of Conservative and Nonconservative Contaminants in Groundwater Systems." *Water, Air and Soil Pollution*, Vol. 16, pp. 339-350.

IAEA. 1997. *Planning and Management Options for Cleanup of Radioactively Contaminated Groundwater.* International Atomic Energy Agency, Vienna, Austria.

ICRP. 1992. *Recommendations of the International Commission on Radiological Protection,* Users edition. ICRP-60, International Commission on Radiological Protection, Pergamon Press, Oxford, United Kingdom.

Il'ichev AV, VD Akhunov, et al. 1992. *Study of Sites of Radioactive Waste Burial, Development of Technologies and Measures for Long-Term Isolation or Excavation of Wastes, Realisation of Radiological and Hydrogeological Monitoring at Waste Burial Sites and Adjoining Territories.* Research Institute of Industrial Technologies. Moscow-Chernobyl, p. 181 (in Russian).

James BR, DD Huff, JR Trabalka, et al. 1996. "Allocation of Environmental Remediation Funds Using Economic Risk-Cost-Benefit Analysis: A Case Study. *Groundwater Monitoring & Remediation.* 16(4)95-105.

Jury WA, DD Focht, and WJ Farmer. 1987. "Evaluation of Pesticide Groundwater Pollution Potential from Standard Indices of Soil-Chemical Adsorption and Biodegradation." *J. Environ. Qual.*, Vol. 16, pp. 422-428.

LeGrand HE. 1982. "Hydrogeologic Issues in Hazardous Waste Management." *Ground Water*, Vol. 20, pp. 488-491.

Massman J and RA Freeze. 1987. "Groundwater Contamination from Waste Management Sites: The Interaction Between Risk-Based Engineering Design and Regulatory Policy. 1. Methodology." *Water Resour. Res.*, Vol. 23, pp. 351-367.

Massmann J, RA Freeze, L Smith, T Sperling, and B James. 1991. "Hydrogeological Decision Analysis: 2. Application to Groundwater Contamination." *Ground Water*, 29(4):536-548.

Mishutina BI, AI Ledenev, and OV Khvesik. 1995. "Evaluation Reliability of Isolation of Radionuclides in Temporary Sites of Radioactive Waste Localisation in Chernobyl NPP Exclusion Zone." *Problems of Chernobyl Exclusion Zone,* № 2, pp. 54-57 (in Russian).

OTA. 1991. *Complex Cleanup: The Environmental Legacy of Nuclear weapons Production*, Appendix B: "Groundwater Contamination." OTA-O-484. Office of Technology Assessment, Washington, DC.

Raucher RL. 1983. "A Conceptual Framework for Measuring the Benefits of Groundwater Protection." *Water Resources Research,* 19(2):320-326.

Rosen L and HE LeGrand. 1997. "An Outline of a Guidance Framework for Assessing Hydrogeological Risks at Early Stages." *Groundwater*, 35(2):195-204.

Skalskiy AS and DA Bugai. 1996. *Hydrogeological Grounding of Urgent Water Protection Measures in Exclusion Zone for Prevention of Radionuclide Transport from Flood-Plain Territories of the 10-km Zone of Chernobyl NPP.* Institute of Geological Sciences, Kiev, p. 162 (in Russian).

Smith L and P Gaganis. 1998. "Strontium-90 Migration to Water Wells at the Chernobyl Nuclear Power Plant: Re-evaluation of Decision Model." *Environmental & Engineering Geoscience*, 4(2):161-174.

Sobotovich EV and GN Bondarenko. 1995. "Problem of Radioactive Substances in Chernobyl NPP Exclusion Zone." *Chernobyl Catastrophe,* VG Baryakhtar, Ed. Naukova Dumka, Kiev, pp. 366-373 (in Russian).

Voitsekhovich OV. 1994. *Scheme of Water Protection Measures to Prevent Radioactive Contamination of Surface Waters and Groundwater in the CNPP Exclusion Zone, Vol. 2, Assessment of Contamination of Geological Medium and Radionuclide Migration Predictions."* Kiev, p. 287 (in Russian).

Voitsekhovich OV (Ed.). 1997. "Radioecology of the Water Bodies Within the Zone of Influence of the Accident at Chernobyl NPP, Vol. 1, Monitoring of Radioactive Contamination of Natural Waters of the Ukraine." *Chernobylinterinform,* Kiev (in Russian).

Chapter 8
Where Do We Go from Here? Construction of the New Safe Confinement

Yasuo Onishi
Washington State University, Richland, WA, USA
Formerly Pacific Northwest National Laboratory, Richland, WA, USA

Mark J. Zheleznyak
Ukrainian Center of Environmental and Water Projects and National Academy of Sciences of Ukraine, Kiev, Ukraine

Oleg V. Voitsekhovich
Ukrainian Hydro-Meteorological Institute, Kiev, Ukraine

8.1 Introduction

The Chernobyl Shelter (formerly called the sarcophagus) consists of the damaged Chernobyl Unit 4 structures and newer structures built after the accident. A 100-m-high, 270-m wide, 144-m-long movable cover called the New Safe Confinement (NSC) will be built over the Shelter (Bechtel et al. 2003, National Geographic 2006). The NSC will lessen the potential for future radionuclide contamination in the environment by reducing or eliminating radionuclide releases to atmosphere and groundwater, provide safer working conditions for dismantling the Shelter and for radioactive waste management, and generate economic and social benefits. It is intended to last for at least 100 years. This chapter examines (1) NSC design requirements of ventilation and insulation to avoid condensation from the air, reduce maintenance needs, and minimize radiation exposure to workers; and (2) potential benefits of the NSC to the Pripyat/Dnieper river system and nearby groundwater through computer simulations. The study indicates that a double roof and a forced-air ventilation rate of 2.5 million cubic meters per day will minimize condensation within the NSC. The evaluation also indicates that the construction of the NSC would further reduce radionuclide contamination in the rivers and nearby groundwater; however, even if the Chernobyl Shelter collapses before the NSC is built, the peak ^{90}Sr and ^{137}Cs concentrations in the Dnieper River would still be below drinking water limits.

Figures 8.1 and 8.2 show Unit 4 before and after construction of the Shelter. Figure 8.3 shows the arch-shaped NSC that will be constructed over the Shelter.

Y. Onishi, O.V. Voitsekhovich and M.J. Zheleznyak (eds.), Chernobyl – What Have We Learned?, 247–278.

Figure 8.1. Chernobyl Unit 4 after the April 26, 1986 accident

Figure 8.2. Chernobyl Shelter that was built after the accident

This chapter describes the NSC conceptual design development and assesses the potential aquatic environment and related health benefits resulting from NSC construction. The funds for designing and constructing the NSC are being provided by the European Bank of Reconstruction and Development, approved by G7 developed and other donor nations.

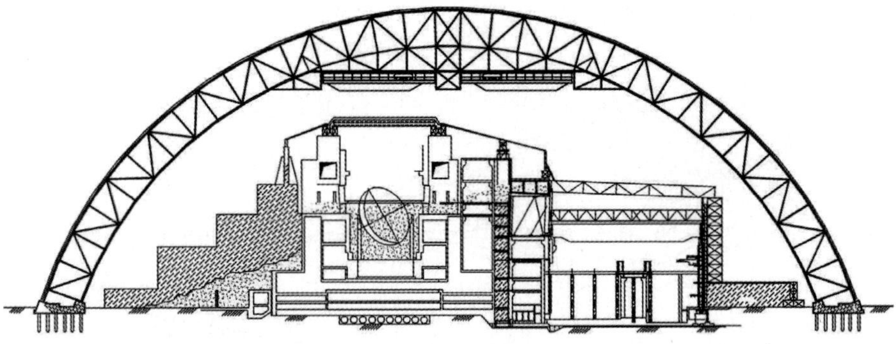

Figure 8.3. New Safe Confinement over the Chernobyl Shelter. North is left on the figure.

8.2 Conceptual Design of the NSC

The arch-shaped NSC will be constructed over the Shelter for Chernobyl Unit 4, as shown in Figure 8.3. The NSC is about 270 m wide, 100 m high, and 166 m long. It is a confinement building, not a containment building like Western nations build over nuclear power plants. To alleviate radiation exposure to construction workers, the NSC will be constructed adjacent to the Shelter and then slid over the Shelter from west to east. It will be the world's largest movable structure. Figure 8.4 shows the NSC construction sequence.

8.2.1 NSC Conceptual Design Evaluation Objective

The NSC is intended to last for about 100 to 400 years. Thus, it is important to design it to minimize required structural maintenance. Reducing air condensation inside the NSC is critical to minimize the maintenance, especially avoiding rusting of NSC's metallic materials, and to reduce seepage of radionuclides to the groundwater environment. The NSC may have the following metal roofs:

- a metal outer roof (if it has two roofs) (see Figure 8.3)
- an inner roof with a metal top straight portion (if it has two roofs)
- an entirely metal roof if it has only one roof.

Approximately 4.5 m^3 of water condenses daily within the Shelter on average, in addition to 5.5 m^3 of precipitation and 0.7 m^3 of dust-control spray when its rooms are not heated, as shown in Figure 8.5. However, once the NSC is constructed, precipitation will be kept out, and all the water, including a pool in Room 001/3 (bottom left of Figure 8.5) will disappear in about 1.5 years due to evaporation and seepage through the concrete floor and walls

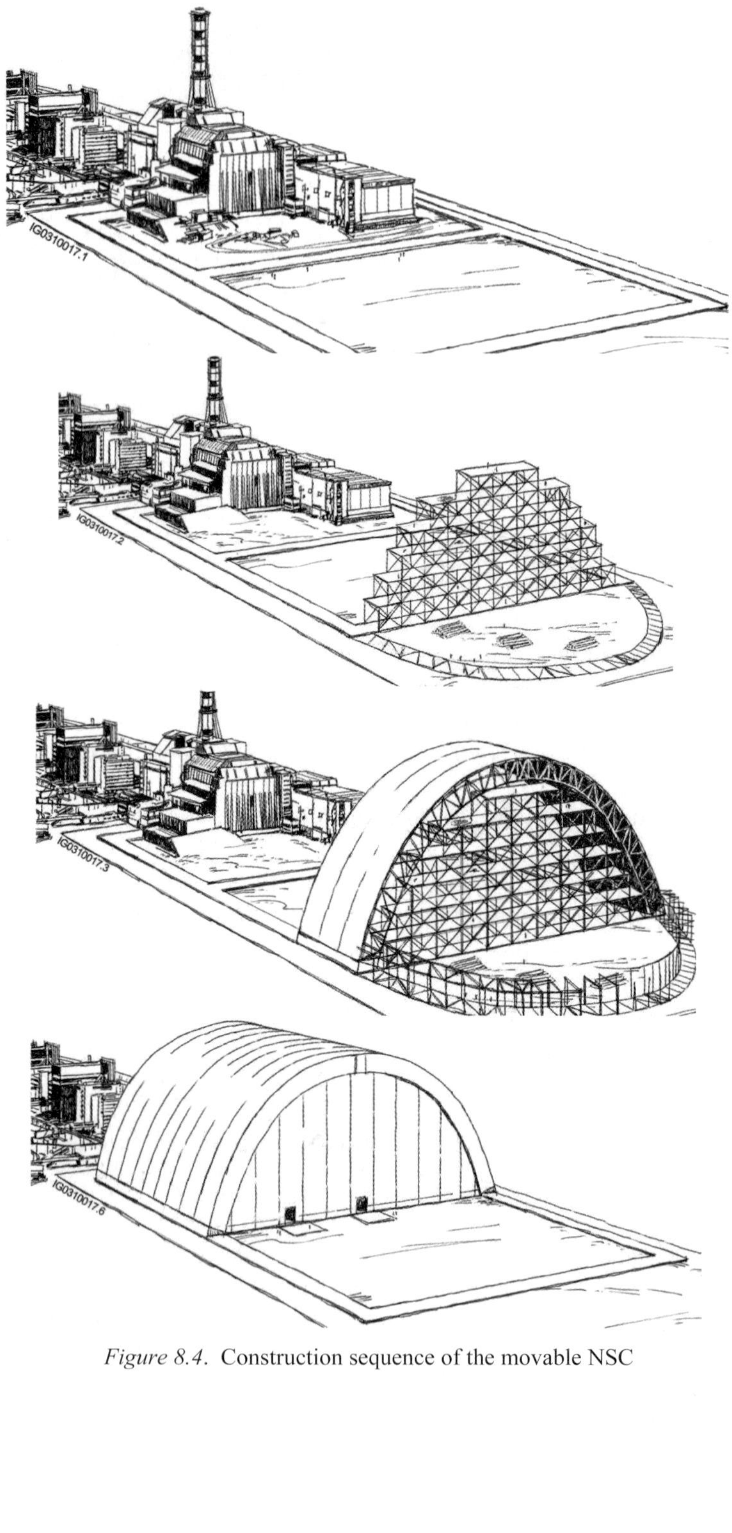

Figure 8.4. Construction sequence of the movable NSC

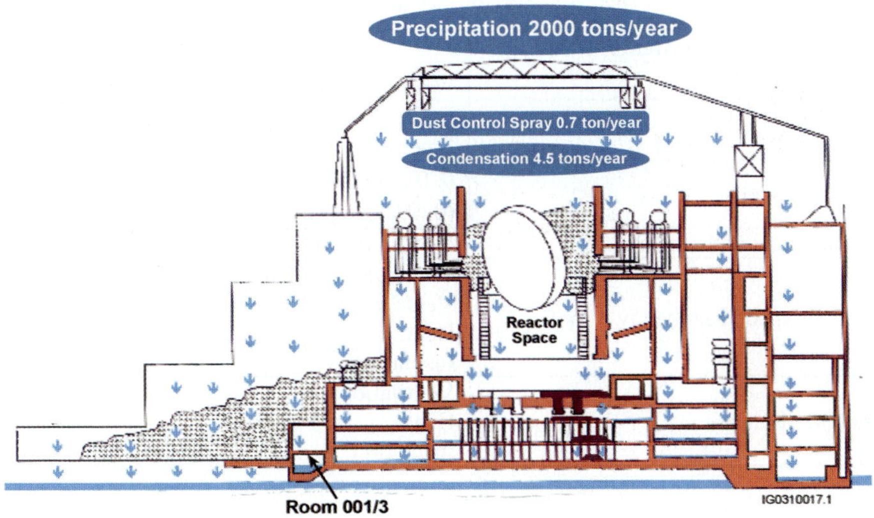

Figure 8.5. Moisture sources and water flow inside the Shelter

(SSE ChNPP 2003). The only expected moisture source will be condensation from the air within the NSC. A key consideration for the conceptual design was reducing condensation from the air inside the NSC to avoid rusting.

The NSC conceptual design evaluation was to determine the possibility of condensation within the NSC (including the Shelter); the amount, type, and location of ventilation; and the amount of insulation needed to avoid condensation, especially on the metal roofs. Specifically, the design questions in Table 8.1 were determined by the modeling. The evaluation also provided information on the radionuclide releases to the environment, which was needed for the environmental impact assessment (Section 8.3). Numerical modeling of air movement and temperature distribution within the NSC was conducted with the commercial computational fluid dynamics code, StarCD®.

8.2.2 Modeling Assumptions and Setup

Two 2-dimensional (2-D) models (one fine-grid, one coarse-grid) and one 3-dimensional (3-D) coarse-grid model were developed for this evaluation. The 2-D models focus on the center of the Shelter (see Figure 8.3), and the 3-D version models the NSC's west side from the outside metal wall to 34 m inside the NSC inside plastic wall (the west side is the front of the NSC; see Figure 8.4). The fine-grid 2-D model has three variations, one with two roofs, one with only the outer roof, and the third with only the inner roof.

Table 8.1. Specific NSC design questions

Design Items	Questions
Ventilation rate	0.25, 1 or 3 NSC volumes per day
Ventilation type	Forced or natural ventilation
Number of NSC roofs	Inner only, outer only, or two roofs
Insulation value, "R"	0, 0.2, or 28.4 hr· ft^2· °F/Btu (0, 0.03, or 5.0 m^2· K/W)
Air inlet and outlet position	Bottom or top inlet
Annular space between two roofs	Connected or segregate Ventilation? Heat? Insulation?
West wall area	Ventilation? Heat? Insulation?

The coarse-grid, 2-D model predicted air movement and temperature distributions similar to those predicted by the fine-grid 2-D model, although the coarse-grid model predicted slightly lower temperatures in some areas. This model comparison indicates that the resolution of the coarse-grid models is fine enough to produce reasonably accurate air flow and temperature distributions. Although there is some east-west air circulation in the NSC (as predicted by the 3-D model), the overall air movement and temperature distributions are 2-D, supporting the current study approach of using three sets of 2-D and 3-D models to evaluate air condensation and ventilation in the NSC in a timely manner.

For a base reference case, the temperature of the solid materials of the Shelter was assigned to be 8° C, the average annual outdoor air temperature. These materials are the 1-m-thick Shelter concrete walls, floors and roofs, soil and concrete piles at the north (left in Figures 8.3 and 8.5) of the reactor building, and a turbine hall south of the reactor building (right in Figure 8.3). In the reactor building, we assigned the surface temperature of the fuel-containing masses to be 37° C for the base case. The temperatures of fuel-containing masses vary both spatially and temporally. The daily temperatures varied from 38° to 43° C in some areas; other areas varied from 5° to 17° C in January through November 2000. Thus we also evaluated a case with fuel-containing masses, wall, and roof surface temperatures of 5° C.

The preliminary assessment indicated that the heat conduction flux between the NSC outer metal roof and outside air would be one to several orders of magnitude greater than solar radiation/cooling heat flux (positive and negative); thus we did not account for solar radiation heating and cooling in the condensation and ventilation assessment. The temperature of the outer surface of the outer roof is assigned to be the same as that of the outdoor air. This maximizes the heat transfer between the NSC roof surface and outdoor air, corresponding to an infinitely large heat transfer coefficient on the outer surface of the roof, and accounts for any strong wind that would enhance the heat transfer of the outer roof.

The measured hourly variations of the outdoor air temperature and moisture content at the Chernobyl plant over one day for each month between 1993 and 2000 were used for the time-varying modeling. July is the warmest month and January the coldest. In July, the highest temperature (23.6° C) typically occurs at noon with relative humidity of 61%. The lowest temperature (14.6° C) usually occurs at about 3 am. When the warm noon air cools to 15.6° C, the excess water vapor starts to condense until the air temperature reaches the minimum of 14.6° C. In January, the highest temperature is –1.7° C with relative humidity of 83%, and the lowest temperature is –4.9° C. The highest-temperature January air is expected to begin condensing its excess water vapor at –4.2° C. The differences between the highest temperatures (12 pm in every month) and corresponding dew points are shown in Table 8.2. This table indicates that January and December have the smallest difference (2.4° C) between the highest temperature and the dew point.

Table 8.2. Difference between highest temperature and dew point

Temp. Diff.	Jan	Feb	Mar	Apr	May	Jun	Jul	Aug	Sept	Oct	Nov	Dec
° C	2.4	3.7	5.5	8.3	10.4	9.2	8.0	8.5	6.2	5.5	2.6	2.4

Although measured data reveal that condensation from air can occur in every month, July conditions produce the greatest amount of water, and January has the smallest temperature drop needed to begin condensation from the air. Thus, we selected these two months as reference case conditions. The coldest recorded air temperatures in the Kiev-Chernobyl area occurred on February 7, 1929 at –32.2° C and January 6, 1935 at –30.8° C. The moisture in the warmest daytime air (–23.7° C) on these two days would have condensed during the night. The highest recorded air temperatures occurred on July 30, 1936 at 39.4° C and August 20, 1946 at 39.3° C. We also examined conditions on these coldest and hottest days.

8.2.3 Evaluation Results

To answer the specific design questions listed in Table 8.1, we predicted air movements and temperature distributions for various combinations of these variables to prevent condensation from inside the NSC, especially on the metal roof surfaces (SSE ChNPP 2003). The model results shown here were obtained after the air temperature reached final periodic diurnal values. The recommended ventilation rate is an example of the modeling procedure.

8.2.3.1 Ventilation rate

An air exchange rate of 0.1 ($2.5x10^5$ m^3/day) or 1 ($2.5x10^6$ m^3/day) NSC volume per day is reasonable for this type of structure based on the U.S. Nuclear Regulatory Commission's air exchange estimate; the ventilation rate may be as high as 1.3 ($3.3x10^6$ m^3/day) NSC volumes per day based on NSC construction information (SSE ChNPP 2003). We tested three ventilation rates using the 2-D fine-grid model with insulation values of R = 0, 0.2, and 28.5 hr· ft^2· °F/Btu (0, 0.03, and 5.0 m^2· °K/W):

- 0.25 NSC volume per day ($6.3x10^5$ m^3/day)
- 1 NSC volume per day ($2.5x10^6$ m^3/day)
- 3 NSC volumes per day ($7.5x10^6$ m^3/day).

Simulations indicate that convection by air movement is the main mechanism controlling air temperature distribution in the NSC. Although all three rates would produce some condensation around the air inlets during the coldest time of day, results show that the ventilation rate of 1 NSC volume per day would produce the least.

At the lower ventilation rate, 0.25 NSC volume per day, a large, very slow air flow would move counter-clockwise above and clockwise to the north of the turbine hall, trapping air as it enters from the bottom inlet. The temperature in these relatively large areas is at or below the dew point, causing condensation. The coldest areas are adjacent to the walls and roofs due to the colder temperature (8° C) of the concrete walls and roofs in July. At this ventilation rate, relatively large areas would experience condensation, especially inside the Shelter and above the turbine hall.

Ventilation of 3 NSC volumes per day was tested with the air inlet at the top and outlet at the bottom for January conditions. The results show that this ventilation rate would bring more cold air into the NSC at night, generating larger cold areas and causing more condensation than 1 NSC volume per day. A ventilation rate of approximately 1 NSC volume per day is recommended.

8.2.3.2 Recommended NSC Conceptual Design

We modeled each of the NSC design questions (Table 8.1) similar to the ventilation rate decision modeling discussed above. The results led to the recommendations for NSC configurations and systems presented in Table 8.3.

Predicted air movement and temperature at the coldest hours of typical January and July days are shown in Figures 8.6 and 8.7 for the recommended NSC configurations and operating systems listed in Table 8.3, except that the ventilation air for the annular space and west wall area was heated 1ºC, not 2ºC, above the outdoor temperature. These results were obtained with the 2-D coarse-grid model with buoyancy effects. The dew points below which condensation occurs are –4.2° C in January and 15.6° C in July. As indicated in the figures, no condensation would occur in the annular space between the roofs and on the metal outer roof and top straight portion of the inner roof. Condensation would occur on the south (right) and north (left) sides inside the inner roof. Simulations indicate that convection by moving air is the main mechanism controlling temperature distribution in the NSC. Air circulation in the annular space, reversing direction at different times of the day, mixes the air and keeps the relatively uniform temperature above the dew point.

A space approximately 1.2 m wide would exist between the outer metal and inner plastic walls on the west side of the NSC. Because the metal wall is exposed to the outdoors, we also evaluated a possible ventilation requirement in this space and the insulation needed to prevent condensation inside the plastic wall and in the annular space between the two roofs. We used the 3-D model with buoyancy effects for this evaluation. The modeling area covered the western portion of the NSC up to 34 m from the west wall.

Insulation of R = 0.2 hr· ft^2· °F/Btu (0.03 m^2· K/W) and ventilation air heated 2° C above the time-varying outdoor air entering the space between the

Table 8.3. Model-guided NSC design decisions

Design Items	**Questions**	**Selections by Model**
Ventilation rate	0.25, 1 or 3 NSC vol. per day	1 NSC vol. per day
Ventilation type	Forced or natural	Forced ventilation
Number of roofs	Inner, outer, or two roofs	Two roofs
Insulation value, "R"	0, 0.2, or 28.4 hr·ft^2·°F/Btu (0, 0.03, or 5.0 m^2·K/W)	R ≥ 0.2 hr·ft^2·°F/Btu (0.03 m^2·K/W)
Air inlet, outlet position	Bottom or top inlet	Bottom inlet, top outlet
Annular space	Connected or separate Ventilation? Heated air? Insulation?	Segregated space 0.1 NSC vol. per day Heated 2°C above outdoor R ≥ 0.2 hr·ft^2·°F/Btu (0.03 m^2·K/W)
West wall area	Ventilation Heated? Insulation?	0.04 NSC vol. per day Heated 2°C above outdoor R ≥ 0.2 hr·ft^2·°F/Btu (0.03 m^2·K/W)

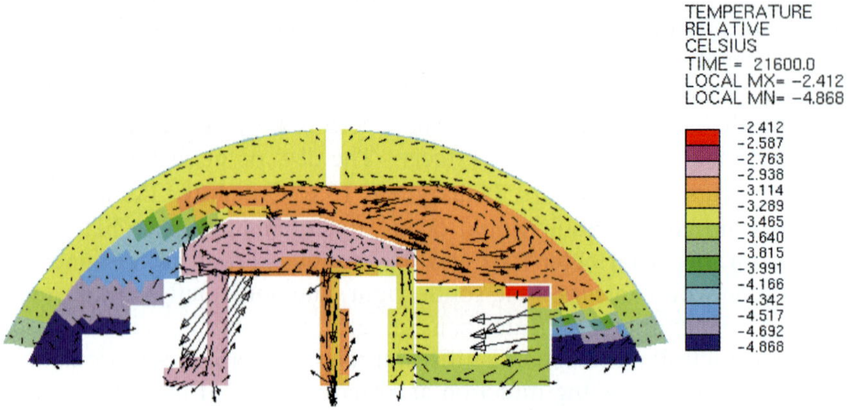

Figure 8.6. Predicted air movement and temperature distribution at 6 am in January

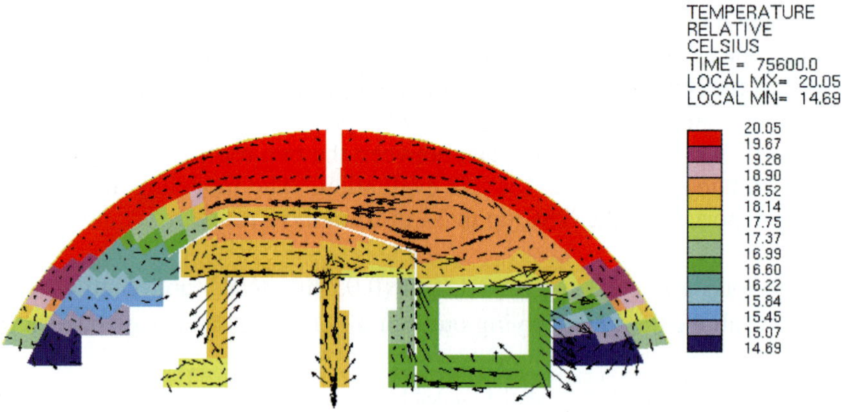

Figure 8.7. Predicted air movement and temperature distribution at 3 am in July

two walls are recommended (see Table 8.3). In this 1.2-m-wide space, ventilation air at 0.04 NSC volume per day ($1.0x10^5$ m^3/d) would enter at the floor and exit at the roof. Figures 8.8 and 8.9 show predicted air flow and temperature distribution at 6 am in January (air heated 1°C above outdoor temperature in this simulation) 0.1 and 1 m, respectively, inside the west wall.

This 3-D modeling at the coldest time of day shows the air circulation. Air flows down the west wall (Figure 8.8), moves east along the floor toward the inner plastic wall, flows up this wall (Figure 8.9), then moves toward the west wall near the top due to the cooling effect of the metal. At 12 pm, this circulation reverses direction, further mixing the air inside the 1.2-m-wide space. Because of this mixing, the air temperature is rather uniform at about –4°C in this space. The model predicted that no condensation would occur in the annular space between the roofs or in this 1.2-m-wide space.

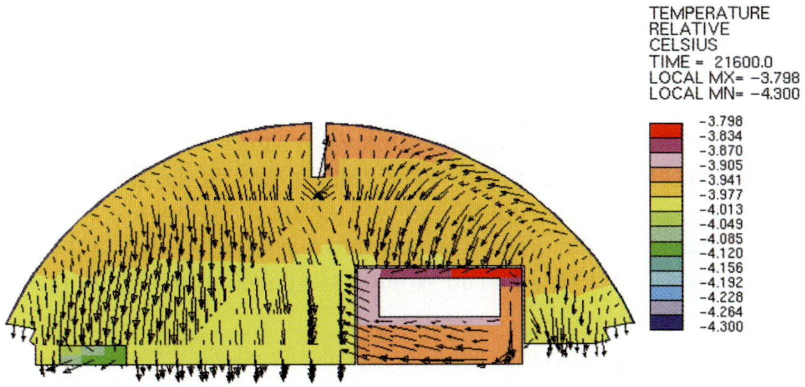

Figure 8.8. Predicted air movement and temperature distribution 0.1 m east of west wall at 6 am in January

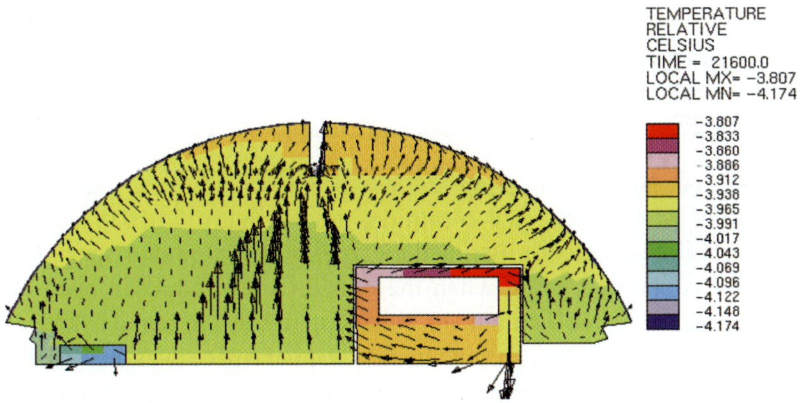

Figure 8.9. Predicted air movement and temperature distribution 1 m east of west wall at 6 am in January

8.2.3.3 Additional test cases

In this section, we discuss the ventilation and insulation assessments under (1) lower surface temperatures of the solid materials of the Shelter, (2) extreme outdoor air temperatures, and (3) no air inflow to the NSC. Except when specifically stated, other modeling conditions are the same as the recommended NSC conceptual design listed in Table 8.3.

The surface temperatures of the fuel-containing masses vary spatially and temporally. Because daily temperatures varied from 38° to 43° C for some measurements and 5° to 17° C for others in Room 305/2 (elevation 9.1 m) between January and November 2000, we conducted additional modeling with the temperature of the Shelter's solid surfaces (fuel-containing masses and walls/floors/roofs) at 5° C as a lower bound. All other conditions matched those in Table 8.3. Simulation results for this case under January conditions indicate that no condensation would occur along the metal roofs.

On average, the air temperatures within a day vary from –4.9 to –1.7° C in January and 14.6 to 23.6° C in July. The coldest days in the Kiev-Chernobyl area for which temperature and humidity data are available are February 7, 1929 and January 6, 1935. The lows on those days, as shown in Table 8.4, were –32.2° C and –30.8° C. The air would have condensed during the night.

Table 8.4. Temperature and humidity on coldest days in the Kiev-Chernobyl area

	February 7, 1929		**January 6, 1935**	
	Temperature, ° C	**Humidity, %**	**Temperature, ° C**	**Humidity, %**
7 am	-31.3	77	-29.5	87
1 pm	-24.4	69	-27.6	89
9 pm	-26.7	77	-28.2	84
Daily avg	-27.4		-28.4	87
Maximum	-23.7		-25.6	
Minimum	-32.2		-30.8	

We conducted modeling to assess the ventilation, condensation, and insulation of the NSC for conditions on February 7, 1929 and January 6, 1935. In both cases, we assigned air inflow to the annular space to be heated 2° C above the dew point when the outside temperature dropped below that. Otherwise, unheated outdoor air was supplied to the annular space. For the February 7, 1929 case, inflow to the annular space was heated to –27.5° C when the outside air temperature dropped below this temperature. For the January 6, 1935 case the air was heated to –25.8° C. We simulated these coldest-day conditions for three days, though they occurred on one day each: February 7, 1929 and January 6, 1935.

On February 7, 1929, the predicted air temperatures in the annular space were –27° to –27.5° C at 6 am, above the dew point of –29.5° C due to active mixing (shown in Figure 8.10). Thus there would be no condensation in the annular space. However, condensation would occur in a significant portion of the southern and lower northern areas of the NSC inside the inner roof because air temperatures there would drop as low as –32.4° C.

On January 6, 1935, the predicted air temperature in the annular space was about –27.2° C, slightly above the dew point of –27.8° C, at 5 am (the coldest time on that day). Thus, no condensation was expected in the annular

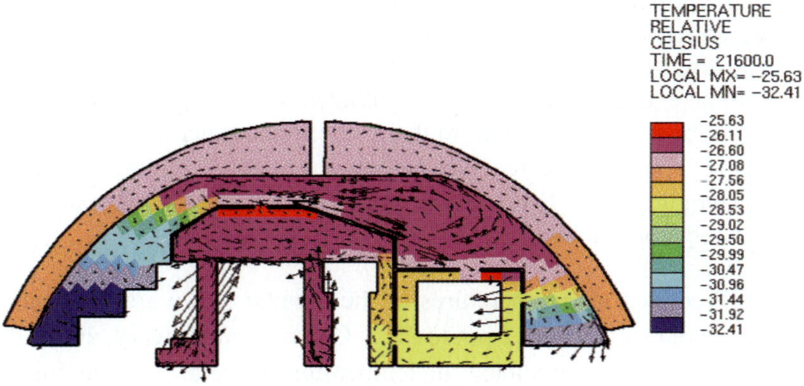

Figure 8.10. Predicted air movement and temperature distributions at 6 a.m. on February 7, 1929 for R = 0.2 hr•ft2•°F/Btu (0.03 m^2•K/W)

space. Similar to the February 7, 1929 case, condensation would occur in both southern and northern areas of the NSC inside the inner roof.

The hottest days in the area were July 30, 1936 and August 20, 1946. The maximum temperatures on these days were 39.4° C and 39.3° C, respectively (Table 8.5). On July 30, 1936, if we conservatively assume that the humidity of the air is 46% (daily average), condensation would occur at 25.6° C, near the lowest temperature of the day. Rain was recorded at 7 pm, when the temperature was 25° C. Because air in the annular space of the NSC is relatively well mixed, no condensation is expected to occur under July 30, 1936 conditions. On August 20, 1946, the lowest temperature (22.4° C) was well above the dew point, so no condensation was expected. Thus, even on the four coldest and hottest days on record, condensation would not occur along the metal outer roof and the straight metal portion of the inner roof.

Table 8.5. Temperature and humidity on the hottest days in the Kiev-Chernobyl area

	July 30, 1936		**August 20, 1946**	
	Temperature, °C	**Humidity, %**	**Temperature, °C**	**Humidity, %**
1 am	27.9	43	26.3	41
7 am	27.9	45	25.9	39
1 pm	38.9	22	37.8	21
7 pm	25.0	73 (rain)	33.4	25
Daily avg	29.9	46	30.8	31.5
Maximum	39.4		39.3	
Minimum	24.8		22.4	

Some adverse weather conditions could cause condensation on these metal roofs. Because the air temperature distributions are strongly affected by the convection induced by the ventilation air and buoyancy-driven airflow, adjusting the timing and amount of air ventilation inflows to the NSC may minimize condensation. We evaluated this situation, assuming that there would be no ventilation air to the inside of the inner roof but the annular space would continue to receive airflows heated 1° C above the outdoor air temperature at 0.1 NSC volume per day. We assessed the case under January and July conditions.

Predicted air temperatures in the annular space are very uniform over a day, about –2.5° C at 12 pm to –2.7° C at 6 am in January, above the due point of –4.3° C. The predicted air temperature there at 3 am in July (the coldest time of day) varies from 18.2° C to 20.6° C, both of which are also above the dew point of 15.6° C. Thus, there is no condensation along the outer roof or the metal portion of the inner roof. However, because the Shelter structures are cooler than the air in summer; the air near the floor at the north and south ends of the NSC and in the reactor building is as cool as 14.2° C. Thus, some condensation would occur in those areas in July if no ventilation air was supplied to the inside of the inner roof for three days or more. These simulation results indicate that the recommended NSC design listed in Table 8.3 avoids condensation along the NSC's metal roof surfaces. The final NSC conceptual design follows these recommendations closely.

8.3 Effects of NSC on Aquatic Environment

The NSC will be constructed over the Shelter to reduce the potential for future radionuclide contamination of the environment, as discussed above. In this section we evaluate the potential benefits of the NSC on the surface- and groundwater environments.

Approximately three million Kiev residents drink water from the Dnieper River, and up to 20 million Ukrainians eat foods irrigated with Dnieper River water. Therefore, the aquatic pathway was analyzed in detail for potential risk in the far field associated with various NSC scenarios. We predicted radionuclide concentrations in (1) the Pripyat and Dnieper Rivers if the Shelter and/or the NSC should collapse, and (2) groundwater seeping into the Pripyat River when the NSC is constructed. Estimated radionuclide concentrations were used to evaluate potential human health impacts through the aquatic pathways with and without the NSC (Ramsdell et al. 2003).

8.3.1 Effects of the NSC on the Pripyat and Dnieper Rivers

8.3.1.1 Model assumptions and setup

The Chernobyl nuclear plant and cooling pond are seen alongside the Pripyat River in Figure 8.11. The plant is the white cluster at the northwest corner (top left) of the cooling pond just below the river. A dike (thick white line) along the east (upper) river bank separates the pond from the river. The Pripyat joins the Dnieper at the upstream end of Kiev Reservoir above the city of Kiev. The Dnieper discharges into the Black Sea after passing through six dams and reservoirs: Kiev, Kanev, Kremenchug, Dneprodzerzhinsk, Dnieprovskoe, and Kahovka, in this sequence (Figure 8.12).

Because the Dnieper River is the main aquatic pathway for radionuclides, the radionuclide concentrations in the river were determined after atmospheric fallout on the surface of the river and floodplain near the NSC. Radionuclides deposited on the floodplain were washed into the river by flooding and rainfall. The following four accident scenarios were evaluated:

- Scenario 1: The Shelter collapses without the NSC
- Scenario 2: The NSC collapses over the Shelter
- Scenario 3: The Shelter collapses inside NSC; NSC ventilation rate is 1 NSC volume per day (see Section 8.2)

Figure 8.11. Satellite image of the Chernobyl plant and the Pripyat River

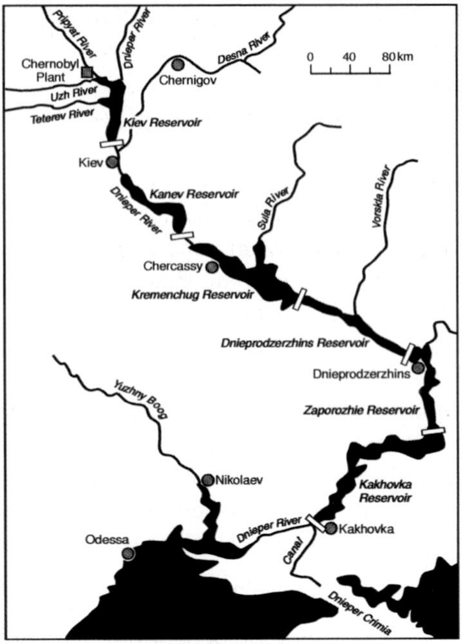

Figure 8.12. Dnieper River and its cascade of reservoirs

- Scenario 4: The Shelter collapses inside NSC; NSC ventilation rate is 0.1 NSC volume per day.

Scenario 1 corresponds to a condition where the Shelter collapses before the NRC is constructed. In Scenario 2, the NSC collapses while being placed over the Shelter, and the impact destroys the Shelter. In Scenarios 3 and 4, the Shelter collapses under the NSC while the NSC remains intact and operating normally with air ventilation rates of 1.0 and 0.1 NSC volume per day ($2.5x10^6$ and $2.5x10^5$ m^3/day), respectively. The ventilation rate for Scenario 3 is the recommended value to minimize condensation inside the NSC, as discussed in Section 8.2.

Key radionuclides for aquatic pathways are ^{90}Sr, ^{137}Cs, ^{238}Pu, $^{239\text{-}240}$Pu, and ^{241}Am, as discussed in previous chapters. Atmospheric dispersion and deposition of these radionuclides under the four scenarios were determined by a Gaussian dispersion model (Hemond and Fechner-Levy 2000) modified to reflect the initial dimension of the cloud and its depletion by radionuclide deposition on the ground (Ramsdell et al. 2003). Solids deposition was estimated by deposition velocities varying with particle size, density, and wind speed. If the Shelter should collapse, the dust is assumed to contain 1-μm radioactive particles accounting for 70% of the total released activity and 10-μm particles for the other 30%. Particle densities were assigned to be 6 and 10 g/mL for 50% each of these two particle-size fractions. The median atmospheric conditions (50th percentile) in the Chernobyl area are wind speed of 3 m/s and Pasquill atmospheric stability category of D (neutral).

Atmospheric stability of F (very stable) was selected for the assessment to limit atmospheric dispersion and maximize fallout within the 10-km zone. The wind direction was selected to maximize radionuclide deposition directly on the surface of the Pripyat River to produce worst-case hydrological conditions. Figure 8.13 shows the predicted ^{90}Sr distribution under Scenarios 1 and 2. Figure 8.14 presents ^{90}Sr surface contamination without the accident.

Table 8.6 shows the calculated ^{90}Sr and ^{137}Cs fallout in the Pripyat River and on its floodplain southwest of (below) the dike (Ramsdell et al. 2003), which is shown as a thick white line in Figure 8.11. Estimated total deposition of ^{238}Pu, $^{239-240}$Pu, and ^{241}Am in the river and floodplain are given in Table 8.7.

As shown in these tables, Scenario 1 (Shelter collapse without NSC) and Scenario 2 (NSC collapse over Shelter) have the same amount of radionuclide fallout. Thus, the Dnieper River assessments under Scenarios 1 and 2 are identical. Scenarios 3 and 4 (the Shelter collapses inside the NSC) have about 15 and 150 times less fallout, respectively, than Scenarios 1 and 2.

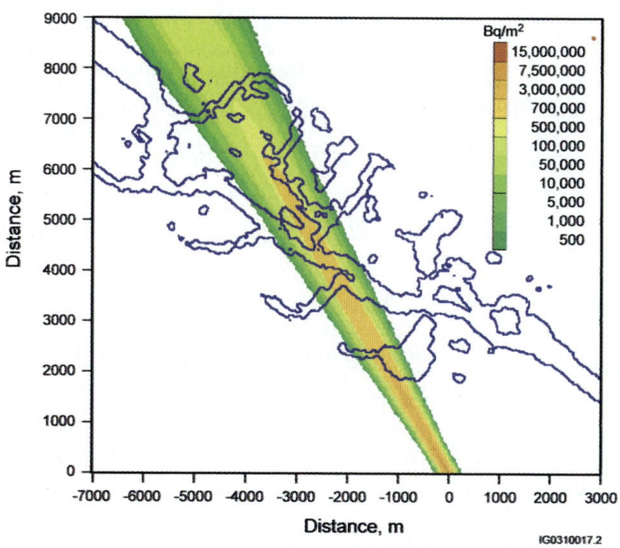

Figure 8.13. Predicted ^{90}Sr deposition for Scenarios 1 and 2

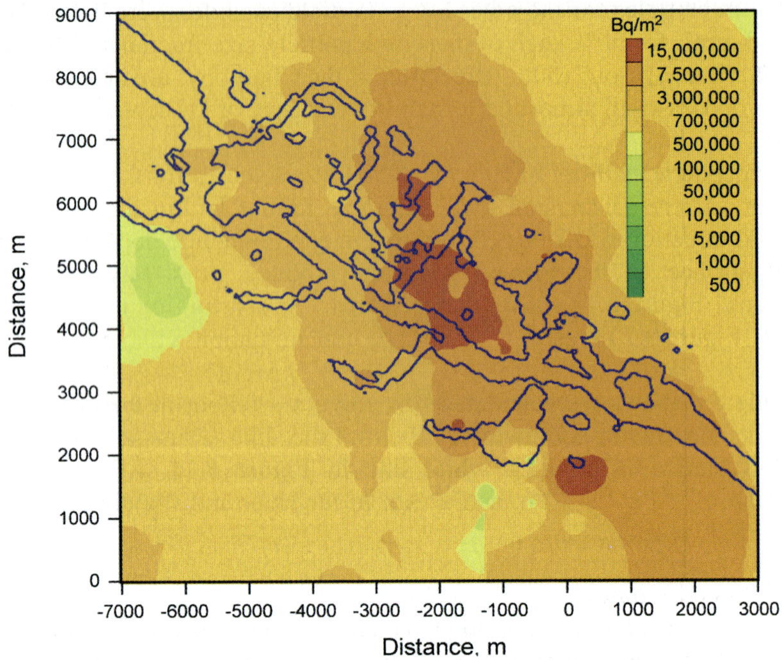

Figure 8.14. ^{90}Sr concentration near the Chernobyl Nuclear Power Plant

Table 8.6. Calculated ^{90}Sr and ^{137}Cs deposited on the Pripyat River and floodplain

Deposition Areas	**Radionuclide Deposition Amount, 10^{12} Bq**							
	Scenario 1		Scenario 2		Scenario 3		Scenario 4	
	^{90}Sr	^{137}Cs	^{90}Sr	^{137}Cs	^{90}Sr	^{137}Cs	^{90}Sr	^{137}Cs
River	0.776	1.74	0.776	1.74	0.053	0.114	0.005	0.011
Floodplain	2.964	6.28	2.964	6.28	0.203	0.414	0.020	0.041
Total	3.74	8.02	3.74	8.02	0.256	0.528	0.025	0.052

Table 8.7. Calculated total ^{238}Pu, $^{239\text{-}240}Pu$, and ^{241}Am deposited in Pripyat River and floodplain

Radionuclide	**Radionuclide Deposition Amount, 10^{10} Bq**			
	Scenario 1	Scenario 2	Scenario 3	Scenario 4
^{238}Pu	1.60	1.60	0.081	0.0081
$^{239\text{-}240}Pu$	4.05	4.06	0.206	0.0206
^{241}Am	5.15	5.15	0.262	0.0262

Shelter collapse was arbitrarily assigned to occur at noon to track the simulation time. We assumed that all radionuclides deposited directly on the Pripyat River were diluted by river water in the first hour after the accident. We conservatively assumed that 10% of the radionuclides deposited on the floodplain were washed into the river by flooding during the next two days and 11 hours (Zheleznyak et al. 1992). We assigned rainfall of 50 mm in the first 2.5 days; average monthly precipitation is 41 mm in this area. Based on measured ^{90}Sr wash-off by rainfall (Voitsekhovich et al. 1990), 1% of the ^{90}Sr was assumed washed into the river by rain. The ^{137}Cs wash-off was also conservatively assigned as 1%. Total radionuclides washed out by flooding (10%) and rainfall (1%) were distributed as 60% during the remaining 11 hours of the first day, 30% the second day, and 10% the third day. Similar assumptions were made for transuranics.

We used the one-dimensional surface water code RIVTOX (Zheleznyak et al. 1992) to simulate radionuclide migration in the Pripyat and Dnieper rivers under Scenarios 1 and 3. This code simulates river flow, sediment migration (transport, deposition, and resuspension), and migration of dissolved and sediment-sorbed radionuclides, accounting for sediment and radionuclide interactions (i.e., adsorption, deposition, and resuspension of contaminated sediments).

8.3.1.2 Model results

There are six reservoirs on the Dnieper River between its confluence with the Pripyat and its discharge to the Black Sea: Kiev, Kanev, Kremenchug, Dneprodzerzhinsk, Dnieprovskoe, and Kahovka, in that order (Figure 8.12). We used 1999–2001 hydrologic conditions for this modeling.

Time-varying simulation results of ^{90}Sr concentrations at the hydropower plant dams of the six reservoirs are presented in Figure 8.15 for Scenarios 1 and 2. These simulation results reflect the combined effects of the accidental

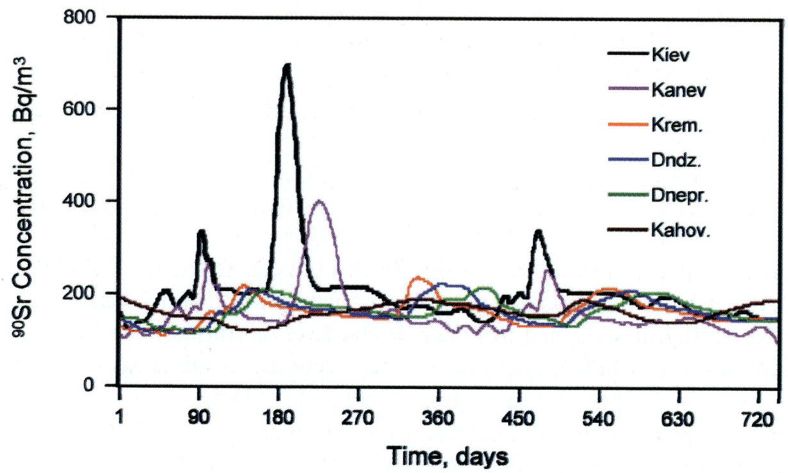

Figure 8.15. Predicted ^{90}Sr concentrations at six hydropower plants under Scenarios 1 and 2

radionuclide release and natural conditions. Higher ^{90}Sr concentrations are often associated with spring floods (Voitsekhovich 1994). As shown in this figure, ^{90}Sr reduces its concentration as it travels toward the Black Sea. The predicted peak ^{90}Sr concentration at Kiev hydropower plant is 685 Bq/m^3 on Day 191. Without the radionuclide influx due to the Shelter accident, the ^{90}Sr concentration would be 190 Bq/m^3. Thus, this hypothetical Shelter collapse would add 495 Bq/m^3 to the natural level at that time. The 685-Bq/m^3 peak concentration is reduced as ^{90}Sr migrates toward the Black Sea.

The peak ^{90}Sr concentration is reduced to 389 Bq/m^3 at Kanev hydropower plant due to the dilution from the large Desna River, which joins the Dnieper just below Kiev Reservoir (see Figure 8.12). The peak concentration of 225 Bq/m^3 arrives at Kremenchug hydropower plant more than 146 days after passing through Kiev Reservoir. The highest ^{90}Sr concentration at the last and largest reservoir, Kakhovka, arrives 330 days after passing through Kiev Reservoir. The peak level is 178 Bq/m^3, only 40 Bq/m^3 greater than the peak concentration under natural conditions. The Ukrainian drinking water limit for ^{90}Sr is 2,000 Bq/m^3 (DU-97), so even at the Kiev hydropower plant the peak concentration under Scenarios 1 and 2 is 35% of the limit.

The ^{137}Cs peak concentrations caused by Shelter collapse under Scenarios 1 and 2 are significant only in the Kiev Reservoir (455 Bq/m^3 in solute and 175 Bq/m^3 in suspended sediments) and Kanev Reservoir (137 Bq/m^3 in solute and 74 Bq/m^3 in suspended sediments). Further downstream at the Kremenchug Reservoir, the predicted peak concentration is only 5 Bq/m^3 above the 1999–2001 natural conditions. This significant ^{137}Cs concentration reduction is due to dilution, interaction between ^{137}Cs and the river sediment, and deposition of suspended sediment-sorbed ^{137}Cs. At the Kakhovka hydropower plant, the difference between Scenarios 1 and 2 and natural conditions is less than 1 Bq/m^3, indicating there is practically no effect of Shelter collapse on this, the largest reservoir. The Ukrainian drinking water standard for ^{137}Cs is 2,000 Bq/m^3 (DU-97). Thus, similar to ^{90}Sr, even at the Kiev and Kanev hydropower plants, the peak ^{137}Cs concentrations under Scenarios 1 and 2 are 32% and 11% of the drinking water limit, respectively.

Based on these simulation results, peak concentrations of ^{238}Pu, $^{239/240}Pu$, and ^{241}Am at Kiev hydropower plant under Scenarios 1 and 2 were estimated to be no higher than 1 Bq/m^3 each. Because Ukrainian limits for these radionuclides are 1,000 Bq/m^3 each, the expected peak concentrations are much less than the maximum permissible levels. Thus, radionuclide fallout from the collapsed Shelter considered under Scenarios 1 and 2 would have no significant effects on the Dnieper River water.

Scenario 3 addresses Shelter collapse inside the NSC with a ventilation rate of 1 NSC volume per day releasing radionuclides to the outside. The radionuclide influx for this case is 5~7% of that under Scenarios 1 and 2 (see Tables 8.6 and 8.7). Thus, Shelter collapse would have much less impact on the river than the accidents in Scenarios 1 and 2. Predicted ^{90}Sr concentrations at Kiev Reservoir with and without the Shelter accident under Scenario 3 are shown in Figure 8.16.

The ^{90}Sr concentration difference at Kiev hydropower plant between Scenario 3 (the purple line) and natural conditions (the blue line) is almost non-existent, as these two concentration lines overlap in this figure. The difference appears only in around the sixth month (180th day). At that time, the predicted concentration is 230 Bq/m^3, 46 Bq/m^3 (25%) greater than natural conditions of 184 Bq/m^3. This concentration is only 11.5% of the Ukrainian drinking water limit. At Kremenchug hydropower plant the maximum influence of the hypothetical accident on ^{90}Sr concentration, 148.8 Bq/m^3, occurs in the 11th month, exceeding natural conditions of 140.2 Bq/m^3 by 8.6 Bq/m^3 (6%). Downstream of the Kremenchug hydropower plant the radionuclide release has no effect on ^{90}Sr concentration in the Dnieper River. The average annual ^{90}Sr concentration in the Kiev Reservoir is 190 Bq/m^3 in the first year in Scenario 3, exceeding natural conditions of 188 Bq/m^3 by only 2 Bq/m^3 (2%).

Shelter collapse under the NSC has some short-term effects on ^{137}Cs concentrations, especially in Kiev Reservoir. The maximum effect from Shelter collapse inside the NSC under Scenario 3 is an increase of 23 Bq/m^3 (91%) in the ^{137}Cs level to 44 Bq/m^3 over the natural condition of 21 Bq/m^3. However, this concentration is only 2% of the Ukrainian drinking water limit.

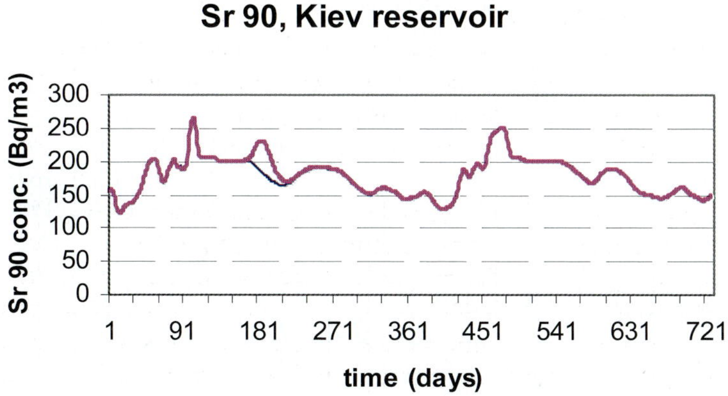

Figure 8.16. Predicted ^{90}Sr concentrations at Kiev hydropower plant under Scenario 3 (pink line) and 1999–2001 natural conditions (blue line)

At the Kremenchug hydropower plant, the maximum influence of the hypothetical accident on ^{137}Cs concentrations occurs in the 11th month. At that time, the ^{137}Cs concentration is 8.35 Bq/m^3 under Scenario 3, exceeding the natural condition of 7.82 Bq/m^3 by 0.53 Bq/m^3 (7%). Similar to ^{90}Sr, there are no effects downstream of the Kremenchug hydropower plant from the radionuclide releases considered under Scenario 3. There would also be no measurable effects of ^{239}Pu, $^{239\text{-}240}$Pu, and ^{241}Am on the Dnieper River.

In Scenario 4 the Shelter collapses inside the NSC with a ventilation rate of 10% NSC volume per day. As shown in Tables 8.6 and 8.7., radionuclide releases under Scenario 4 are about 0.5% of Scenarios 1 and 2 (without the NSC) and 10% of Scenario 3. Expected ^{90}Sr and ^{137}Cs concentrations would be just 1% and 9% above natural conditions. There would be no measurable impacts to the river from the transuranics in Scenario 4.

8.3.1.3 Aquatic pathways

Radionuclide dose estimates are based on average annual concentrations. Table 8.8 shows the predicted ^{90}Sr and ^{137}Cs concentrations at the hydropower plants of six reservoirs averaged over each year. As shown in this table, both ^{90}Sr and ^{137}Cs concentrations at all six reservoirs are well below the drinking water limits of 2,000 Bq/m^3.

The Shelter collapse considered under Scenarios 1 and 2 would increase the average annual ^{90}Sr concentration at Kiev hydropower plant by 24 Bq/m^3 (or 12.5 percent) over natural conditions in the first year and by 17 to 20 Bq/m^3 (12–14 percent) at Dnieprovskoe and Kakhovka hydropower plants

Table 8.8. Predicted average annual ^{90}Sr and ^{137}Cs concentrations in the Dnieper reservoirs

Reservoirs	**Year**	**^{90}Sr Concentrations (Bq/m^3)**				**^{137}Cs Concentrations (Bq/m^3)**			
		Scen. 1&2	**Scen. 3**	**Scen. 4**	**Nat. Cond.**	**Scen. 1&2**	**Scen. 3**	**Scen. 4**	**Nat. Cond.**
Kiev	1	211	190	188	188	42.1	23.6	22.5	22.5
	2	190	184	184	184	31.7	21.5	20.8	20.8
Kanev	1	162	149	147	147	24.2	17.3	16.9	16.9
	2	148	142	141	141	25.3	16.4	15.8	15.8
Kremenchug	1	153	147	146	146	9.0	8.3	8.3	8.3
	2	156	143	142	142	13.9	10.0	9.7	9.7
Dnieprodzer	1	150	146	146	146	7.5	7.1	7.1	7.1
	2	159	142	142	142	11.5	8.5	8.5	8.5
Zaporozhie	1	146	146	146	146	5.8	5.8	5.8	5.8
	2	162	142	142	142	10.1	7.5	7.5	7.5
Kakhovka	1	145	145	145	145	1.5	1.5	1.5	1.5
	2	159	143	143	143	3.2	2.8	2.8	2.8

in the second year. For ^{137}Cs, significant effects of Shelter collapse appear only in the Kiev and Kanev reservoirs. At Kiev Reservoir, the average annual ^{137}Cs concentration is 19.6 Bq/m^3 (87 percent) higher than natural conditions in the first simulation year, while at Kanev Reservoir the concentration exceeds 7.3 Bq/m^3 (73 percent) over natural conditions. In the second simulation year, due to radionuclide exchanges (adsorption and desorption) with contaminated sediment at the river bottom, the annual average ^{137}Cs concentration is still 50–60 percent greater than natural conditions in Kiev and Kanev reservoirs.

Under Scenario 3, the annual average of ^{90}Sr concentrations in Kiev Reservoir is 190 Bq/m^3 in the first year, exceeding natural conditions by only 2 Bq/m^3 (or 2 percent). For ^{137}Cs, the largest difference between Scenario 3 and natural conditions occurs in Kiev Reservoir in the first year. The average annual ^{137}Cs concentration in Kiev Reservoir is 23.6 Bq/m^3 in the first year under this scenario, exceeding natural conditions by only 1.1 Bq/m^3 (5 percent). Thus, Table 8.8 indicates that there is no measurable increase in ^{90}Sr and ^{137}Cs concentrations under Scenario 4, even in Kiev Reservoir.

As indicated above, the Dnieper River provides aquatic pathways for these radionuclides to reach Ukrainians mostly through

- drinking water
- water sports and recreation
- consumption of fish caught in these six reservoirs
- consumption of produce irrigated with Dnieper River water.

Ramsdell et al. (2003) estimated radiation doses from the Dnieper River to determine potential health impacts. They selected three groups: professional fishermen, local residents who occasionally eat fish, and farmers who grow and eat irrigated produce. Their estimates indicate that a professional fisherman would receive the highest dose, mostly from eating fish, especially those caught in Kiev Reservoir. Each of these fishermen would receive up to 47 μSv of radiation effective dose in the first year, resulting in 2.4×10^{-6} deaths from cancer. In the second year, fisherman would receive a smaller dose of 24 μSv. Fishermen living by the other five reservoirs would receive a higher radiation dose in the second year but still smaller than near Kiev Reservoir.

The average local resident would receive a smaller radiation dose than a fisherman but more than the farmer. People living near Kiev Reservoir would get the highest dose, mostly from drinking Dnieper River water. These local residents would receive the maximum dose of 1.1 μSv in the first year. This translates to a 6×10^{-8} risk of cancer death. Similar to fishermen, those living near the other five reservoirs would get higher doses in the second year but still less than Kiev residents. Among irrigation farmers, those living near Kremenchug Reservoir would get the maximum dose in the second year, up to 0.021 μSv. This dose would cause a risk of cancer death of $1x10^{-9}$. These dose levels indicate that the aquatic pathways from Shelter collapse before or during NSC construction (under Scenarios 1 and 2) would not be a significant link to the majority of the population.

8.3.2 Effects of the NSC on Groundwater

8.3.2.1 Model assumptions and setup

The force driving radionuclides from the Shelter to the subsurface is the water inside the Shelter. The amount of water from precipitation, condensation, and spraying for dust control exceeds evaporation. Much of this excess water flows into Room 001/3 at 2.3~3.6 m^3/day on average. This room, shown near the bottom left corner of the Shelter in Figure 8.5, has a drainage hole in the wall, 1.4 m above the floor, that maintains the water depth at 1.4 m. The pool contains $1.0x10^9$, $5.2x10^9$, and 360 Bq/m^3 of ^{90}Sr, ^{137}Cs, and ^{239}Pu, respectively. The water filters through the 1-m-thick concrete floor to the vadose zone beneath and eventually reaches groundwater, carrying radionuclides toward the Pripyat River. One of the main benefits of the NSC is eliminating this excess water inside the Shelter, removing the driving force transporting the radionuclides from the Shelter to the subsurface environment.

The migration of ^{90}Sr, ^{137}Cs, and ^{239}Pu in the subsurface water (vadose zone and groundwater) from the Shelter (through the concrete floor and walls) to the subsurface was simulated. The groundwater moves toward the Pripyat River, 2.5 km away. The time-varying, 2-D, vertical profile code SUSTOX (Kivva 1997) was used for the evaluation by accounting for nonequilibrium radionuclide adsorption/desorption processes. The 2-D modeling is conservative (predicts higher radionuclide concentrations) because it does not account for lateral dispersion and dilution that occur in actual conditions.

The hydraulic head difference between the Chernobyl cooling pond and the Pripyat River is about 6 m and is the main driving force of groundwater in the area (see Figure 8.11). The water level in the cooling pond will be reduced to river level after the power plant discharge is terminated. This would slow groundwater flow toward the river significantly. Our analysis conservatively used the current water level in the Chernobyl cooling pond.

8.3.2.2 Scenarios 5 and 6

In Scenario 5 (Shelter without NSC) the radionuclide contamination of the subsurface environment is due to water infiltration from Room 001/3 under current hydrological conditions. The water was fixed at 1.4 m deep, or 112.8 m above Baltic Sea level. Water in the incoming channel and Pripyat River were assigned as 109.82 and 104.2 m above Baltic Sea level, respectively. Scenario 6 considers the Shelter with the NSC. After NSC construction, precipitation influx to the Shelter would stop, and evaporation would be greater than influx from dust suppression and condensation. The water level in Room 001/3 would diminish because of evaporation and seepage through the walls and floor. Seepage was simulated as a part of the subsurface water and radionuclide transport modeling; results indicate the pool in Room 001/3 would disappear in 1½ years.

The effect of NSC excavation and construction was simulated by imposing a more permeable barrier and the NSC pilings. The NSC was assumed to last 100 years. By then all radionuclides would be removed from the Shelter. Figure 8.17 shows the vertical profile of the region modeled. The Shelter and Pripyat River are at 1,000 m and 3,500 m, respectively. Table 8.9 lists subsurface model parameters.

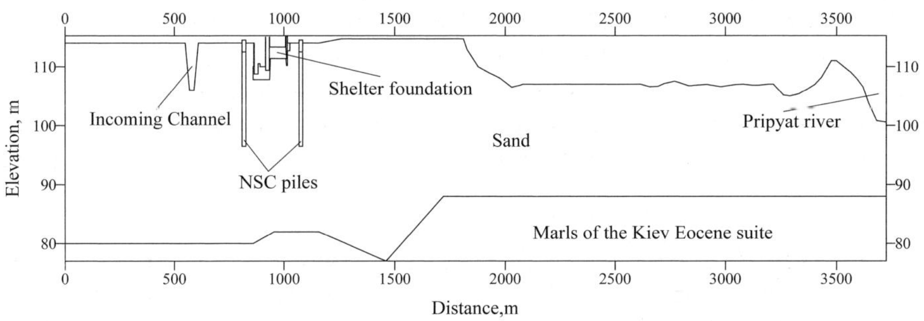

Figure 8.17. Scenario 6 model region with Shelter and NSC foundation

Table 8.9. Subsurface model parameters(a)

Types	n	ρ_S kg/L	α_L m	α_T m	K m/day	^{137}Cs			^{90}Sr	^{239}Pu
						K_d L/kg	α_{SF} 1/day	α_{FS} 1/day	K_d L/kg	K_d L/kg
Sand	0.40	2.66	1.2	0.1	10.0	2.0	$1x10^{-3}$	$5x10^{-5}$	1.0	30
Concrete	0.08	2.67	1.0	0.1	$8.65x10^{-4}$	2.0	$1x10^{-3}$	$5x10^{-5}$	1.0	30
Pilings	0.24	2.66	1.2	0.1	3.0	2.0	$1x10^{-3}$	$5x10^{-5}$	1.0	30

(a) n = porosity; ρ_S = solid density; α_L = longitudinal dispersivity; α_T = transverse dispersivity; K = hydraulic conductivity; K_d = distribution coefficient; α_{SF} = irreversible adsorption rate, and α_{FS} = desorption rate.

8.3.2.3 Model results

Simulation of ^{90}Sr for Scenario 5 (Shelter without NSC) reveals its slow movement. After 100 years, ^{90}Sr, with its concentration above $4x10^9$ Bq/m^3, has moved less than 100 m from the Shelter (Figure 8.18). The 100 Bq/m^3 level is reached 600 m from the Shelter, a fraction of the 2.5-km distance to the Pripyat River. Figure 8.19 shows the sensitivity analysis on the radionuclide migration by depicting a series of breakthrough curves at the Pripyat River with longitudinal dispersivity values ranging from 0.2 to 10 m.

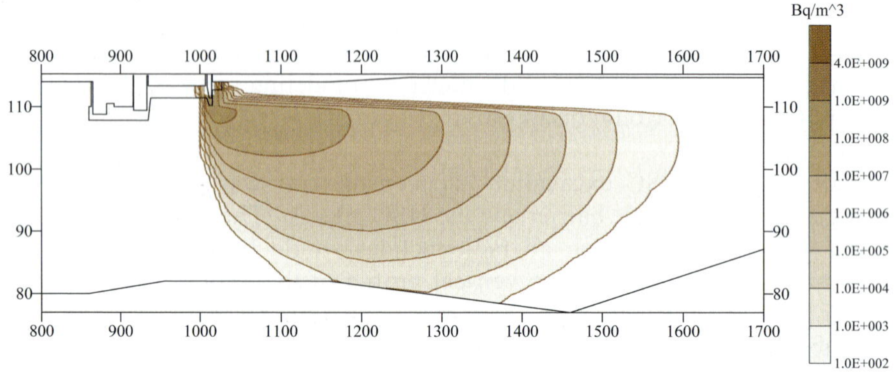

Figure 8.18. Predicted ^{90}Sr concentration in aqueous phase without the NSC after100 years

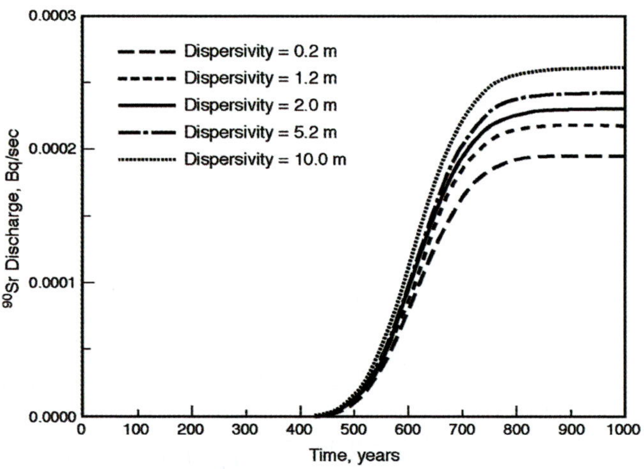

Figure 8.19. Sensitivity of ^{90}Sr seepage into the Pripyat River from the Shelter on the longitudinal hydraulic dispersivity

As shown in Table 8.9, a longitudinal dispersivity value of 1.2 m was used. The sensitivity analysis results show that in spite of the large variation in dispersivity tested the arrival time of ^{90}Sr for all five dispersivity values is about 800 years, and its concentration at that time varies by only one order of magnitude. With a half-life of 29.12 years, the ^{90}Sr level would be reduced to 5.7×10^{-9} of the original concentration due to radionuclide decay. This is reflected in the very small ^{90}Sr influx (about 0.0002 Bq/s) to the river at that time. When this ^{90}Sr influx is fully mixed with the average Pripyat River discharge of 404 m^3/s, the ^{90}Sr concentration would be $1x10^{-6}$ Bq/m^3; the current level is 100 Bq/m^3, and the drinking water limit is 2,000 Bq/m^3. Moreover, the simulation results are conservative because the 2-D model (as opposed to actual 3-D phenomena) and the current water level of the Chernobyl cooling pond were used. Thus, infiltration of ^{90}Sr from the Shelter will have no harmful effects on the Pripyat River, even without the NSC.

Construction of the NSC will stop precipitation in the Shelter. Water in Room 001/3 will dry out in 1½ years. Comparing Figures 8.18 (Scenario 5: Shelter without NSC) and 8.20 (Scenario 6: Shelter with NSC) reveals that the NSC will decrease ^{90}Sr concentration because less will filter through the floor with little or no water in Room 001/3. This further reduces ^{90}Sr concentration and influx to the Pripyat River. Thus, constructing the NSC would reduce radionuclide flux to subsurface and river environments. But even without the NSC there would be no harmful effects from ^{90}Sr to the Pripyat River.

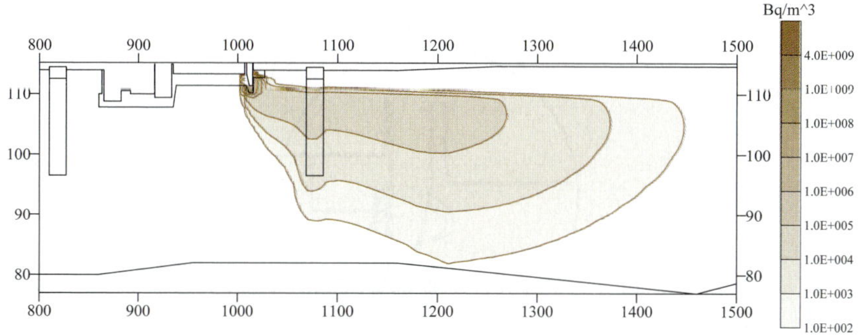

Figure 8.20. Predicted ^{90}Sr concentration in aqueous phase with NSC after100 years

Predicted ^{137}Cs concentration distributions indicate that it moves slower than ^{90}Sr; its larger distribution coefficient produces a larger retardation factor. After 2,000 years, its plume would still be within 200 m of the Shelter even without the NSC. With the NSC, it would be even slower. With a half-life of 30 years, radionuclide decay over 2,000 years would reduce ^{137}Cs concentrations by a factor of $1x10^{-20}$. Thus, ^{137}Cs would not harm the Pripyat River or humans through the aquatic pathway with or without the NSC.

Unlike ^{90}Sr and ^{137}Cs, ^{239}Pu has a half-life of 24,065 years; thus radionuclide decay cannot solve ^{239}Pu contamination. Predicted ^{239}Pu concentrations after 100 years with and without the NSC are shown in Figures 8.21 and 8.22. Because of high adsorption to the soil matrix (high Kd value, 30 L/kg),

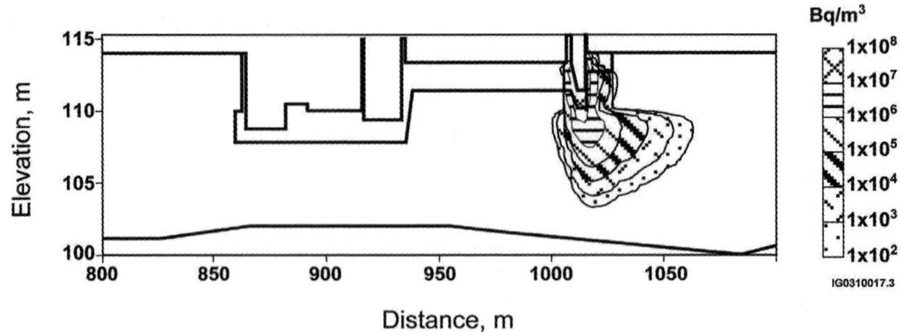

Figure 8.21. Predicted ^{239}Pu concentration in aqueous phase without the NSC after 100 years

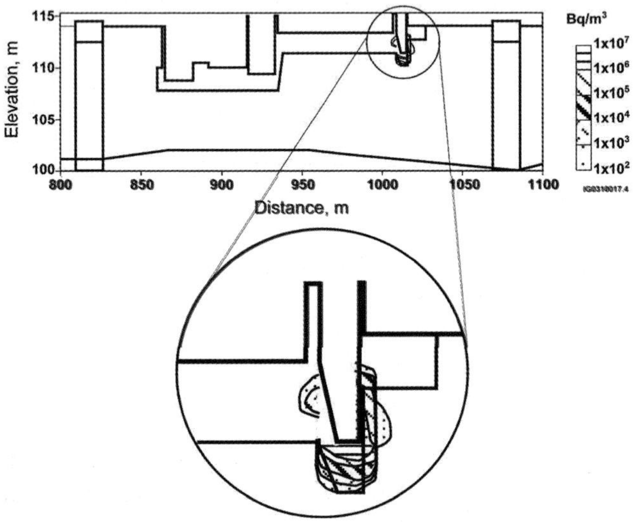

Figure 8.22. Predicted ^{239}Pu concentrations in the aqueous phase with the NSC after 100 years

^{239}Pu barely migrates from the Shelter. The beneficial effect of the NSC on ^{239}Pu migration is seen by comparing the figures.

Figure 8.23 depicts a breakthrough curve at the Pripyat River, showing it would take ^{239}Pu about 25,000 years to reach the river without the NSC. At that time, the ^{239}Pu influx would be about 0.6 Bq/s. When fully mixed with the average Pripyat River discharge of 404 m^3/s, the resulting concentration would be only 0.002 Bq/m^3. The current level is 0.25 Bq/m^3. Both are much less than 1,000 Bq/m^3, the Ukrainian regulatory limit. Thus, infiltration of ^{239}Pu from the Shelter, even without the NSC, will not harm the Pripyat River. The NSC would reduce ^{239}Pu levels already too small to be of concern.

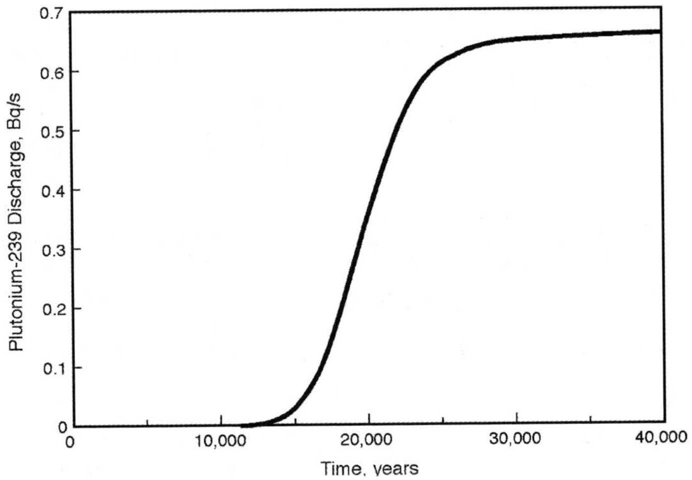

Figure 8.23. Predicted ^{239}Pu seepage into Pripyat River from the Shelter without the NSC

8.4 Conclusions

The NSC will be built over the Chernobyl Shelter to (1) reduce radionuclide releases to the environment, (2) provide safer working conditions for the Shelter dismantling and waste management, and (3) generate economic and social benefits. This chapter discussed the NSC conceptual design and its benefits on the aquatic environment.

The modeling determined the possibility of condensation within the NSC and the amount and type of ventilation and insulation needed to avoid condensation on the metal roofs. Numerical modeling of air movement and temperature distribution within the NSC was conducted with two 2-D models and one 3-D model. The modeling evaluation concluded the following to avoid condensation on the outer roof and metal portion of the inner roof:

- Place two roofs on the NSC.
- Install at least R = 0.2 hr· ft^2· °F/Btu (0.03 m^2· K/W) insulation along the outer roof.
- Use forced air ventilation with flow rate controlled at both inlets and outlets.
- Provide ventilation air at the bottom and remove it at the top.
- Supply 0.9 NSC volume per day (2.3x10^6 m^3/day) unheated air to the inside of the NSC.

- Supply 0.1 NSC volume per day ($2.5x10^5$ m^3/day) heated air (at least 2° C above outdoor temperature) to the annular space between roofs.
- Supply 0.04 NSC volume per day ($1.x10^5$ m^3/day) heated air (at least 2° C above outdoor temperature) to the 1.2-m-wide space between the outer and inner west walls.

It is likely that deviating from the conditions used in this assessment would produce condensation along the metal roofs. NSC operators may control the timing and rate of the ventilation and amount of heating for the air inflow to mitigate adverse weather and accident conditions.

During normal operation of the NSC, radionuclides would be released to the environment at the recommended rate of 1 NSC volume per day. This release rate was used as one of the scenarios for the radionuclide transport modeling for atmospheric and surface water environments.

The impact of Shelter collapse on the Dnieper River was evaluated for the following four scenarios:

- Scenario 1. Shelter collapse before the NSC construction
- Scenario 2. Shelter collapse caused by NSC collapse
- Scenario 3. Shelter collapse inside the NSC with a ventilation rate of 1 NSC volume per day
- Scenario 4. Shelter collapse inside the NSC with a ventilation rate of 0.1 NSC volume per day.

The NSC would reduce atmospheric fallout from Shelter collapse by 15~20 times under Scenario 3 and 150~200 times under Scenario 4. Without the NSC standing over the Shelter, the fallout from the collapsed Shelter would have some impacts on ^{90}Sr and ^{137}Cs concentrations in the Dnieper River. The Kiev Reservoir would experience the highest radionuclide concentrations with and without the NSC, as shown in Table 8.10.

Table 8.10. Predicted ^{90}Sr and ^{137}Cs concentrations at hydropower plant dam on Kiev Reservoir with and without the NSC

Res.	^{90}Sr Concentrations Bq/m^3				^{137}Cs Concentrations Bq/m^3			
	Scen. 12	**Scen. 3**	**Scen. 4**	**Nat. Cond.**	**Scen. 12**	**Scen. 3**	**Scen. 4**	**Nat. Con.**
Kiev	211	190	188	188	42.1	23.6	22.5	22.5

Without the NSC, the maximum increase to the annual average ^{90}Sr and ^{137}Cs concentrations at Kiev hydropower plant would be 27 Bq/m^3 (15 percent) over natural conditions of 188 Bq/m^3 for ^{90}Sr and 19.6 Bq/m^3 (or 87 percent) over natural conditions of 22.5 Bq/m^3 for ^{137}Cs. The concentrations are still below Ukrainian drinking water limits of 2,000 Bq/m^3. Shelter collapse under Scenarios 1 and 2 does not affect ^{238}Pu, $^{239/240}Pu$, and ^{241}Am concentrations in the Dnieper River to any measurable degree. With the NSC in place (Scenarios 3 and 4), impacts to the river from Shelter collapse are very small.

The Dnieper River provides aquatic pathways for radionuclides to reach Ukrainians, mostly through drinking water, water sports and recreation, consumption of fish caught in the Dnieper reservoirs, and consumption of produce irrigated with Dnieper River water. A professional fisherman would receive the highest radiation dose, mostly from eating fish, especially those caught in Kiev Reservoir. This fisherman would have 2.4×10^{-6} risk of death from cancer without the benefit of the NSC shielding the collapsed Shelter.

The subsurface water (vadose zone and groundwater) simulations were performed to determine potential benefits of constructing the NSC over the Shelter to reduce or prevent ^{90}Sr, ^{137}Cs, and ^{239}Pu in the Shelter from reaching the Pripyat River through subsurface water. The major benefit of the NSC is blocking precipitation, thereby eliminating a water pool in the Shelter that drives the radionuclides to the subsurface though the concrete floor.

It would take about 800 years for ^{90}Sr to reach the Pripyat River even without the NSC, and decay would reduce ^{90}Sr to 5.7×10^{-9} of the original concentration. The concentration in the river would be only 1×10^{-6} Bq/m^3; the current level is 100 Bq/m^3, and the drinking water limit is 2,000 Bq/m^3. Because ^{137}Cs migrates much slower than ^{90}Sr and has a half-life of 30 years, it would cause no harmful effects to the river. It would take ^{239}Pu about 25,000 years to reach the Pripyat River due to its higher affinity for sediment sorption. By then the ^{239}Pu concentration in the river would be 0.002 Bq/m^3; current level is 0.25 Bq/m^3 and the regulatory limit 1,000 Bq/m^3.

Comparing predicted radionuclide concentrations with and without the NSC indicates that concentrations will be smaller with the NSC than without it because there will be less infiltration through the Shelter's concrete floor with a smaller or no driving force of water. This would further reduce the ^{90}Sr concentrations and influx to the Pripyat River.

Thus, building the NSC will reduce radionuclide flux to the subsurface and river environments, even though without the NSC there will be no measurable harmful effects from ^{90}Sr, ^{137}Cs, and ^{239}Pu on the Pripyat River environment. The biggest public health problem caused by the Chernobyl accident is mental health—feelings of helplessness, anxiety for health and life expectancy, lack of community, and hopelessness (the Chernobyl Forum 2005). A major socio-economic problem is the lack of a vital local economy, lower living standards, high unemployment, and increased poverty. Thus, regardless of NSC's specific remedial benefits, NSC construction will provide employment, higher self esteem, and a high-quality labor force and will develop long-term manufacturing and construction capabilities. These elements will revitalize the economy, thus addressing the most critical Chernobyl problems: mental health and socio-economic issues.

8.5 References

Bechtel, Electricity DeFrance, and Battelle. 2003. "Environmental Impact Assessment, Site Characterization", San Francisco.

Hemond HF and EJ Fechner-Levy. 2000. *Chemical Fate and Transport in the Environment.* Academic Press, Boston.

IGS. 1995. *Report on the Study, Investigation of Radionuclide Contamination of Groundwater at Promploshadka and Unit Shelter*. Kiev, p. 118 (in Ukrainian).

Kivva SL. 1997. "SUSTOX: A Numerical Simulator for Fluid and Species Transport in the Subsurface." RODOS(WG4)-TN(99)06, Institute of Mathematical Machines & System Problems, Kiev, Ukraine.

National Geographic. April 2006. "The Long Shadow of Chernobyl," pp. 32-53.

Ramsdell JV Jar, BA Napier, and D Timmins. July 2003. "Off-Site Radiation Accident Analysis for Chernobyl New Safe Confinement." *48th Annual Meeting of the Health Physics Society*, San Diego, California.

Shelter Implementation Plan (SIP). 2000. "Shelter Water Management Water Characterization and Analysis Report." SIP report TN/00139, Kiev, Ukraine.

State Specialized Enterprise Chernobyl Nuclear Power Plant (SSE ChNPP). 2003. "Environmental Impact Assessment, New Safe Confinement Conceptual Design Chernobyl Nuclear Power Plant–Unit 4." Kiev, Ukraine.

Voitsekhovich OV, VV Demchuk, and GV Laptev. 1990. "Analysis of Secondary Process of Pripyat River Contamination Following Floodplain Area Flooding of New Closed-in Zone." *Proceedings of the Ukrainian Hydrometeorological Institute.* Kiev, Ukraine, Vol. 214, pp. 54-76 (in Russian).

Voitsekhovich OV, MJ Zheleznyak, and Y Onishi. 1994. *Chernobyl Nuclear Accident Hydrologic Analysis and Emergency Evaluation of Radionuclide Distributions in the Dnieper River, Ukraine, During the 1993 Summer Flood.* PNL-9980, Pacific Northwest National Laboratory, Richland, Washington.

Zheleznyak MJ, R Demchenko, S Khursin, Y Kuzmenko, P Tkalich, and N Vitjuk. 1992. "Mathematical Modeling of Radionuclide Dispersion in the Pripyat-Dnieper Aquatic System after the Chernobyl Accident." *The Science of the Total Environment,* Vol. 112, pp. 89-114.

Chapter 9
Summary and Conclusions

Yasuo Onishi
Washington State University, Richland, WA, USA
Formerly of Pacific Northwest National Laboratory, Richland, WA, USA

Oleg V. Voitsekhovich
Ukrainian Hydrometeorological Institute, Kiev, Ukraine

Mark J. Zheleznyak
Ukrainian Center of Environmental and Water Projects and National Academy of Sciences of Ukraine, Kiev, Ukraine

9.1 Summary

Twenty years ago, on April 26, 1986, Unit 4 of the Chernobyl nuclear power plant suffered a core meltdown and released 12×10^{18} Bq of radioactivity to the environment. The accident also caused the deaths of 50 emergency workers. The accident had significant environmental impacts that continue to this day in the affected ecosystems, and as many as 4,000 deaths from cancer and leukemia are expected to result (Chernobyl Forum 2005).

The Chernobyl nuclear plant is situated on the Pripyat River, a tributary of the Dnieper River, which flows into the Black Sea. Annual releases of ^{137}Cs and ^{90}Sr from contaminated floodplains and watersheds to the Dnieper River are 2 to 4 trillion Bq and 10 to 20 trillion Bq, respectively. Thus radionuclides have been carried from the heavily contaminated watersheds to the less contaminated downstream areas through the six Dnieper reservoirs. Cesium-137 in the river has mostly accumulated in the reservoirs, whereas a soluble form (a majority) of ^{90}Sr has reached the Black Sea.

Water pathways make up a small part of the total dose received by the population from all dose-forming factors. However, the Dnieper River system delivered the radionuclides to people living in areas that were not originally contaminated. Thus, radionuclide mobility in the river system deserves special attention. The river is the main aquatic pathway for the radionuclides to reach 20 million Ukrainians through drinking water; fishing; swimming and other water recreation; consuming irrigated agricultural produce; and drinking milk and eating the meat of cattle fed with contaminated feed and water. Many research and remediation efforts have been focused on minimizing radionuclide transfer from the Chernobyl Exclusion

Y. Onishi, O.V. Voitsekhovich and M.J. Zheleznyak (eds.), Chernobyl – What Have We Learned?, 279–286.

Zone (CEZ) to the Dnieper River to reduce radiological risks. A significant effort was focused on reducing possible radionuclide transfer to the human body through consumption of irrigated crops and fish. Irrigation affects the radionuclide distribution in soil and irrigated crops, especially in rice paddies. In the first few years, radionuclide concentrations increased by five to ten times in the topsoil layer. However, due to the significant drop of the radionuclide concentrations in the Dnieper River, the low additional radionuclide influx into the soil of irrigated lands is now balanced by washout into drainages and radionuclide decay. Based on a large amount of monitoring data, radionuclide transfer processes were quantified and better agricultural practices developed to reduce radionuclide transfer to the irrigated crops. Studies of fish contamination identified a clear dimensional (size) effect on ^{137}Cs accumulation in fish. However, a large number of field and experimental data have not been comprehensively assessed to quantify the major mechanisms of radionuclide accumulation in fish.

Radiation doses could reach 3,000 man-sv, due mainly to ^{137}Cs and ^{90}Sr in the water, if all current protective activities implemented before 1991 are terminated. The contribution of cesium to the exposure dose is mostly restricted to Ukraine's northern regions. But due to the relatively uniform strontium distribution in the Dnieper cascade of reservoirs, the strontium contribution to the expected collective dose of the population is relatively uniform. Protective measures (e.g., dike construction) implemented in 1992 and 1993 along the Pripyat River's northeast (left)-bank floodplain decreased exposure doses by about 700 man-sv. The construction of similar protective measures on the southwest (right) bank in 1998 to 2003 would decrease the dose by another 200 to 300 man-sv. These dose estimates from aquatic pathways are crucial for developing optimal water protection measures in the areas affected by the Chernobyl accident. Such estimates must consider the aquatic system as a whole, incorporating radionuclide migration, water use, and population dose.

The modeling results for the dose estimates are based on the best available information on radionuclide concentrations in rivers and reservoirs, radiological information on soil, agricultural products from irrigated lands, and fish contamination. However, the reliability of dose estimates is determined by the accuracy of the required information, such as long-term changes in radionuclide contents of the rivers and reliability of forecasting radiological contamination levels of soil, agricultural products, and fish, as well as future population data. These studies should be continued to improve the predictive methodology and databases.

Conceptual approaches and health, economic, social and psychological criteria are also presented to justify and optimize specific countermeasures to protect Ukrainians from radiation. The best-developed application for water protection optimization is MOIRA (Model-Based Computerized System for Management Support to Identify Optimal Remedial Strategies for Restoring Radionuclide Contaminated Aquatic Ecosystems and Drainage Areas). MOIRA uses a GIS database and reliable validated models to predict temporal behavior of radionuclides in the fresh water environment and their ecological, social, and economic impacts. It was implemented in a personal computer-based decision support system with all relevant information. The strategy and technologies applied to the CEZ during the last 15 years were carefully evaluated for fresh water environmental management. This evaluation is important to understand the MOIRA philosophy as a tool for the decision support system. Actions taken for various water bodies and aquatic ecosystems are considered in their historical context and examples evaluated for their effectiveness. In the future, MOIRA may be widely used in decision-making for water quality management at radioactively contaminated sites.

Stochastic groundwater simulations of ^{90}Sr migration to the Pripyat Town water wells show the low probability of significant radioactive contamination, mostly because of radionuclide decay during the long travel time. Conservatively estimated risks for hypothetical residents of the CEZ from the groundwater used as a drinking water supply would be less than $1x10^{-5}$/yr under the average radiological and hydrogeologic conditions in the CEZ. For the Red Forest radioactive waste dump site, risk from radioactive contamination of groundwater is up to 10^{-3}/yr and 20 times the drinking water standard. It will take 150 and 250 years, respectively, for the risk caused by using groundwater for drinking water to be reduced to $1x10^{-6}$/yr. Soil contamination in the CEZ by ^{137}Cs is a priority dose-forming factor compared with groundwater contamination.

To determine groundwater protection criteria and countermeasures related to ^{90}Sr migration into Pripyat Town's water supply wells, a cost-benefit analysis methodology was used, accounting for uncertainty in groundwater flow parameters of contaminant transport modeling. A low economic risk of water supply contamination indicates that expensive countermeasures are not justified. Modeling of ^{90}Sr migration from the RAWTLS sites and other radionuclide sources in the CEZ shows that ^{90}Sr transport from these sources to surface waters would be relatively low. The expected travel time of maximum subsurface transport ranges from 33 to 145 years. However, ^{90}Sr migration assessment suggests that subsurface water migration from the contaminated watersheds of small rivers and water reclamation systems of the CEZ under certain conditions may be an essential mechanism for radioactive contamination of surface waters during low-flow periods.

In general, because of the dilution of contaminated groundwater seepage from watersheds of the CEZ by the Pripyat River and the reservoirs of the Dnieper cascade, subsurface migration would not expected to cause catastrophic transport of radioactivity from the CEZ. However, these assessments are subject to significant uncertainty in radionuclide transport model parameters, necessitating continued monitoring of groundwater migration in the CEZ.

A movable cover called the NSC will be built over the Chernobyl Shelter. The NSC will be built to reduce radionuclide releases to the environment, to provide safer conditions for workers dismantling the Shelter and performing waste management, and to generate economic and social benefits. The computer modeling determined the possibility of condensation within the NSC and the amount and type of ventilation and insulation needed to avoid condensation on the metal roofs in order to reduce maintenance needs over its 100-year design life and to minimize radiation exposure to workers.

The NSC will reduce atmospheric fallout from Shelter collapse by 15–20 times under its normal operation, further reducing radionuclide contamination in the Pripyat and Dnieper rivers. However, even if the Chernobyl Shelter collapses before the NSC is built, the peak ^{90}Sr and ^{137}Cs concentrations in the Dnieper River would still be below drinking water limits. Under this condition, the maximum increase to the annual average ^{90}Sr and ^{137}Cs concentrations at Kiev hydropower plant would be 15% for ^{90}Sr and 87% for ^{137}Cs, respectively, over the natural conditions. Shelter collapse with or without the NSC does not affect ^{238}Pu, $^{239/240}$Pu, and ^{241}Am concentrations in the Dnieper River to any measurable degree. It would take about 800 years for ^{90}Sr to reach the Pripyat River even without the NSC, and radionuclide decay would reduce ^{90}Sr to 5.7×10^{-9} of the original concentration. Because ^{137}Cs migrates much slower than ^{90}Sr and has a similar half-life, 30 years, it would not cause any harmful effects to the river. Comparing predicted radionuclide migration with and without the NSC indicates that concentrations in the groundwater will also be smaller with the NSC, because less water will filter through the Shelter floor with little or no condensation. This would further reduce radionuclide concentrations and influxes to the Pripyat River.

Thus, building the NSC would reduce the radionuclide flux to the subsurface and river environment, though even without the NSC there would be no measurable harmful effects from ^{90}Sr, ^{137}Cs, and ^{239}Pu to the Pripyat River environment. The Chernobyl Forum (2005) states that the biggest public health problem caused by the Chernobyl accident is psychological stress (e.g., helplessness and anxiety). A major socio-economic problem is the lack of a vital local economy, lower living standards, high unemployment, and increased poverty. Thus, regardless of the remedial benefits of the NSC, its construction would provide employment, higher self esteem, and a high-quality labor force and develop long-term manufacturing and construction capabilities. These elements will revitalize the economy and potentially address the most critical Chernobyl problem: mental health and socio-economic issues.

9.2 Conclusions

The experience gained from the water protection strategies used in the CEZ and the affected areas after the Chernobyl accident proved that, as never before, the procedure for justifying countermeasures requires in-depth knowledge of monitoring data, field experiments on the parameters of aquatic pathways of radionuclide transfer, radioecological processes predicted by comprehensive mathematical models to describe radionuclide migration in water-soil media, analysis of radionuclide accumulation in food products and their effects on the human organism considering various elements of water use and consumption.

It was impossible to use the whole set of knowledge without making adaptations and improvements to the regulatory constraints on radiation protection. Efforts to justify water protection activities in the CEZ showed the importance of the psychological response of society to radioactive contamination. This psychological factor cannot be ignored in justifying any mitigative activity.

It is important to reduce the uncertainties in decisions made on the basis of social and economic benefits. The experience gained by developing and applying the models of radionuclide migration in surface water is now used in developing tools for decision making (e.g., the Decision Support System for Off-Site Emergency Management in Europe).

The analysis of the basic parameters of radionuclide migration via aquatic pathways has been used to prepare recommendations for radioecological monitoring and optimizing countermeasures. The methods developed for assessing radiation risks made it possible to perform an objective analysis of the real level of radioactive danger. Thus, it was possible to approach the issues of water protection and the expediency of the countermeasures in the CEZ more objectively on the basis of not only dose but also social and economic criteria. A practical application of a new Ukrainian water standard regulation, NRBU-97, to address water protection problems was demonstrated. The experience of protecting surface water and groundwater during the years after the accident should be considered in its historical context when applying technologies to protect the environment. The

Chernobyl experience gained from the comprehensive and systematic approach is unique and should be studied as an example when solving man-caused contamination of groundwater and large water systems. A vast amount of experience in planning and managing radioactively contaminated environments has been accumulated in Ukraine. The experience refers to both justification of strategies and development of technologies that can be applied in many areas of environmental protection.

Reviewing important archived materials provides reliable information on the atmosphere and style of decision making regarding water protection in the first years after the accident. It shows that even with the unprecedented attention to the contamination of the water resources, it was impossible for management to be effective without sufficient scientific justification in selecting appropriate countermeasures. The decisions were often based only on the personal experience of the experts, and decision makers acted on information and forecasts without accounting for the effects of the protective measures.

The normative and legislative basis of environmental protection in the CEZ was improved, including water protective interference. Discussions will go on for many years as to whether it is possible to consider the near zone of the Chernobyl nuclear plant and the contaminated territories adjacent to it as analogous to the wider area of the plant's industrial site in the later stages of accident prevention. It is possible that at some time activities in the contaminated territories can return to normal. Then the maximum permitted discharges may be applied to the Pripyat River where it crosses the boundaries of the CEZ, and water-protection activity will reach its planned completion. It then may be possible to minimize radioactive discharges from the CEZ. The rehabilitative measures will continue, as will the costs for rehabilitative measures. It will probably be necessary to maintain the water protection facilities in the CEZ until the theoretical basis for their justification is developed, incorporating modern ideas of radiation hazards for human health and ecosystem as a whole.

The recently developed "Optimization of Countermeasures in the Practice of Radiological Protection" became the most important tool for supporting water protection projects. However, the application of such an approach may be effective only if the decision makers are provided with reliable data on monitoring, accurate modeling, and comprehensive assessment of effects of various proposed strategies for water protection.

Thus, it is necessary to preserve and support the systems set up for the CEZ water bodies during the post-accident years and the scientific experts who can analyze and forecast the consequences of radiation accidents on the environment. This experience is impossible to be over-estimated. It must be studied and be more widely introduced into the practice of ecological monitoring and forecasting of water ecosystems where nuclear plants and other hazardous industrial plants are operated.

New technologies can significantly reduce the uncertainties regarding the anticipated effects of a water protective activity. In the years after the accident, the computer systems for decision making and support and the GIS technologies that were introduced allowed scientists to analyze the spatial extent of radionuclide migration and radionuclide fluxes through surface water and groundwater. Natural sorbents were used for cleaning radioactive flows and have been used as possible phyto-rehabilitative technologies to restore ecosystems in the contaminated territories.

At present, a cost-benefit analysis is used to assess the economic efficacy of radiation protection measures. The high water protection costs encourage proposals of useful technologies to reduce the costs of implementing these measures. However, the methods are often not sufficiently practical for applications in the CEZ.

Tens of technical engineering projects proposed by scientists and experts during the first years after accident have not been applied and unfortunately will never be applied; they have been archived and likely lost. For example, many volumes of proposals and technologies for decontamination and localization of bottom contamination of the Chernobyl cooling pond were available in 1989, but they have been lost. Organizing and filing the accumulated materials remains to be done.

An important lesson that has not yet been completely realized is that water protection in zones of radioactive accidents cannot be focused only on radiological protection. In spite of the fact that the areas suffered from radioactive contamination and the goal of radiation protection will always be a priority, the optimal balance of various aspects of water management in the contaminated territories will always be important to officials.

The CEZ with its rivers and lakes, watersheds and floodplains, bogs and forests will not always be considered exclusively as the source of radioactive contamination of the Dnieper water system. The ecological state of the Dnieper reservoirs is controlled, to a great extent, by processes occurring in the watersheds and floodplains of many rivers, abandoned and swampy lands, and many other factors. Thus, limited water protection measures in the CEZ will continue to protect the populace from exposure to radiation.

The authors hope that the studies described in this book will be continued in the affected areas. In particular, the model parameterization studies based on experimental data must be developed to improve the predictive capabilities of the mathematical models. Studies must be continued on the ecological consequences of chemical and radioactive substances on human health and ecosystems, development of monitoring and assessment methods for radioactive contamination in the environment, improvement of a standardization base, and development of more complete technologies for purification and management of radioactive flows in the affected ecosystems.

As stated above, the most important conclusions are (1) scientifically defensive assessment tools and required data must be developed and applied, (2) countermeasures and remediation selections must be based on a cost-risk analysis that directly connects the main physical and chemical processes to environmental/human health risks and costs, (3) decision makers must be knowledgeable on the phenomena being evaluated, and (4) decision makers must communicate facts quickly and honestly to the affected public.

9.3 References

Chernobyl Forum. 2005. "Chernobyl Legacy: Health, Environmental, and Socio-economic Impacts: and Recommendations to the Governments of Belarus, the Russian Federation and Ukraine." International Atomic Energy Agency, Vienna.

Subject Index

ENVIRONMENTAL POLLUTION

1. J. Fenger, O. Hertel and F. Palmgren (eds.): *Urban Air Pollution – European Aspects.* 1998 ISBN 0-7923-5502-4
2. D. Cormack: *Response to Marine Oil Pollution – Review and Assessment.* 1999 ISBN 0-7923-5674-8
3. S.J. Langan (ed.): *The Impact of Nitrogen Deposition on Natural and Semi-Natural Ecosystems.* 1999 ISBN 0-412-81040-9
4. C. Kennes and M.C. Veiga (eds.): *Bioreactors for Waste Gas Treatment.* 2001 ISBN 0-7923-7190-9
5. P.L. Younger, S.A. Banwart and R.S. Hedin: *Mine Water: Hydrology, Pollution, Remediation.* 2002 ISBN 1-4020-0137-1; Pb 1-4020-0138-X
6. K. Asante-Duah: *Public Health Risk Assessment for Human Exposure to Chemicals.* 2002 ISBN 1-4020-0920-8; Pb 1-4020-0921-6
7. R. Tykva and D. Berg (eds.): *Man-Made and Natural Radioactivity in Environmental Pollution and Radiochronology.* 2004 ISBN 1-4020-1860-6
8. L. Landner and R. Reuther (eds.): *Metals in Society and in the Environment.* A Critical Review of Current Knowledge on Fluxes, Speciation, Bioavailability and Risk for Adverse Effects of Copper, Chromium, Nickel and Zinc. 2004 ISBN 1-4020-2740-0; Pb 1-4020-2741-9
9. P.F. Ricci: *Environmental and Health Risk Assessment and Management.* Principles and Practices. 2006 ISBN 1-4020-3775-9
10. J. Davenport and J.L. Davenport: *The Ecology of Transportation: Managing Mobility for the Environment.* 2006 ISBN 1-4020-4503-4
11. M.B. Adams, D.R. DeWalle and J.L. Hom (eds.): *The Fernow Watershed Acidification Study.* 2006 ISBN 1-4020-4614-6
12. Y. Onishi, O.V. Voitsekhovich and M.J. Zheleznyak (eds.): *Chernobyl – What Have We Learned? The Successes and Failures to Mitigate Water Contamination Over 20 Years.* 2007 ISBN 1-4020-5348-7